THE STRUCTURE OF THE LIFE COURSE: STANDARDIZED? INDIVIDUALIZED? DIFFERENTIATED?

ADVANCES IN LIFE COURSE RESEARCH

Series Editor: Timothy Owens

Recent volumes:

Volume 1: Work, Retirement and Social Policy, 1986

Volume 2: Family Relations in Life Course, 1986

Volume 3: Personal History Through Life Course, 1990

Volume 4: Delinquency and Disrepute Life Course, 1995

Volume 5: Self and Identity Through the Life Course in Cross-Cultural Perspective, 2000

Volume 6: Children at the Millennium: Where Have We Come From, Where Are We Going?
Edited by Sandra L. Hofferth and Timothy J. Owens, 2001

Volume 7: New Frontiers in Socialization
Edited by Richard A. Settersten, JR. and Timothy J. Owens, 2002

Volume 8: Changing Life Patterns in Western Industrial Societies
Edited by Janet Zollinger Giele and Elke Holst, 2004

ADVANCES IN LIFE COURSE RESEARCH VOLUME 9

THE STRUCTURE OF THE LIFE COURSE: STANDARDIZED? INDIVIDUALIZED? DIFFERENTIATED?

EDITED BY

ROSS MACMILLAN
Department of Sociology, University of Minnesota, Minneapolis, MN, USA

2005

ELSEVIER
JAI

Amsterdam – Boston – Heidelberg – London – New York – Oxford
Paris – San Diego – San Francisco – Singapore – Sydney – Tokyo

ELSEVIER B.V.
Radarweg 29
P.O. Box 211
1000 AE Amsterdam
The Netherlands

ELSEVIER Inc.
525 B Street, Suite 1900
San Diego
CA 92101-4495
USA

ELSEVIER Ltd
The Boulevard, Langford
Lane, Kidlington
Oxford OX5 1GB
UK

ELSEVIER Ltd
84 Theobalds Road
London
WC1X 8RR
UK

First edition 2005

British Library Cataloguing in Publication Data
A catalogue record is available from the British Library.

ISBN: 0-7623-1193-2
ISSN: 1040-2608 (Series)

∞ The paper used in this publication meets the requirements of ANSI/NISO Z39.48-1992 (Permanence of Paper).
Printed in The Netherlands.

CONTENTS

PART III: CONTEXTS AND CONTINGENCIES

PART IV: CONSEQUENCES

LIST OF CONTRIBUTORS

Alexandra Berkowitz	Department of Sociology, University of Indiana, Bloomington, IN, USA
Pamela Braboy Jackson	Department of Sociology, University of Indiana, Bloomington, IN, USA
Hannah Brückner	Department of Sociology, Center for Research on Inequalities and the Life Course, Yale University, New Haven, CT, USA
Philippa Clarke	Center for Aging and Human Development, Duke University, Durham, NC, USA
Regina Deil-Amen	Department of Education Policy Studies, Pennsylvania State University, University Park, PA, USA
Patricia Drentea	Department of Sociology, University of Alabama, Birmingham, AL, USA
Elizabeth Fussell	Department of Sociology, Tulane University, New Orleans, LA, USA
Shin-Kap Han	Department of Sociology, University of Illinois, Urbana, IL, USA and Department of Sociology, Yonsei University, Seoul, Korea
Steffen Hillmert	Chair of Sociology I, University of Bamberg, Germany
Helga Krüger	University of Bremen, Bremen, Germany
Ross Macmillan	Department of Sociology, Life Course Center, University of Minnesota, Minneapolis, MN, USA

Karl Ulrich Mayer	Department of Sociology, Center for Research on Inequalities and the Life Course, Yale University, New Haven, CT, USA
Phyllis Moen	Department of Sociology, Life Course Center, University of Minnesota, Minneapolis, MN, USA
Jeylan T. Mortimer	Department of Sociology, Life Course Center, University of Minnesota, Minneapolis, MN, USA
Sabrina Oesterle	Social Development Research Group, University of Washington, Seattle, WA, USA
Ann E. Person	Department of Human Development and Social Policy, Institute for Policy Research, Northwestern University, Evanston, IL, USA
James E. Rosenbaum	Department of Human Development and Social Policy and Department of Sociology, Institute for Policy Research, Northwestern University, Evanston, IL, USA
Victor Shaw	Department of Sociology, California State University, Northridge, CA, USA
Stephen Sweet	Department of Sociology, Ithaca College, Ithaca, NY, USA
Raymond Swisher	Policy Analysis and Management, College of Human Ecology, Cornell University, Ithaca, NY, USA
Blair Wheaton	Department of Sociology, Center for Aging and the Life Course, University of Toronto, Toronto, ON, Canada

PREFACE

While dormant over the last couple of decades, the structure of the life course is re-emerging as a central issue in life course social science. Much of this interest reflects on-going speculation and debate about the way in which life courses are changing over time. For example, a recent contribution to the *Annual Review of Sociology* by Michael Shanahan provides an overview of an emergent theoretical discussion over the structure of the life course. This debate focused on the changing nature of the life course in modern society and the degree to which life course structures were evolving in a particular direction. The chapter juxtaposes in tight fashion arguments about the "standardization," "individualization," and "new individualization" of the life course and poses some provocative questions about what any of these trends might mean for theory and research. The chapter also serves a second, likely unintended, function. Although the thorough literature review brought together work from demography, social psychology, and sociology and included contributions from various nations across the globe, it still highlights the lack of serious systematic research on the structure of the life course, how it has changed over time and differentiated across groups, and its consequences for social and personal development.

Against this backdrop, Tim Owens agreed to have the 2005 volume of *Advances in Life Course Research* focus on the structure of the life course. The hope was to bring together theoretical and empirical works that tackled the issue in creative ways and marshalling empirical evidence that addressed key theoretical questions of what the structure of the life course means for individuals and society. In this regard, I could not be more pleased with the contributions. They are of exceptionally high quality, are diverse in their theoretical and empirical content, and speak about a wide variety of issues. In the end, I am particularly grateful to the authors for the thought-provoking work they produced and the opportunity they provided me to learn about the myriad ways in which the life course operates and the various and creative ways in which we can study it.

In addition to the contributors, I am also grateful for the support of the series editor, Tim Owens. Michael Shanahan at the Department of Sociology, University of North Carolina at Chapel Hill contributed to my thinking and re-thinking the modern life course, as well as provided an initial spark for the idea of a thematic compilation. Jeylan Mortimer and the Life Course Center at the University of Minnesota also provided intellectual stimulation and a wide variety of support. The general support of the College of Liberal Arts and the University of Minnesota provided research assistance at various points that freed up time to concentrate on this project. Ann Corney and the editorial staff at Elsevier were also helpful. Finally, I want to express my appreciation to Annette Nierobisz, Department of Sociology and Anthropology, Carleton College, for partnering with me in the on-going discussion of the life course and what it all means.

Ross Macmillan
Volume Editor

ABSTRACT

Studies of the structure of the life course attempt to consider the life span in a holistic manner. This involves attention to the multidimension unfolding of social roles of school, work, marriage, and parenthood and the ways in which their timing and ordering give rise to unique life course forms. Such issues are increasingly important in the light of existing debates over the transformation of the life course and questions of its 'standardization,' 'individualization,' and 'differentiation' across social groups, societies, and time periods. This book brings together research which consider the characteristics of life course structures, the contexts and contingencies that influence their formation, and the consequences they have for personal and social development. Such research plays an important role in understanding the nature of modern lives and how they are connected to broader patterns of social change. The first chapter provides an overview of how questions of the structure of the life course are connected to classic issues, as well as current controversies. Succeeding chapters in Part II cover conceptual issues and empirical efforts to characterize the life course in a multidimension, dynamic manner. Part III includes chapters that show the important influences of social contexts and life course contingencies on the structuring of individual lives. The closing chapters in Part IV consider consequences for psychological well-being, the nature of activities over the life span, and role-related experiences and affect. Overall, the various chapters make important contributions by considering the implication of the structuring and re-structuring of the life course for individuals and society.

PART I:
INTRODUCTION

THE STRUCTURE OF THE LIFE COURSE: CLASSIC ISSUES AND CURRENT CONTROVERSIES

Ross Macmillan

ABSTRACT

The structure of the life course occupies a central, yet often unacknowledged, position in life course research. On the one hand, life course theory emphasizes in various and often disparate ways life stages, role trajectories and transitions, and the timing and ordering of events in the life course. All these coalesce to produce a "normative" life course that corresponds to social timetables of given times and given places. At the same time, empirical efforts seldom describe the life course in a sufficiently holistic manner that adequately reflects existing theory. The gap between theory and research is all the more significant with the emergence of current debates over the problematic nature of the life course in the late 20th and early 21st centuries. Here, scholars argue that the overall structure of the life course has changed in profound ways, becoming "destandardized," "de-institutionalized," and increasingly "individualized." Equally important, there is increased recognition of social differentiation in the unfolding of human lives based on time, place, and social position. Both the changing nature of the life course in general and differences across social groups are further connected to increases in and the reproduction of inequalities through the undermining of life

The Structure of the Life Course: Standardized? Individualized? Differentiated?
Advances in Life Course Research, Volume 9, 3–24

ISSN: 1040-2608/doi:10.1016/S1040-2608(04)09001-X

chances in a wide number of domains. This chapter reviews some of the key issues and debates and discusses the relevant contributions of the various chapters included in this volume.

INTRODUCTION

The structure of the life course involves both the timing and ordering of events in the life span and occupies a central, yet often unacknowledged, position in life course research. On the one hand, it is central in life course theory. It is tied to general principles of life history and biography and is closely connected to key concepts such as transitions and trajectories. It also bears a sharp affinity to the emerging notion of "social pathways" that focuses squarely on "life patterns and their dynamics in time (Elder, Johnson, & Crosnoe, 2003, p. 7). Studies of modern societies emphasize the "three boxes of life" (Riley, Kahn, & Foner, 1994) and the tripartite life course (Kohli, 1987), both of which suggest that the unfolding of lives over time is structured in uniform ways. Most generally, groups both within and across societies are differentiated in terms of (modal) structures of the life course and the study of structures of the life course, their antecedents, and their consequences is the foundation for a wide variety of research.

At the same time, the incorporation of life course structures into empirical work is typically partial and fragmentary. The vast majority of life course research imports life course concepts, such as timing, into existing theoretical frameworks and substantively focused research (George, 2003). The life course as a holistic experience, a purposeful creation, and social formation is much less visible. While this gap between theory and research is problematic in and of itself, it may be increasingly so. Contemporary scholarship is currently confronting debates over the restructuring of the life course. Such work considers long-term trends in the structure of the life course and contrasts images of "standardization," "individualization," and "new individualization" (Shanahan, 2000). A key thread that links the various arguments is the idea that life courses are changing in multi-dimensional ways that are not easily apprehended in conventional research. At the heart of current discussions is the idea that a life course that is structured in an orderly or normative manner has given way to some new and potentially problematic formations. These involve the decompression of markers of adulthood, increased overlap of social roles, increased reversibility and instability of roles, and the decoupling of role trajectories over the life span (Buchmann, 1989; Kohli, 1986; Shanahan, 2000).

The objective of this volume is to bring together works that extend our understanding of the structure of the life course and demonstrate its relevance to both classic life course issues and contemporary controversies. In particular, the various chapters advance life course theory by addressing issues of conceptualization, empirical characterization of the life course, the importance of contexts and contingencies, and investigation of consequences for life span development and experience. The chapters are broad in scope, involving comparisons both within and across nations and time periods and incorporating both objective and subjective elements. As such, they provide a variety of lenses on questions of the individualization, standardization, and differentiation of the life course and their implications for personal and social well-being. The purpose of this introductory chapter is to situate theory and research on the structure of the life course with respect to both classic life course issues and contemporary controversies and to discuss the various chapters included in this volume as they relate to these broader issues.

CLASSIC ISSUES

The Order and Timing of Social Roles in the Life Span

The concepts of trajectories and transitions are central and by now familiar themes in life course studies (Elder, 1985). They are used as central descriptors of the life course and characterize it in both the short and long term. Trajectories are life course dynamics that take place over an extended period of time. Traditionally they reference time spent in specific social roles. One speaks of trajectories of schooling, of work (i.e., careers), of marriage, and of parenthood. In recent years, the term is used more generally to index a temporal progression. Here, one speaks of trajectories of offending (Sampson & Laub, 1993) or trajectories of mental health (McLeod & Shanahan, 1996). Trajectories are marked at the beginning and at the end by transitions. Transitions are shorter in duration. They index change as people move from one role to another, begin or cease a course of activity, experience a particular state, or stop doing so. Importantly, transitions are always embedded in trajectories that give them discrete form and meaning (Elder, 1985). From this conceptual foundation, much life course inquiry has examined issues such as timing, duration, and cessation (or desistance), as well as ordering, sequencing, and occurrence.

Yet, the recognition of life course dynamics as rooted in trajectories and transitions yields a more complicated theoretical scenario than is often recognized. Current theory suggests conceptual limits in examining trajectories and transitions in and of themselves and instead highlights the interlock of role trajectories as the central descriptor of the life course. As Elder (1985, pp. 32–33) notes,

> Life course dynamics arise in part from the *interplay* of trajectories and transitions, an interdependence played out over time and in relation to others. Interdependence emerges from the socially differentiated life course of individuals, its multiple trajectories, and their synchronization. The interdependence among earnings, childbearing and marriage is also expressed in the concurrence and overlap of transitions along different pathways, such as the transitions to adulthood (education, work, and marriage) and retirement (emphasis added).

From Elder's description, the life course is not found in trajectories and transitions themselves, but in the dynamic, interconnected unfolding of trajectories and transitions over time. When one conceptualizes the interlock of role trajectories over time, one confronts the general structure of the life course. Life courses are structured by virtue of the order *and* timing of multiple social roles over the life span. How we understand lives is predicated upon our ability to effectively delineate how lives unfold in multidimensional ways. The authors in this volume contribute to such discussions in their attempts to consider the life course as a multifaceted process and to understand its sociohistorical origins, its immediate antecedents, and its consequences and implications.

The conceptual shift from trajectories and transitions to the interlock of trajectories and transitions and the structure of the life course is not without costs. In particular, immense heterogeneity is typically apparent when one attempts to characterize the structure of the life course via the order and timing of multiple events in the life span. For example, Rindfuss, Swicegood, and Rosenfeld (1987) analysis of data from the *High School Class of 1972* found that over 1,100 sequences of roles were necessary to describe the transition to adulthood for 6,700 males, while more than 1,800 sequences were required to describe the transitions among the 7,000 females. Likewise, Hogan's (1978) earlier effort to classify the temporal ordering of school, work and marriage in a large sample of American males revealed that only half (54%) experienced a life course characterized by schooling followed by a first job or schooling followed by a first job followed by marriage. A quarter of the sample experienced a variety of atypical patterns and one in five respondents was "non-classifiable." Ironically, many have argued that white males during this period, the middle of the 20th century (cohorts born between 1907 and 1952), were particularly likely to experience an "orderly" or "normative" life course (Moen, 1985; Bradburn, Moen, & Dempster-McClain, 1995). A similar study by Marini (1984b) echoes this argument, finding that a "normative" sequence of school, work, marriage, and parenthood characterized only 40% of females and 37% of males. While such heterogeneity has likely influenced the popularity of studies that incorporate life course concepts rather than conceptualize and model the life course (see discussion in George, 2003), the overall structure of the life course is also an important object of inquiry. Indeed, Rindfuss and colleagues concluded their seminal study with the hope that their research would "not stimulate more work on the determinants of first birth, first marriage, or educational attainment, but rather, encourages a more careful look at the life course as it is actually lived, not as we wish it to be for the sake of order in research" (Rindfuss et al., 1987, p. 799).

Several of the chapters included in this volume reflect Rindfuss and colleagues call and in doing so make important conceptual, methodological, and empirical contributions. Jackson and Berkowitz provide a detailed assessment of the sequencing of roles in the transition to adulthood using data from the National Survey of Families and Households (NSFH). While employing an analytic strategy similar to Hogan's and Marini's earlier work, Jackson and Berkowitz extend earlier their work by incorporating role exits (i.e., unemployment, separation, divorce), as well as role entries, to describe the structure of the life course. Fussell's study of the transition to adulthood in Mexico makes use of a measure of entropy to summarize multifaceted role sequences. Han innovatively combines sequence analytic techniques and log-linear models to examine the multifaceted temporal dynamics of careers in work, marriage, and parenthood. Hillmert adapts goodness of fit statistics to measure the interdependencies between transition events of household formation, employment, marriage, and parenthood across cohorts in Germany through much of the 20th century. Clarke and Wheaton make use of hierarchical linear models to examine the joint impact of role entries and role exits as a possible explanation of the age-conditional effects of neighborhood context on mental health. All of these contributions articulate different approaches by which researchers can tap into the broader structure of the life course and incorporate it into various research endeavors. They showcase the complex ways in which biographies can be understood and measured and highlight some very real gains that accrue by focusing on multiple, interconnected transition events.

Lives in Sociohistorical Contexts

Consistent with the goal of understanding human development within sociohistorical contexts (Elder, 1994), much of the research on transitions and trajectories has sought to map out the broad contours of the life course as a means of understanding how individuals construct different types of life courses within the opportunities and constraints of society and history. Two general strands of research exist. The first considers the impact of historical events in the structuring and restructuring of the life course. Elder's (1999 [1974]) *Children of the Great Depression* stands as the seminal text, delineating the ways in which exposure to the great depression shaped when and how individuals moved into adult social roles, including work, marriage, and parenthood. The insights of this work were many, but of particular importance was the explicit linking of history and biography as a means of showing how life chances were implicitly branded by the sociohistorical origins of individuals. Subsequent studies have considered other significant historical events including both World War II (Sampson & Laub, 1996), the Civil Rights Movement (McAdam, 1989), and the Vietnam War (Hagan, 2001).

A second strand of research is more concerned with broad patterns of historical change. In recognizing that life span development occurs against a backdrop of sociohistorical opportunity and constraint, large-scale cultural and structural change are important determinants of the structure of the life course. Considerable life course research, stretching from Elder (1999 [1974]) to Hogan (1981) to Modell (1989), can be read as efforts to understand the unfolding of the life course against a backdrop of expanding socioeconomic opportunity in the post World War II era (see also Blau & Duncan, 1967). Likewise, the large-scale movement of women into the paid labor force (England & Farkas, 1986; Huber, 1990) reconfigured the landscape of economic opportunities with consequences for both male and female lives. Other notable trends such as the removal of formal, legal impediments to school and work in the post Civil Rights era and the general expansion of higher education (Rosenbaum, 2001) also altered the structure of modern lives.

The significance of sociohistorical context and social change is an important theme in many of the chapters included in this volume. Bruckner and Mayer marshal data from the German Life History Study to examine changes in the timing of key life course transitions across the varying economic contexts of the 20th century. Hillmert further draws upon two distinct subsets of these data, an East German and a West German, to further show the importance of the Second World War and the different political–economic contexts that subsequently emerged in the structuring of the life course from the late 1920s through to the early 1970s. The work of Moen, Sweet, and Swisher sits against a backdrop of increased female labor force participation and the concomitant rise of dual career households in examining its role in shaping retirement decisions and planning in the latter years of the life span. Han in his chapter also focuses on the dual career household by modeling the inter-connected dynamics of husbands and wives careers both within and beyond the workforce. Attention to the socio-historical complexities of work–family linkages also is seen in Hillmert's emphasis on variation in social welfare policies between East and West Germany and how this enhanced the ability of women to incorporate paid employment into their adult lives.

Expansion and transformation of higher education and its implications is also prevalent in many of the contributions. In short, the decades since the 1960s have seen tremendous growth in higher education with both an expansion of the number of institutions available (Rosenbaum, 2001) and the proportion of the population involved (Pallas, 2002). While implications of this for the structure of the life course are many, its impact on the school-to-work transition (Booth, Crouter, & Shanahan, 1999; Kerckhoff, 2002; Rosenbaum, 2001) and the timing of marriage and parenthood, particularly among females (Goldstein & Kenney, 2001; Marini, 1978, 1984c) is particularly significant. In her contribution to this volume, Fussell maps out the changing nature of the life course in Mexico against a backdrop of massive educational expansion. The expansion of higher education and its implications is also central to Mortimer, Oesterle, and Kruger's examination of the ways in which institutional context shapes the relationship between post-secondary education and the timing of parenthood in emphasis in Germany and the United States. At a more meso level, Person, Rosenbaum, and Diel-Amen use a multimethod, multidimensional approach to examine the role of institutional context for understanding how higher education impacts the life course. At a microlevel, Shaw usefully delineates the structure of an academic career and discusses its institutional dimensions, as well as its consequences of subjective well-being.

Social Differentiation and Inequality

While cultures may construct the life course differently by regulating the pace and process of human development in any given society, it is also the case that economic realities and social constraints may alter cultural connections and cause certain individuals to depart from the social timetables (Furstenberg, 2003). It is here that difficulties in conforming to social scripts of the life course may reveal cultural and structural contradictions. Such contradictions may reflect the disjuncture between norms about the life course and the ability to actualize such norms in everyday life. This disjuncture, as well as speculation about subcultural values and their importance, provides the foundation for a third key issue in life course research, the link between the structure of the life course and broader patterns of stratification, inequality, and differentiation. Variation by gender, ethnicity and race, and social class has been the focus on considerable attention (see reviews in Furstenberg, 2003; Fussel & Furstenberg, 2004; Shanahan, 2000).

Still, it is important to recognize that the majority of work focuses on specific types of roles or specific role domains. For example, there is a wealth of research examining the stratification in educational attainment (Mare, 1981), employment (Browne, 1997), marriage (Espenshade, 1985), and parenthood (Ruggles, 1994). Further work considers relationships between specific types of role trajectories, including the school to work transition (Kerckhoff, 2002; Mortimer, 2003; Schneider & Stevenson, 1999), the relationship between occupational attainment and marriage (Drobnic, Blossfeld, & Rohwer, 1999; Oppenheimer, Kalmijn, & Lim, 1997), the relationship between marriage and parenting (Bumpass & Lu, 2000; Manning, 1995), as well as other role dynamics. All this work focuses on a specific role or pair of roles with an assumption, often quite arbitrary, of causal order. While there is little question that the study of roles and role pairings has contributed greatly to our understanding of life course dynamics, such work downplays the fact that interdependencies between roles, both within and across time, is almost always multidimensional (Elder, 1985). Understandings of the life course in general, as well as the antecedents and consequences of discrete social roles, are thus enhanced by a consideration of more general matrices of roles, what might be called role configurations, their timing in the life span, and the pathways through life that their dynamic unfolding reveals (Macmillan & Eliason, 2003). The key implication of this is that stratification and social differentiation in the life course occurs not just in terms of discrete statuses or roles, but in the role configurations and pathways that make up the general structure of the life course.

An equally important consideration is that understanding the structuring of the life course often requires simultaneous consideration of multiple dimensions of stratification. Gender and race do not operate independently of one another in the life course (Anderson & Collins, 2001) and social class and age merely adds a further element of contingency. For example, Goldstein and Kenney (2001) find that race differences in marriage among women diverge at both the ends of the class perspective. African-American females who did not graduate from college were particularly unlikely to marry, while white females with college degrees had particularly high odds of marrying and this had increased over time. Likewise, Oppenheimer et al. (1997) showed that race differences in the timing of marriage among males were relatively small in the general population but grew substantially when educational attainment and career success were taken into account. Different dimensions of social stratification interact in important ways in the structuring of the life course.

Both life course scholarship and theory and research on stratification and inequality clearly benefit from consideration of both the multidimensionality of roles over the life span and the multifaceted nature of stratification. Several of the contributors to this volume reflect this. Fussell's study of the structuring of the life course in Mexico, for example, is careful to examine variation by cohort, gender, and urban–rural residence, noting that these factors combine to produce unique social locations in light of the transformation to Mexican society in the latter half of the 20th century. Brayboy Jackson and Berkowitz directly consider the intersection of racioethnicity and gender in the American context. Their research highlights important variation in the sequencing of social roles that would likely be obscured when examining broader categories. Clarke and Wheaton demonstrate the interaction between neighborhood poverty, age, and nexuses of role entries and role exits characterizing the life course in shaping mental health. This work is unique in showcasing the importance of both "time and space," neighborhood stratification and the life course, in the production of psychological well-being. While the joint study of contingencies in life course dynamics and contingencies in social statuses is only beginning, the contributions to this volume suggest great promise and sharpened insights that accompany such work.

The Normative Basis of the Life Course

Studies of differentiation in the structure of the life course often reference a general thesis of, or debate over, the cultural basis of the life course. At the heart of this issue is the question of whether there exist clear "norms" governing the order and timing of events and the general stability of life course structures across time and place (Marini, 1984a). The claim to a normative structure of the life course has its basis in two facts. First, there is clear statistical regularity in the patterning of social roles in the life course (Hogan, 1978). Accompanying this is evidence of specific values and beliefs regarding the appropriate order and timing of roles in the life span. Most of the evidence pertains to individual views about the "best" or "ideal" age at which to leave school, work, marry, leave the family home, or have children. Neugarten, Moore and Lowe (1965) seminal contribution noted that

> Expectations regarding age-appropriate behavior form an elaborated and pervasive system of norms governing behavior and interaction, a network of expectations that is imbedded throughout the cultural fabric of adult life. There exists what might be called a prescriptive timetable for the ordering of major life events: a time in the life span when men and women are expected to marry, a time to raise children, a time to retire. This normative pattern is adhered to, more or less consistently, by most persons in society (p. 711).

Buttressing this view is considerable evidence concerning the ideal age to marry or have children. Neugarten et al. (1965) themselves showed that large segments of the population, typically 80% and higher, concurred that different life events should occur within specific, well-defined age bands (e.g., 19–24 for the "best age for a woman to marry"). Likewise, public opinion polls consistently reveal that American's have relatively concrete ideas about the optimal age to start a family (Modell, 1980; Neugarten & Datan, 1973; Rindfuss & Bumpass, 1976) and that the perceived "best" age clusters around the modal ages at which these events actually occur. While evidence on sequencing norms is less apparent, there is certainly anecdotal evidence that some roles are seen as inappropriate to particular stages of the life course (i.e., "teen parents") and that others occur out of appropriate order (i.e., "unmarried parents") (see discussion in George, 1993).

Against this backdrop, Marini (1984a) offers an elaborate critique. Specifically, she argues that there is insufficient evidence that social norms explain behavior during the transition to adulthood or in later life, that the use of social norms has been inconsistent, used primarily to explain role transitions that do not convey material status (i.e., marriage and parenthood), and that the use of "life course norms" has been misleading because of the way the concept has been operationalized. Importantly, much of the evidence used to support normative arguments has been found in studies of the structure of the life course, including those by Hogan (1978) and Rindfuss et al. (1987), as well as Marini's own research (Marini 1984b). Yet, as Marini notes, such work is not evidence of norms and instead may reference a wide array of structural and institutional processes that produce population homogeneity in life course patterns. Equally important, we are only beginning to fully theorize the nature of life course norms and produce research that actually assesses such things. Of particular note, Settersten (1998; see also Settersten & Mayer, 1997) argues that "age norms" reference "statistical age norms" which describe statistical regularities in the timing of life course transitions, "optimal age norms" that describe collective notions about the "best," "ideal," or "preferred" ages at which to make particular transitions, as well as "prescriptive" or "proscriptive" age norms that describe when certain transitions should or should not occur. Ultimately, the existence and significance of norms in the life course remains an important aspect of life course research and are intimately connected to research on the structure of the life course and key debates over social change in contemporary societies (Buchmann, 1989; Settersten, 1998; Shanahan, 2000).

The contributors to this volume extend discussions of life course norms and the general cultural foundation of the life course in both theoretical and empirical ways. Bruckner and Mayer begin such efforts by drawing attention to the large scale "value changes" that characterized the 1960s and 1970s. Characteristics of youth and student rebellions, such changes are seen as organizing or reorganizing the way in which individuals envision and construct their private lives. Such value change impacts the life course by fostering delays in marriage and parenthood and the rise of non-marital unions, divorce, and remarriage. Concomitant values of autonomy and self-realization also sped up departures from parental homes, independent of the acquisition of traditional markers of adulthood and produced new structures of the life course. Social norms may also produce consistency in life course structures across different sociohistorical contexts. For example, Fussell's examination of the structure of the life course in a changing Mexican society shows remarkable stability across cohorts. Ultimately, she concludes that structural changes in education and employment over-ride and perhaps even enhance gender norms, which tightly structure the prevalence and timing of marriage and childbearing and the broader life course.

Mortimer, Oesterle and Kruger further differentiate an age-norm hypothesis with institutional determinisms in examining cross-national differences in the ordering of school completion and childbearing among women. Emphasizing a "lag" in which changes in cultural norms trail institutional change, their work highlights the importance of institutional environment (which in turn is connected to national context) for establishing the efficacy of norms for structuring the life course. In the context of unfolding academic careers, Shaw's theorizing suggests that changes in both individual and institutional position produce different sets of expectations for work. Such expectations can be viewed as "normative" in the sense that they characterize individuals at a give career state (i.e., are statistically normative) and that they are proscriptive of action. They are grounded in the rank-specific cultural environment and are adopted by individual academics. This perspective is complemented by Drentea's detailed consideration of changes in the nature of activities across the life span. For Drentea, a comparison of variation in activities across the age span reveals the importance of "life stage" and the specific, sequenced social roles that people occupy. Such roles (and role configurations) are imbued with social norms that shape expectations of institutionally based activities and how these vary over the life span. With respect to the latter stages of the life span, Moen and colleagues demonstrate that individuals in organizations with older work forces are more apt to alter their retirement plans. They suggest that such organizations are likely to foster (sub)cultural values that both value older workers and place issues of retirement "in the air." In either case, organizations serve as a locus of norms that serve to structure both the planning and timing of retirement. As in most life course research, the works included in this volume tend to infer or reflect social norms than actually measure them. Yet, the corpus of work provides some important theoretical observations and empirical evidence of a cultural foundation of the life course.

CONTEMPORARY CONTROVERSIES

Characterizations of the life course and studies of sociohistorical context, social differentiation, and life course norms continue to be important in life course research. Yet, their significance and connection to studies of the structure of the life course has renewed importance with increasing debate over the changing nature of the life course in the latter half of the 20th century. Bruckner and Mayer's contribution to this volume provides a conceptual road map to these debates, delineating the key concepts and their theoretical and empirical implications. As their chapter is both comprehensive and thought-provoking, I provide only a cursory discussion of the salient issues and a brief description of the contributions of the various works in this volume.

Standardization of the Life Course

Several commentators argue that the life course became increasingly standardized through the 19th and 20th centuries. Tied to general patterns of modernization (Hogan, 1981), variation in life course patterns were increasingly eroded through the expansion of state activities and the ever-increasing institutionalization of social life (Meyer, 1986). Kohli (1986), for example, argues that the organization of public services and the age structuring of employment through law rendered the life course more orderly. More generally, Buchmann (1989) emphasizes the general rationalization of economy and polity and the ties between school, work, marriage, and parenthood that established institutional timetables for various life course transitions. The resulting standardization of the life course is shown through decreased variation in the order and timing of key markers of adulthood including marriage and parenthood, leaving the family home and establishing one's own residence (Modell, Furstenberg, & Hershberg, 1976), and the transition from school to work (Hogan, 1978). Shanahan (2000) argues that a key aspect of standardization is the general "compression" of transition markers.

The idea that the life course is increasingly standardized has important implications. From a theoretical standpoint, it draws our attention to macro-level social change and the role of institutions in the structuring of the life course. Importantly, large-scale patterns of social change are not simply sources of opportunity, as often proposed, but are also sources of constraint. Institutional environments may increasingly impinge upon the efforts of agents to construct their own lives, echoing Weber's (1952) long-standing imagery of the "iron cage." Empirically, a standardization thesis suggests homogeneity in both the overall structure of the life course, as well as the implications that it has for individual experience. Both dimensions are important in illuminating both the nature of the life course in modern society and the meaning it has for social and psychological well-being. Several chapters in this volume contribute to this understanding both directly and indirectly.

At the outset, a blunt hypothesis that can be derived from standardization arguments is that the life course should be reasonably homogeneous across time, across cultures, and across social groups. Bruckner and Mayer analyses of several cohorts from Germany suggest that the institutional environment has a pervasive and significant impact on the structure of the life course. They further conclude that the life courses of males and females increasingly resemble one another over time. Such continuity is echoed in Hillmert's and Fussell's historical comparisons. Homogeneity across social groups is equally prominent. Jackson and Berkowitz detailed analyses of American data show that a small number of role sequences typically characterize the vast majority of the population and that sequences are often quite similar across race–sex groups despite very different social locations and opportunity structures. Consistent with this, Fussell's entropy index reveals a tight clustering of values over much of the age span for males and females regardless of urban–rural residence. Han similarly shows that a small number of sequences usefully summarize the patterning of employment, marriage, and parenthood. Drentea's research reveals that the patterning of social roles over the age span plays a key role in explaining age-differentiation in activities both within and beyond the workplace. Likewise, Clarke and Wheaton demonstrate a distinct age patterning of role entries and role exits, revealing how explicit age-span differentiation characteristics of contemporary U.S. society. Finally, Shaw's chapter suggests that the institutional context provides a general framework of academic careers that is productive of particular experiences and particular psychological orientations that ebb and flow over the life span. In general, these chapters contribute to our understanding of the standardization of the life course by illuminating the role of structural, cultural, and institutional factors in structuring the life course and revealing the general pattern of social roles and life course experiences in modern society.

Individualization and 'New' Individualization

In commenting on the influence of modernization on the life course, a variety of theorists have suggested either a complementary or competing hypothesis of increased "individualization." Giddens in a variety of projects (Giddens, 1990, 1991, 2003) argues that modernization has increasing loosened the social and cultural ties that embedded individuals in familial and other local contexts. A key consequence of this is that domains of agency increasingly expand and the life course increasingly becomes a particular or deliberate project. While individuals are still subject to cultural and institutional constraints, such constraints are more varied, more diffuse, and less universal. Within a context of less constraint and increased choice, the life course becomes a more variable and less uniform phenomenon (Shanahan, 2000).

Although the notion of less constraint or greater choice is powerful imagery for thinking about modern lives, others suggest a less optimistic picture. In particular, a variety of scholars have drawn attention to the global transformation of modern economy and its implications for the structuring of the life course. For example, Booth et al. (1999) *Transitions to adulthood in a changing economy* includes the provocative sub-title "no work, no family, no future?" A key feature of their argument is that changing economic conditions make it harder to transition to adulthood. The transition to adulthood through much of the 20th century was organized around movement into work roles (Hogan, 1978, 1980, 1981) and the ways in which these provided an economic base for family formation. The transformation of work, particularly following the 1973 Oil Crisis, increasingly involves the prevalence of 'bad' rather than 'good' jobs with poorer career prospects, decreased opportunity for mobility, lower wages and fewer benefits (Booth et al., 1999; Kalleberg, Reskin, & Hudson, 2000; Smith, 2001). The new economic context also exacerbates problems in the life course progress through the increased frequency of recessionary conditions that produce spikes in unemployment rates (Macmillan, 2005). Equally important are implications for later life and the ways in which the new economy undermines the stability and viability of occupational careers through mid-life (Buchmann, 1989) and even the retirement process in the later life span (Farkas & O'Rand, 1998).

Whether a consequence of changing cultural conventions or changing economic prospects, increased individualization is typically located in increased variability in the sequencing and overlap of life course transitions (Shanahan, 2000). In other words, the lock step sequence of school to work to marriage to parenthood that would have been regarded as the "normative" or "orderly" sequence through much of the 20th century has given way to multiple, less orderly patterns. There are four relevant dimensions. First, there is a general decompression of markers of adulthood. This typically involves increased time between life course transitions and appears to be particularly acute in the context of family roles (Stevens, 1990). Second, individuals increasingly occupy multiple social roles simultaneously, particularly roles those that were previously seen as characteristic of disparate life stages. Schneider and Stevenson (1999), for example, describe the increased overlap between school and (full-time) work (see also Mortimer, 2003; Rosenbaum, 2001). Third, there is increased reversibility and instability of roles. While marriage, divorce, and remarriage are commonly referenced (Espenshade, 1985; Popenoe, 1993), others note the increasing phenomenon of "returning" to school following time in full-time work (Pallas, 2002). Likewise, Goldscheider and Goldscheider (1999) carefully examine "renesting" in American families as adult children return after leaving the family home.

A final dimension is the decoupling of social roles that may reflect some combination of certain roles being in decline, roles being delayed, or a general lack of synchronization or 'scrambling' of the roles in the life course. Popenoe (1993), for example, suggests that marriage is in significant decline in that fewer people are choosing to get married, people are spending increasingly smaller amount of their life courses in marital relationships, and that marriages are increasingly unstable unions. Others draw attention to unclear connections between schooling and work (Kerckhoff, 2002; Mortimer, 2003; Schneider & Stevenson, 1999). In general, the increased individualization of the life course is seen in the decoupling of social roles (Shanahan, 2000), increased deviations away from social norms (Buchmann, 1989), and increasing disorder in the life course (Hogan, 1978; Rindfuss et al., 1987).

The contributions to this volume tackle the individualization thesis in many different ways. Bruckner and Mayer begin with a thorough conceptual overview that differentiates the terms that scholars have used to characterize the (perceived) changes in life course structure over the past 40 years. They further bring longitudinal data from a number of cohorts of Germans to examine stability and change in the unfolding life course. Hillmert extends the scope of their work by further considering individualization in the context of the founding and development of East and West Germany and the eventual reunification in the 1980s. Jackson and Berkowitz focus on the American context and consider similarity and difference in patterns of role entries and role exits across racioethnic–gender groupings. Fussell provides a further point of observation by examining whether the structure of the life course is substantively different for more recent cohorts in Mexico in light of expansion of education and increasing economic crisis during the latter decades of the 20th century. Mortimer and colleagues explicitly consider both institutional environment and economic instability as increasing variability in pathways into adulthood. The importance of institutional context is further highlighted by Person and colleagues who show that educational environment plays a key role in shaping the link between individual decision-making and educational outcomes. In this regard, they highlight 2-year colleges as institutions that are implicated in the individualization of the life course. Considering life course dynamics at mid-life, Han's emphasis on the work–family interface reveals the importance of dynamics into and out of work and how these are connected to patterns of marriage, marital dissolution, and parenthood. Moen, Sweet, and Swisher focus on the latter stage of the life span and conclude that the coupling of work and retirement is increasingly complicated by the rise of dual career families and the lack of taken-for-granted scripts about how and when retirement should occur in the life course. The work of both Han and Moen and colleagues extends our understanding of individualization by focusing attention beyond the transition to adult which has to date been the main focus of inquiry.

Clarke and Wheaton's research is firmly situated in the increasing 'geography' of disadvantage which is further tied to large scale transformations of economy (Massey 1998; Massey & Eggers, 1990; Wilson, 1987) and its links to increases in role entries and role exits. That these characterize the structure of the life course in contemporary society and are sources of personal stress suggests that increased variability in the life course may be intimately connected to diminished psychological well-being (cf. Arnett, 2000). Drentea's consideration of activities over the life span rests against a backdrop of unfolding social roles that, as Shanahan notes, may be increasingly overlapping and temporally undifferentiated. Finally, Shaw concludes his essay with a plea for accommodation, an effort to balance the various life roles that sit alongside an academic career and may be sources of both strain and resilience in the unfolding life course.

CONCLUSION

It is hoped that this chapter has served two purposes. First, it has attempted to situate theoretical and empirical consideration of the structure of the life course within the broader landscape of life course research. The goal here is to describe some central themes that organize the field, as well as indicate their importance for a wide variety of social science questions. Second, it has attempted to situate each of the contributions to this volume in relation to the different issues they address. It is testament to the quality of the works included that they typically make multiple contributions and address a wide variety of issues. At the same time, they all share an important attribute. They each consider the life course, theoretically or methodologically, in a dynamic way. It is this attribute that unites the works and set them apart from much that is done in the field. As some have questioned the utility of broad conceptualizations and operationalizations of the life course for social science inquiry (see for e.g., George, 2003), it seems important to recognize the potentialities of holistic attempts to conceptualize and study the life course and the limitations of fragmentary incorporation of concepts and measures into extant theory and substantive research. The hallmark of life course inquiry is its attention to interdependence (Elder, 1985). Human lives are characterized by implicit and important links between earlier and later events with the former conditioning the meaning and implications of the latter (Wheaton, 1990). They are also characterized by the interlock of social roles over time which unfolds in unique and interesting ways indicative of pathways over the life span. These both characterize and constitute stages of the life course that have social meaning. They conform (or do not conform) to cultural scripts with wide recognition and ultimately determine the social aspect of aging. The various contributions to this volume, while adopting a varying theoretical perspectives and diverse methodological approaches, all contribute to this multidimensional, dynamic understanding of the life course.

A second feature of the various contributions is that they speak in very direct ways to debates over the structure of the life course and questions of standardization, individualization, and differentiation. There is little question that the modern life course is transforming in significant ways. Yet, whether this involves increasing standardization, individualization, or long-standing social differentiation is open to question. In the end, the contributions present convincing evidence that the contemporary life course is highly structured, but also variable in the context of sociohistorical conditions and social location. In a variety of ways, the chapters highlight the importance of continued inquiry into the social foundations of the life course and the important ties between life course and extra-life course structures. Equally important, the implications of these large-scale changes are potentially great and are in need of further study. Here, the contributors to this volume have done an admirable job laying a foundation for further inquiry. They have answered some questions and raised others. Probably most important, they have laid out theoretical puzzles and offered methodological suggestions for future research. It is hoped that this volume and the contributions within will foster further work on the structure of the life course and its implications for social and individual development.

ACKNOWLEDGMENTS

This volume was completed with support from the Life Course Center at the University of Minnesota and from the College of Liberal Arts. I also thank Jeylan Mortimer, Phyllis Moen, Annette Nierobisz, and Jennifer Schultz for discussions on various aspects of the work.

REFERENCES

Anderson, M. L., & Collins, P. H. (2001). *Race, class and gender: An anthology*. Belmont: Wadsworth.

Arnett, J. J. (2000). Emerging adulthood: A theory of development from the late teens through the twenties. *American Psychologist*, *55*, 469–480.

Blau, P. M., & Duncan, O. D. (1967). *The American occupational structure*. New York: Wiley.

Booth, A., Crouter, A. C., & Shanahan, M. J. (1999). *Transitions to adulthood in a changing economy: No work, no family, no future?* Westport: Praeger.

Bradburn, E. M., Moen, P., & Dempster-McClain, D. (1995). Women's return to school following the transition to motherhood. *Social Forces*, *73*, 1517–1551.

Browne, I. (1997). Explaining the black–white gap in labor force participation among women heading households. *American Sociological Review*, *62*, 236–252.

Buchmann, M. (1989). *The script of life in modern society: Entry into adulthood in a changing world*. Chicago: University of Chicago Press.

Bumpass, L., & Lu, H. (2000). Trends in cohabitation and implications for children's family contexts in the United States. *Population Studies*, *54*, 29–41.

Drobnic, S., Blossfeld, H., & Rohwer, G. (1999). Dynamics of women's employment patterns over the family life course: A comparison of the United States and Germany. *Journal of Marriage and the Family*, *61*, 133–146.

Elder, G. H. (1994). Time, human agency, and social change: Perspectives on the life course. *Social Psychology Quarterly*, *57*, 4–15.

Elder, G. H. (1999 [1974]). *Children of the great depression: Social change in life experience*. Chicago: The University of Chicago Press.

Elder, G. H., Jr. (1985). *Life course dynamics: Trajectories and transitions, 1968–1980*. Ithaca: Cornell University Press.

Elder, G. H., Johnson, M. K., & Crosnoe, R. (2003). The emergence and development of life course theory. In: J. T. Mortimer & M. J. Shanahan (Eds), *Handbook of the life course* (pp. 3–19). New York: Plenum.

England, P., & Farkas, G. (1986). *Households, employment, and gender: A social, economic, and demographic view*. New York: Aldine Pub. Co.

Espenshade, T. J. (1985). Marriage trends in America: Estimates, implications, and underlying causes. *Population and Development Review*, *11*, 193–245.

Farkas, J., & O'Rand, A. (1998). The pension mix for women in middle and late life: The changing employment relationships. *Social Forces*, *76*, 1007–1032.

Furstenberg, F. F. (2003). The future of the life course: On a path toward inequality. In: J. T. Mortimer & M. J. Shanahan (Eds), *Handbook of the life course* (pp. 661–670). New York: Plenum.

Fussel, E., & Furstenberg, F. F., Jr. (2004). Race, nativity, and gender differences in the transition to adulthood in the 20th century. In: R. A. Settersten, F. F. Furstenberg & R. Rumbault (Eds), *On the frontier of adulthood: Theory, research, and public policy* (pp. 29–75). Chicago: University of Chicago Press.

George, L. K. (1993). Sociological perspectives on life transitions. *Annual Review of Sociology*, *19*, 353–373.

George, L. K. (2003). The future of the life course: Late modernity and life course risks. In: J. T. Mortimer & M. J. Shanahan (Eds), *Handbook of the life course* (pp. 671–680). New York: Plenum.

Giddens, A. (1990). *The consequences of modernity*. Stanford: Stanford University Press.

Giddens, A. (1991). *Modernity and self-identity: Self and society in the late modern age*. Stanford: Stanford University Press.

Giddens, A. (2003). *Runaway world: How globalization is reshaping our lives*. New York: Routledge.

Goldscheider, F. K., & Goldscheider, C. (1999). *The changing transition to adulthood: Leaving and returning home*. Thousand Oaks: Sage.

Goldstein, J. R., & Kenney, C. T. (2001). Marriage delayed or marriage forgone? New cohort forecasts of first marriage for U.S. women. *American Sociological Review*, *66*, 506–519.

Hagan, J. (2001). *Northern passage: American Vietnam War resisters in Canada*. Cambridge: Harvard University Press.

Hogan, D. P. (1978). The variable order of events in the life course. *American Sociological Review*, *43*, 573–586.

Hogan, D. P. (1980). The transition to adulthood as a career contingency. *American Sociological Review*, *45*, 261–276.

Hogan, D. P. (1981). *Transitions and social change: The early lives of American men*. New York: Academic Press.

Huber, J. (1990). Macro-micro links in gender stratification. *American Sociological Review*, *55*, 1–10.

Kalleberg, A., Reskin, B. F., & Hudson, K. (2000). Bad jobs in America: Standard and nonstandard employment relations and job quality in the United States. *American Sociological Review*, *65*, 256–278.

Kerckhoff, A. C. (2002). The transition from school to work. In: J. Mortimer & R. W. Larson (Eds), *The changing adolescent experience. Societal trends and the transition to adulthood* (pp. 52–87). New York: Cambridge University Press.

Kohli, M. (1986). The world we forgot: A historical review of the life course. In: V. Marshall (Ed.), *Later life* (pp. 271–332). Beverley Hills: Sage.

Kohli, M. (1987). Retirement and the moral economy: An historical interpretation of the German case. *Journal of Aging Studies*, *1*, 125–144.

Macmillan, R. (2005). Transition to adulthood and the 'new' economy: Economic recessions and the structure of the life course. A paper presented at the annual meeting of the population association of America, Philadelphia, PA.

Macmillan, R., & Eliason, S. (2003). Characterizing the life course as role configurations and pathways: A latent structure approach. In: J. T. Mortimer & M. J. Shanahan (Eds), *Handbook of the life course* (pp. 529–554). New York: Plenum.

Manning, W. D. (1995). Cohabitation, marriage, and entry into motherhood. *Journal of Marriage and the Family*, *57*, 191–200.

Mare, R. D. (1981). Trends in schooling: Demography, performance, and organization. *The Annals of the American Academy of Political and Social Science*, *453*, 96–122.

Marini, M. M. (1978). The transition to adulthood: Sex differences in educational attainment and age at marriage. *American Sociological Review*, *43*, 483–507.

Marini, M. M. (1984a). Age and sequencing norms in the transition to adulthood. *Social Forces*, *63*, 229–244.

Marini, M. M. (1984b). The order of events in the transition to adulthood. *Sociology of Education*, *57*, 63–84.

Marini, M. M. (1984c). Women's educational attainment and the timing of entry into parenthood. *American Sociological Review*, *49*, 491–511.

Massey, D. (1998). Back to the future: The rediscovery of neighborhood context. *Contemporary Sociology*, *27*, 570–572.

Massey, D., & Eggers, M. (1990). The ecology of inequality: Minorities and the concentration of poverty, 1970–1980. *American Journal of Sociology*, *95*, 1153–1188.

McAdam, D. (1989). The biographical consequences of activism. *American Sociological Review*, *54*, 744–760.

McLeod, J. D., & Shanahan, M. J. (1996). Trajectories of poverty and children's mental health. *Journal of Health and Social Behavior*, *37*, 207–220.

Meyer, J. (1986). The self and the life course: Institutionalization and its effects. In: A. Sorenson, F. Weinert & L. Sherrod (Eds), *Human development and the life course: Multidisciplinary perspectives* (pp. 199–216). Hillsdale: Lawrence Erlbaum Associates.

Modell, J. (1980). Normative aspects of American marriage timing since World War II. *Journal of Family History*, *5*, 210–234.

Modell, J. (1989). *Into one's own: From youth to adulthood in the United States, 1920–1975*. Berkeley: University of California Press.

Modell, J., Furstenberg, F. F., Jr., & Hershberg, T. (1976). Social change and transitions to adulthood in historical perspective. *Journal of Family History*, *1*(1), 7–32.

Moen, P. (1985). Continuities and discontinuities in women's labor force activity. In: G. H. Elder Jr. (Ed.), *Life course dynamics: Trajectories and transitions, 1968–1980* (pp. 113–155). Ithaca: Cornell University Press.

Mortimer, J. T. (2003). *Working and growing up in America*. Cambridge: Harvard University Press.

Neugarten, B. L., & Datan, N. (1973). Sociological perspectives on the life cycle. In: P. B. Baltes & K. W. Schaie (Eds), *Life span developmental psychology: Personality and socialization* (pp. 53–69). New York: Academic Press.

Neugarten, B. L., Moore, J. W., & Lowe, J. C. (1965). Age norms, age constraints, and adult socialization. *American Journal of Sociology*, *70*, 710–717.

Oppenheimer, V., Kalmijn, M., & Lim, N. (1997). Men's career development and marriage timing during a period of rising inequality. *Demography*, *34*, 311–330.

Pallas, A. (2002). Educational participation across the life course: Do the rich get richer? *Advances in Life Course Research*, *7*, 327–354.

Popenoe, D. (1993). American family decline, 1960–1990: A review and appraisal. *Journal of Marriage and Family*, *53*, 527–542.

Riley, M. W., Kahn, R., & Foner, A. (1994). *Age and structural lag: Society's failure to provide meaningful opportunities in work, family, and leisure*. New York: Wiley.

Rindfuss, R. R., & Bumpass, L. L. (1976). How old is too old? Age and the sociology of fertility. *Family Planning Perspectives*, *8*, 226–230.

Rindfuss, R. R., Swicegood, C. G., & Rosenfeld, R. A. (1987). Disorder in the life course: How common and does it matter? *American Sociological Review*, *52*, 785–801.

Rosenbaum, J. (2001). *Beyond college for all*. New York: Russell Sage.

Ruggles, S. (1994). The origins of African-American family structure. *American Sociological Review*, *59*, 136–151.

Sampson, R. J., & Laub, J. H. (1996). Socioeconomic achievement in the life course of disadvantaged men: Military service as a turning point, circa 1940–1965. *American Sociological Review*, *61*, 347–367.

Sampson, R. J., & Laub, J. L. (1993). *Crime in the making: Pathways and turning points through life*. Cambridge: Harvard University Press.

Schneider, B. L., & Stevenson, D. (1999). *The ambitious generation: America's teenagers, motivated but directionless*. New Haven: Yale University Press.

Settersten, R. A., Jr. (1998). A time to leave home and a time never to return? Age constraints on the living arrangements of young adults. *Social Forces*, *76*, 1373–1400.

Settersten, R. A., Jr., & Mayer, K. U. (1997). The measurement of age, age structuring, and the life course. *Annual Review of Sociology*, *23*, 233–261.

Shanahan, M. J. (2000). Pathways to adulthood in changing societies: Variability and mechanisms in life course perspective. *Annual Review of Sociology*, *26*, 667–692.

Smith, V. (2001). *Crossing the great divide: Worker risk and opportunity in the new economy*. Ithaca: ILR Press.

Stevens, D. A. (1990). New evidence on the timing of early life course transitions: The United States 1900–1980. *Journal of Family History*, *15*, 163–178.

Weber, M. (1952). *The Protestant ethic and the spirit of capitalism*. New York: Scribner.

Wheaton, B. (1990). Life transitions, role histories, and mental health. *American Sociological Review*, *55*, 209–223.

Wilson, W. J. (1987). *The truly disadvantaged: The innercity, the underclass, and public policy*. Chicago: University of Chicago Press.

PART II: CONCEPTS AND CHARACTERISTICS

DE-STANDARDIZATION OF THE LIFE COURSE: WHAT IT MIGHT MEAN? AND IF IT MEANS ANYTHING, WHETHER IT ACTUALLY TOOK PLACE?

Hannah Brückner and Karl Ulrich Mayer

ABSTRACT

We explore both conceptually and empirically whether and how precise meanings and measures can be attached to recent ideas about the transformation of the life course. With data from the German Life History Study (GLHS), we assess social change in the transition to adulthood for birth cohorts born between 1921 and 1971, focusing on the de-standardization hypothesis. While we see increasing de-coupling of events in the connections between the school–training–work nexus and family formation, the institutional environment continues to structure the school–training–work nexus and not much change was seen in the way in which cohort members undergo these transitions. On the contrary, there is actually a homogenization as women's and men's life courses converge in terms of education and labor force participation. It is the family formation nexus that shows the most pronounced changes. This is also the realm in which gender differences persist across cohorts. While we find strong

The Structure of the Life Course: Standardized? Individualized? Differentiated?
Advances in Life Course Research, Volume 9, 27–53

ISSN: 1040-2608/doi:10.1016/S1040-2608(04)09002-1

evidence for period effects that produce inter-cohort differences in life course patterns, taken as a whole our indicators do not point to a general process of a de-standardization of the life course.

That lives have become less predictable, less collectively determined, less stable, less orderly, more flexible, and more individualized has become one of the most commonly accepted perceptions of advanced societies (Held, 1986; Buchmann, 1989; Myles, 1993). Private lives and family forms are said to have become pluralized, and working lives more unstable, including increased firm and occupational mobility (Beck, 1986; Heinz, 2003). Education and training are perceived to have become more extended and intermittent. More generally, the traditional sequence of life stages into education, work and retirement is said to have given way to a de-standardized life course where people go back to education after periods of work, take sabbaticals, change occupations in midlife and combine work and leisure in prolonged transitions to final retirement (Riley, Kahn, & Foner, 1994). Demographers and family sociologists see the emergence of patterns of sequential promiscuity and the rise of an "autistic society" (Hoffmann-Nowotny, 1980) where singles are only loosely tied to families. Beck (1999) claims that more than half of the (German) workforce is subject to non-standardized labor contracts and predicts that labor markets in advanced societies will soon resemble the widespread marginality of third world countries. Likewise, Sennett (2000) sees the increasing flexibility of both private and working lives as the hallmark of globalized capitalism. De-standardized life courses, then, are fundamental tenets of both theories of Post-Fordism and theories of post-industrialism.

In this paper, we first examine hypotheses concerning de-standardized life courses in different, recent historical contexts. We then explore both conceptually and empirically whether and how more precise meanings and measures can be attached to ideas of the transformation of the life course. Third, we use data from the West German part of the GLHS covering cohorts born between 1919 and 1971 to assess some of the implicit claims in the de-standardization thesis.

THREE HISTORICAL CONTEXTS OF THE DEBATE: MODERNIZATION, POST-INDUSTRIALISM AND GLOBALIZATION

In the 1970s and early 1980s, the social and temporal organization of human lives was developed as a topic within a distinctly historical framework

(Mayer, 2004). The question implicitly or explicitly put forward was a very broad one: What distinguishes "modern" life courses from more "traditional" ones? Some of the answers given were that life courses had become "institutionalized" as part and parcel of the life discipline imposed by industrial work (Hareven, 1981, 1986; Anderson, 1985; Kohli, 1985) or that lives had become more predictable because cycles of poverty (Rowntree, 1914; Myles, 1993) – which characterized manual worker's lives until the last mid-century – had gradually vanished with rising standards of living and increasing protection by the welfare state (Mayer & Müller, 1986; Mayer & Schoepflin, 1989). The long-term historical cohort comparison by Modell, Furstenberg, and Hershberg (1976) showed that events making up the transitions to adulthood had become more universal, were more temporally connected, and showed less age variation. Conceptually, these developments were framed either in terms of a dichotomy between "traditional" and "modern" or in terms of a linear and fairly universal historical trend. In this sense, it is not a mistake to locate this line of thinking within the more general framework of modernization theory (Zapf, 1991; Schelkle, Krauth, Kohli, & Elwert, 2000). Many different processes were thought to contribute to more institutionalized, more predictable and more "standardized" lives (Mayer & Müller, 1986). First, the expansion of secondary and tertiary education and training created career paths within and between educational institutions. It also moved young adults to labor market entry positions at different levels, thereby minimizing or decreasing initial search mobility and more securely launching them on employment trajectories. Second, larger work organizations, strong trade unions, and an increased prevalence of white-collar jobs enlarged the prevalence and length of working lives characterized as "careers." This enhanced occupational stability over the life span. Economic growth and the upgrading of the occupational structure favored voluntary job shifts and upward mobility in contrast to involuntary, horizontal or downward shifts. Third, the provisions of the welfare state institutionalized new kinds of statuses and events, like sick leave, maternity leave or child leave, and fostered continuity in lives by buffering the impact of income loss due to adverse events like unemployment or illness or old age. Finally, the relative security of income, employment and career advancement supported early marriage and a larger number of children (Esping-Andersen, 1999).

While ideas of standardization and institutionalization had a powerful sociohistorical logic, the latter years of the 1980s saw the accumulation of observations which ran counter to the standardization thesis (Held, 1986; Buchmann, 1989; Kohli, 1986, 1989). In fact, as early as the late 1960s and early 1970s, pervasive value changes (Inglehart, 1977) in connection with youth and student rebellions across a number of advanced Western societies were seen as restructuring the way people organized their private lives. Such value changes manifested themselves in delayed marriage and childbirth and even beyond that, in the rise of non-marital unions, divorce and remarriage. This was coupled with increasing claims to autonomy and self-realization which further mirrored earlier exits from the parental home independent of the event of marriage. This increasing "disorderliness" in young adult lives was believed to result from educational expansion and the inflation of educational credentials, which in turn delayed or jeopardized labor market integration. Some went so far as to describe educational institutions as "waiting halls" (Boudon, 1974; Freeman, 1976; Büchel, Grip, & Mertens, 2003). The women's movement provided an additional, independent and strong force in reshaping the lives of both women and men by encouraging women to pursue education and careers partially independent of family formation. The 1980s also saw the full consequences of the 1973 oil shock hit the training and labor markets for young adults with the result of making the attainment of vocational and professional credentials, as well as the transition to gainful work being more prolonged and more complicated. All of this, however, occurred in a context of increasing welfare state provisions, continued increases in real incomes, and improvements in the material circumstances of parents, which allowed children to invest more in education, to experiment with private living arrangements, and to pursue self-defined goals. It is this curious mixture of value changes, opportunities and adaptive constraints, which fostered the topoi of the "postponed generation" (Mayer, 1994, 1995), and "individualized" or "patchwork" biographies (Beck, 1986).

In the early 1990s, belief in these tendencies continued but was gradually supplanted by perceptions of massive impacts of the forces of globalization, particularly increased international competition, labor market de-regulation and structural unemployment (Mills & Blossfeld, 2003). What some viewed in the 1970s and 1980s as a widening of life pathways due to new options were reinterpreted in the 1990s as difficult adaptations to external constraints (Becker & Hermkens, 1993). These more pessimistic outlooks were interrupted for a short period by the information-technology "bubble" which appeared to foster unconventional lifestyles and careers of almost unlimited opportunities. Another distinctive causal condition which potentially affected life courses in the 1990s were reforms and cuts in the provisions of social security (Esping-Andersen, 1999). These included cuts in the levels of unemployment benefits, health insurance, pension entitlements and benefits to unemployed or underemployed youth. Thus it could be said that the 1990s increased both the incidence of exposure to risks, especially on the labor market but also in family life, and the level of welfare assistance given such risks (DiPrete, 2002).

In sum, the course of more than 30 years gave rise to a number of different macro-social and macro-economic conditions that are widely believed to have had strong impacts in making life courses less conventional, less standardized, less collectively patterned, less predictable and more exposed to risks both in the public and in the private sphere. In retrospect, the pattern of stable employment, progressive income levels and careers, welfare protection, early marriages and stable family lives, as well as high degrees of social integration at the work place and in the community appear to have been constricted to a fairly short and exceptional historical period, a "Golden Age," that was both preceded and succeeded by more turbulent times and turbulent lives (Mayer, 2001; Myles, 1993). Both academic and journalistic observers, then, have developed a broad consensus and a multitude of personal experiences and illustrative evidence on "post-modern" patterns of a greater variety of partly freely chosen, partly imposed life trajectories.

CONCEPTS FOR CHANGES IN LIFE COURSE PATTERNS

Such sweeping generalizations about changes in life course patterns as sketched above suffer from both a lack of precision in the concepts employed and a lack of systematic data in lieu of casual observations. In this section, we review a number of terms used in the attempt to capture both more historical and more recent transformations of life trajectories. We also offer a set of definitions of these terms and some illustrative examples to explicate their intended meanings. In this way we hope to contribute to a more sophisticated, multi-dimensional set of categories denoting social changes in the area of life course research.

Several concepts are often used interchangeably and indiscriminately to capture salient historical changes in life course patterns. These include "(de-) institutionalization" (Mayer & Müller, 1986), "(de-) standardization" (Modell et al., 1976), "individualization" (Beck, 1986), "pluralization" (versus "homogenization") (Zapf, 1987), and "(de-) differentiation (Mayer, 1991). We suggest that this list of concepts actually taps dimensions that can vary independently from each other and should therefore be distinguished from each other.

The *institutionalization* of life courses refers to the process by which normative, legal or organizational rules define the social and temporal organization of human lives. It can refer to stages or states in lives which can be formally or informally decreed like marriage, education, and retirement. It can also refer to events and transitions like leaving school, entry into and exits from labor contracts, or ages of pension entitlements. Kohli (1985) maintains that modern life courses have become a modus of "Vergesellschaftung", that is, of socialization, social integration and social control, and that they have become as, if not more, important than class structures. The welfare state now recognizes not only a number of broader life course-related states: periods with neither paying nor receiving contributions (childhood), periods paying but not receiving (employment), periods not contributing, but receiving (old age, illness, unemployment), but also more fine-graded episodes such as maternity leave, child leave, rehabilitation and the like (Mayer & Müller, 1986).

Conversely, *de-institutionalization* would then mean that states, stages, events, and transitions, which at earlier times were clearly differentiated, are being reintegrated or fused. One case in point is education and work, which in some countries including the U.S., have come to be highly synchronous activities rather than clearly separate life stages (Shanahan, 2000). Another example is the rise of non-marital unions where the establishment of a joint household is no longer tied to marriage (Bumpass & Lu, 2000; Meyer & Schulze, 1983). This is an example where "marriage" becomes partially de-institutionalized, while at the same time the new sequence of first non-marital union and then marriage (with the same or another partner) leads to an increase in the differentiation of the life course by changing into two qualitatively different episodes, whereby it used to be one. The recent increase in home schooling, notable in the U.S., would be a further instance of de-institutionalization of the early phase of education.

The *standardization* of life courses refers to processes by which specific states or events and the sequences in which they occur become more universal for given populations or that their timing becomes more uniform. An example of a highly standardized life course pattern would be, for instance, if all workers retire and all retire at age 65. Apart from the retirement processes increasing standardization has been claimed for gainful employment for women, for secondary education, and, specifically in Germany, for vocational/professional training as well as for maternity leave.

Conversely, *de-standardization* would mean that life states, events and their sequences can become experiences which either characterize an increasingly smaller part of a population or occur at more dispersed ages and with more dispersed durations. Both marriage and parenthood in this sense first (until the 1970s) became more standardized and then became more de-standardized (Espenshade, 1985). "Stop gap jobs" at career entry in the U.S. (Oppenheimer & Kalmijn, 1995) or the increasing fuzziness of age at retirement in Germany (Ebbinghaus, 2002) would be other cases in point.

Differentiation refers to the process where the number of distinct states or stages across the life time increases. For instance, Mayer (1991) has claimed that early life courses become more and more institutionally differentiated. They are increasingly divided up in publicly defined and recognized periods like pre-school, kindergarten, elementary school, secondary education and tertiary education. Likewise, a process where firm tenure with only one or very few employers has been gradually replaced by frequent shifts between firms could be called a differentiation of the work life (Mertens, 1998). The splitting of a single training period into several ones (Jacob, 2003) would be another instance. Gerontologists have made similar claims for the later life, distinguishing between partial or pre-retirement and retirement as well as between being a "young" old and being an "old" old (Baltes & Mayer, 1999). *De-differentiation* would then refer to processes where formerly split life periods become joined. Still, it is hard to think of examples for the latter process, suggesting that irreversibility of the underlying trend is the more salient phenomenon. While the concept of differentiation of the life course refers to the diachronous dimension of a sequence of life states, the idea of *pluralization* (Zapf, 1987) is usually used to refer to an increase in the synchronous number of states or forms of life activity in a given population or even a given person. Pluralization has mostly been applied to family forms. This refers to the rise of non-marital unions, the increase of persons becoming divorced, increases in the number of single mothers or persons living alone due to divorce or widowhood. A parallel example for a given person would be the increasing frequency of holding multiple jobs.

Finally, the term *individualization* (Beck, 1986; Junge, 2002; Schimank, 2002; Wohlrab-Sahr, 1992) is frequently invoked to refer to many of the changes mentioned above. It is a more interpretative concept according to which individuals are assumed to gain greater control over their lives, thus pursuing a wider variety of life designs and life trajectories. Many of the processes referred to above are then assumed to be the result of increasing individualization. Such a more positive meaning of the concept has in recent years become mixed with notions of involuntary "individualization," of being condemned to pursue and experience trajectories, which are not collectively well-trodden pathways (Buchmann, 1989; Shanahan, 2000).

Obviously, all these different dimensions of general changes in life course patterns could be applied one for one to various domains of life to generate more specific hypotheses. But it may suffice here that we have drawn our illustrative examples from life domains such as education, work, family and social security. However, given its central role in the debate on de-standardization, we want to especially highlight changes postulated for the area of work: the declining employment stability due to generally increasing unemployment, decreases in firm tenure due to layoffs and downsizing, declining durations of staying in a given occupation and declining residential stability. This perception is also connected to hypotheses about increasing rates of downward mobility (Newman, 1999) and income fluctuation across the working life. Such tendencies are frequently generalized under the heading of increasing "flexibility" enforced by employers and made possible by weakened trade unions (Rhodes & van Apeldoorn, 1997; Sennett, 2000; Heinz, 2003). The implicit idea here is that the overall stability and continuity of life courses are seriously undermined by these changes in labor markets.

Two important implications follow from this discussion. First, the processes outlined above do not need to go all in the same direction. For instance, it is possible that institutionalization is coupled with de-standardization as in the case of early retirement policies (Ebbinghaus, 2002) or that pluralization might go hand in hand with standardization, as would be the case if a period of non-marital cohabitation almost universally precedes marriage. Therefore, one cannot assume one overarching unidirectional process and should be careful when making grand generalizations. Going even further, one might question whether the invocation of an overall trend does not lead to more confusion than clarity. Second, the empirical charting of diverse changes in life course patterns cannot simply rely on anecdotal and illustrative evidence. Rather one has to look for systematic data which both proponents and opponents of the de-standardization debate would accept as relevant evidence. Given this situation in the debate, the patient and painstaking measurement of empirical developments becomes crucial.

EMPIRICAL TESTS AND HYPOTHESES

A proper empirical examination of changes in life course patterns requires detailed life histories in various life domains across a larger series of birth cohorts. To be useful, data should be drawn from representative samples, register or census data for a given population. Ideally, one would like to be able to map all the transitions of a given birth cohort through a changing institutional structure of life states. Of particular interest are the following aspects of these transition processes: (a) prevalence, the extent to which a given transition or state occurs; (b) age variance, the degree to which transitions occur at specific ages; (c) duration variance, the extent to which people stay in a given state or stage; (d) inter-event dependency, the extent to which the occurrence of one event or state is associated with the occurrence of another event or state, and (e) sequence uniformity, the extent to which the temporal sequence and ordering of events and states are uniformly distributed.

Based on these dimensions, we can draw some hypotheses. First, if *de-standardization* is such a pervasive process, we should expect: (a) that prevalence of certain events or life stages has decreased over time; (b) that variances of ages at given transitions have increased; (c) that the variances of given durations have increased; (d) that the inter-event and inter-state dependencies have decreased; and (e) that the sequences of events or states have decreased in their predictability. To corroborate a shift towards *higher flexibility* we should also expect: (f) increased shifts between jobs; (g) decreased firm tenure; and (h) decreased occupational stability. Moreover, according to most of the scenarios we should expect not just very gradual trends, but fairly sudden discontinuities or trend reversals. In the following analyses, we focus on the processes that may result in a de-standardization of life courses. Although we occasionally invoke institutionalization or differentiation in interpreting our findings, we do not attempt to systematically produce evidence related to these concepts, or the related processes of individualization and pluralization. Finally, it is important to note that each of these indicators taken alone can be indicative of multiple processes of social change that in principle may have nothing to do with de-(standardization) of life courses. For example, a delay in family formation could be expected due to changes in educational participation alone, without being indicative of changes in the overall patterning of life courses. Similarly, decreasing job stability may occur during economic up- or downturns, without indicating a societal trend towards de-standardization. Evidence for de-standardization in the conceptualization employed in this paper would require changes in all or at least most of the indicators mentioned above.

DATA

The data used to examine these issues are drawn from the West German part of the GLHS, directed by Karl Ulrich Mayer. This research program collected detailed retrospective life course information for various birth cohorts of West German Nationals (an overview is given in Brückner & Mayer, 1998; Wagner, 1996; Hillmert, 2003). From 1981 to 1983, 2,171 life histories of a representative sample of people born in 1929–1931, 1939–1941 and 1949–1951 were collected (Mayer & Brückner, 1989). In the years 1985–1987, the cohort group from 1919 to 1921 was added with 1,412 cases, 407 of whom were collected via face-to-face interview, while the remaining 1,005 were collected with a computer-assisted telephone interview (Brückner, 1993). Data for the birth cohorts from 1954–1956 and 1959–1961 were collected in 1989, resulting in 2,008 computer-assisted telephone interviews (Brückner & Mayer, 1995). Finally, the birth cohorts from 1964 and 1971 were interviewed in 1998–1999 with a sample size of 2,909 respondents (Hillmert & Mayer, 2004). For these cohorts, foreign nationals were included in the sample for the first time, reflecting the changing composition of the residential population at that age and time. Our analyses exclude the latter group in order to ensure full comparability with the earlier series. The samples for all parts of the GLHS were carefully evaluated against official statistics (census and micro-census) to ensure that they are representative of their respective populations (documented in the above sources).

For all cohorts, the survey instrument contained detailed questions about family of origin, residential history, education, work life, work interruptions, and family formation, including the formation and dissolution of marital and (for younger cohorts) non-marital unions, as well as children and grandchildren (where applicable). All interviews were carefully checked for errors and chronological consistency. Many respondents were contacted again to clear up inconsistencies in the data, and helped the researchers to reconstruct life histories to the fullest extent possible. The observation window varies for each cohort in the GLHS, from 64–68 years for those born around 1920 to 27 years for those born in 1971.

FINDINGS

In the following, we report findings regarding the experiences of West German men and women born between 1919 and 1971. We explore changes in the timing of selected important life course transitions and in the variation therein. The median age of these events by gender and cohort is reported in Table 1. Fig. 1 shows the inter-quartile range of transition ages. The inter-quartile range is the difference between the age at which 75% of a cohort had experienced an event, and the age at which 25% had experienced

Table 1. Median Age at Selected Life Course Transitions, by Cohort and Gender.

Birth cohort	1920	1930	1940	1950	1955	1960	1964	1971
Men								
Leaving school[a]	14.3	14.3	14.7	15.3	16.0	16.5	16.8	17.2
Completing training[b]	19.3	20.0	19.3	20.3	24.8	24.6	22.3	23.5
1st job[c]	18.1	17.8	18.0	18.8	19.5	19.8	20.2	20.3
Leaving home[d]	29.7	25.7	25.3	24.2	23.9	23.3	23.8	24.2
1st marriage	27.7	25.8	25.4	25.8	26.3	29.5	29.4	—[e]
1st child	29.7	28.3	27.3	30.0	30.6	—[e]	32.6	—[e]
Women								
Leaving school[a]	14.4	14.6	14.7	14.9	16.0	16.5	17.1	17.6
Completing training[b]	18.3	18.9	18.3	18.5	19.5	20.4	20.7	21.6
1st job[c]	16.9	16.5	17.5	18.1	18.9	19.6	20.3	20.4
Leaving home[d]	28.3	23.7	22.2	21.0	20.8	21.0	21.4	21.8
1st marriage	23.3	23.7	22.2	21.5	23.0	25.3	25.5	—[e]
1st child	25.4	25.9	23.8	25.0	27.0	28.2	28.1	—[e]

Source: German Life History Study. Kaplan–Meier estimates.
[a]Primary and secondary schooling.
[b]Includes only those who began a training spell during the observation period; includes vocational apprenticeships, vocational schools, and post-secondary training.
[c]First job with a duration of 2 months and longer.
[d]Age at which first own household was formed.
[e]Less than 50% of cohort had experienced event during the observation period.

the event, and serves as a measure of intra-cohort variation. For marriage and first childbirth, some cohorts had not yet reached the 75th percentile or even the median age, and we will introduce other measures throughout the text to compare the experiences of these cohorts to the older cohorts.

The School–Training–Work Nexus

Table 1 clearly shows the well-documented trend towards delay in major life course transitions. During the observation window, the age of mandatory schooling was raised to age 15, but more importantly, educational expansion led to longer periods of schooling as Germans increasingly graduated from middle school and Gymnasium[1]. Thus, the median age at leaving school rose from 14.3 for men born around 1920 to 17.2 for men born in 1971. For women, it similarly increased from 14.4 for women born around 1920 to 17.6 for women born in 1971. Intra-cohort variation in age at

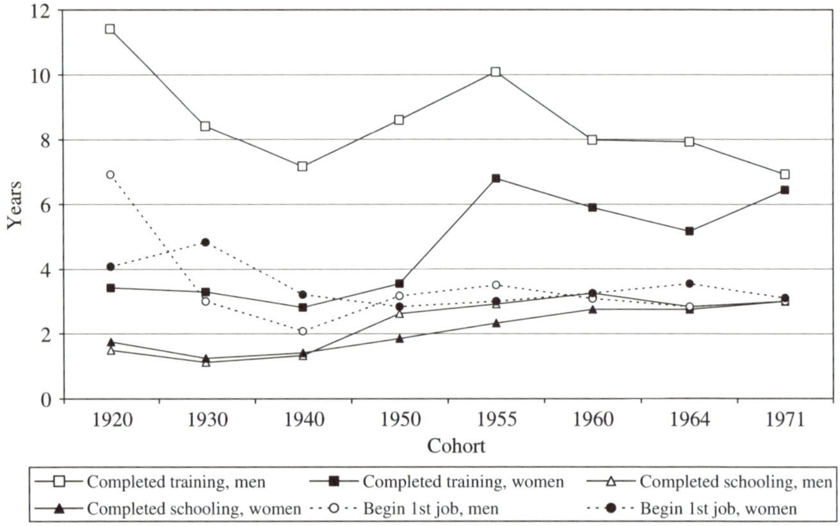

Fig. 1. Interquartile Ranges for Age at Selected Life Course Transitions.

leaving school increased moderately, from 1.5 years for men of the 1920 cohort to 3.3 years for men born around 1960 and 3.0 years for men born in 1971 (see Fig. 1, line marked with white triangles). For women, the increase was more gradual but remarkably similar.

There are more dramatic changes with respect to the age at which cohort members completed their training. Among men who had begun any training during the observation window, completion age rose from 19.3 for those born around 1920 to 24.8 for those born around 1955 (see Table 1). Most of the change happened between the 1950 and 1955 cohorts, when completion age increased by $4\frac{1}{2}$ years. It fell slightly for subsequent cohorts, to 22.3 and 23.5 for those born in 1964 and 1971, respectively. For women, the changes were much more gradual. Completion age rose from 18.3 for women born around 1920 to 21.6 for women born in 1971. Fig. 1 further shows that from the 1950 cohort on, the variation in age at completion of training increased strongly for women, while it declined for men. The inter-quartile range was 3.4 for women born around 1920, 3.5 for women born around 1950, before reaching a peak of 6.8 for women of the 1955 cohort. For men, dispersion was greatest among the 1920 cohort, due to delayed entry into training because of the war and the opportunities for re-entering training after the war. Dispersion then declined steeply for men born around 1930 and 1940, only to rise again with the greater educational opportunities for men born in the 1950s. The 75th percentile for the 1955 cohort is 29.5, compared to 26.8 for the 1950 cohort and 27.7 for the 1960 cohort. Thus, a substantial minority of cohort members delayed completion of training considerably. The difficult labor market in the mid-1970s and mid-1980s (when these cohorts were in their 20s) may have contributed to a tendency to spend more time in the educational system. Here, the institutions of the educational system served as 'waiting halls' for young people who had few prospects in the labor market. These cohorts have been named the 'postponed' generation for this very reason (Mayer, 1994).

Table 2 shows that the cohorts born after 1950 were also increasingly likely to start and complete several training episodes. For example, more than half of the men born around 1955 and 1960 began more than two training episodes, compared to less than a third for the earlier cohorts.

Table 2. Training Spells[a] Started by Age 27, by Cohort and Gender (%).

Birth cohort	1920	1930	1940	1950	1955	1960	1964	1971
Started training spells, men								
None	13.4	16.4	9.1	3.6	1.5	2.0	2.1	0.8
One	59.7	55.0	58.7	57.1	48.7	43.0	55.7	51.7
Two	20.8	21.0	25.6	28.8	31.0	36.1	33.4	36.9
Three or more	6.1	7.5	6.7	10.4	18.8	18.9	8.8	10.6
Completed training spells, men								
None	27.1	26.6	18.9	17.0	14.4	15.4	12.3	13.8
One	63.6	59.0	57.6	59.1	53.3	56.6	69.6	69.4
Two	7.7	12.7	20.0	19.5	22.8	21.9	16.8	15.2
Three or more	1.6	1.7	3.5	4.4	9.6	6.1	1.4	1.5
Started training spells, women								
None	33.7	53.2	34.9	16.8	7.2	4.3	5.4	2.4
One	46.8	38.5	45.6	55.6	51.5	50.9	58.7	53.8
Two	14.8	7.2	16.3	21.1	29.3	32.1	24.1	29.9
Three or more	4.7	1.1	3.1	6.5	12.0	12.7	11.7	13.9
Completed training spells, women								
None	54.8	65.4	44.5	26.8	16.3	15.5	12.9	13.4
One	39.4	31.0	46.2	56.9	59.6	61.8	70.6	71.2
Two	5.3	3.3	8.5	14.9	19.6	17.8	14.9	13.6
Three or more	0.6	0.3	0.8	1.4	4.5	4.9	1.7	1.8

Source: German Life History Study.
[a]Includes vocational apprenticeships, vocational schools, and post-secondary training.

Similarly, the proportion of men born around 1955 who completed more than three training episodes is more than five times greater than that for those born around 1920, and more than twice that for men born around 1940 and 1950. Almost a third of the men born around 1955 completed two or more training episodes, compared to 9.3% for the 1920 cohort and 14.4% for the 1930 cohort. For the two youngest cohorts, however, the proportion that completed two or more training episodes fell again to 18.2% for men born in 1964 and 16.7% for men born in 1971. Finally, the proportion of men who never completed training fell from 27.1% for the 1920 cohort to 14.4% for the 1955 cohort.

The decline of the proportion that never completed any training is even more dramatic for women. More than half of the women born in 1920 and 65.4% of those born around 1930 did not complete training, compared to 16.3% for the 1955 cohort and 12.9% for the 1964 cohort. Similarly, having two or more training episodes was quite rare for women born before 1940 but became more common for women born around 1950 and later. Overall, the marked gender differences in the distribution of the number of started and completed training that are typical for the older cohorts had pretty much disappeared by the time the 1955 cohort left the educational system, although women still completed their training about 2–4 years earlier than men.

Despite such changes, the median age at labor market entry is comparatively stable. For men, it rose from around ages 18 to 20 across cohorts (see Table 1). For women, it increased more, from under 17 to about 20. For the three youngest cohorts, earlier gender differences in age at labor market entry have virtually disappeared. Cohort members entered the labor market within a relatively narrow range of ages, with the exception of the 1920 cohort, which entered the labor market during and after World War II (see Fig. 1). Men in this cohort reached the 75th percentile only at age 24, compared to 19 for the 1930 and 1940 cohort. For later cohorts, the ages at which cohort members reached the 25th and 75th percentiles rose modestly, leading to a stable inter-quartile ratio across cohorts. The same is true for women. It is noteworthy that for men, the median age of first job is always and increasingly younger than that at completion of training, although the proportion of cohort members who never entered training fell across cohorts. Thus, transitions from labor force participation back into the educational system are increasingly common across cohorts.

Employment mobility is shown in Table 3. A full comparison of life courses across the eight cohort groups is possible only until age 27, the age at which the youngest cohort (1971) was interviewed. Table 3 therefore

Table 3. Experiences by Age 27, by Cohort and Gender.

Birth cohort	1920	1930	1940	1950	1955	1960	1964	1971
Men								
Number of jobs[a]	2.2	2.4	2.4	2.3	2.3	2.3	2.1	3.0
Years worked[b]	3.2	8.3	8.1	6.2	5.4	4.8	5.1	5.1
Jobs per year[c]	1.0	0.4	0.3	0.4	0.6	0.7	0.7	1.8
Number of residences	5.2	3.0	2.8	2.2	2.9	3.0	2.2	2.3
Proportion married	0.41	0.56	0.64	0.57	0.51	0.36	0.32	0.17
Proportion with child	0.28	0.39	0.43	0.30	0.27	0.20	0.19	0.11
Women								
Number of jobs[a]	3.0	2.0	2.0	1.9	2.0	2.1	1.9	2.6
Years worked[b]	7.0	6.9	6.3	5.9	5.4	5.1	4.9	5.1
Jobs per year[c]	0.5	0.4	0.4	0.5	0.5	0.7	0.5	1.0
Number of residences	3.9	2.6	3.0	2.3	2.9	2.9	2.6	2.6
Proportion married	0.67	0.74	0.85	0.81	0.66	0.58	0.57	0.39
Proportion with child	0.56	0.56	0.68	0.57	0.47	0.39	0.38	0.27

Source: German Life History Study.
[a]Excludes persons who never worked for longer than 3 months.
[b]Includes persons who never worked.
[c]Job shifts per year of labor force participation until age 27. This excludes those who worked less than 2 months.

reports the number of residences and jobs reported up to that age. The average number of jobs held until age 27 is remarkably stable across cohorts with around 2.3 for men and around 2.0 for women. Only the 1971 cohort for men and women and the 1920 cohort for women show a higher mobility: approximately three jobs. The dispersion of number of jobs is similarly stable over time (data not shown). However, because cohort members entered the labor market later and spent more time in the educational system, the average number of jobs per year changes across cohorts. Aside from the 1920 cohort, there is an increase in job changes for cohorts born after 1950. Thus, the same number of jobs is distributed over a shorter period of time for cohorts born after 1950. The rate of job shifts more than doubled for men born in 1971. Fig. 2 shows the dramatic increase in both median number of job shifts per year and in the inter-quartile range for the 1971 cohort and that this trend is more pronounced for men than for women. One source of this may in part be the dramatic rise in unemployment for this cohort (Hillmert & Mayer, 2004)[2]. Mertens and Mayer (2004) further show that, compared to the 1940 cohort, the risks of involuntary firm switches and downward wage mobility have strongly increased for the cohorts born

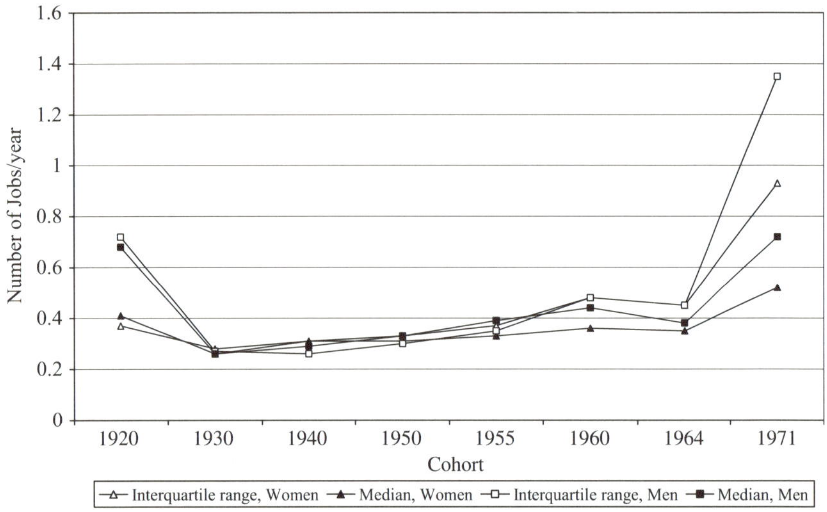

Fig. 2. Job Shifts until Age 27 per Year of Labor Force Participation, by Cohort and Gender.

around 1960 and in 1971, but to a lesser extent for the 1964 cohort. Hillmert (2002), however, has found no decrease in occupational stability for the cohorts born between 1960 and 1971, although their risk of unemployment is comparatively higher (see also Kurz, Hillmert, & Grunow, 2002).

Household and Family Formation

The delay of the transition to adulthood in the school–training–work nexus shown in the previous section did not keep cohort members from moving out and forming their own household. Median age at household formation declined somewhat across cohorts from the late 20s to the early 20s (see Table 1). Fig. 3 shows the inter-quartile range for this transition. There was significant variability in age of household formation for people born around 1920 but this declined steeply until the 1950 cohort. There is a modest increase of variability for the younger cohorts of men (Konietzka & Huinink, 2003).

For residential mobility up to age 27, the 1920 cohort stands out. They reported an average of 5.2 places of residence for men and 3.9 for women (see Table 3). For the other cohorts, the number fluctuates between 2 and 3

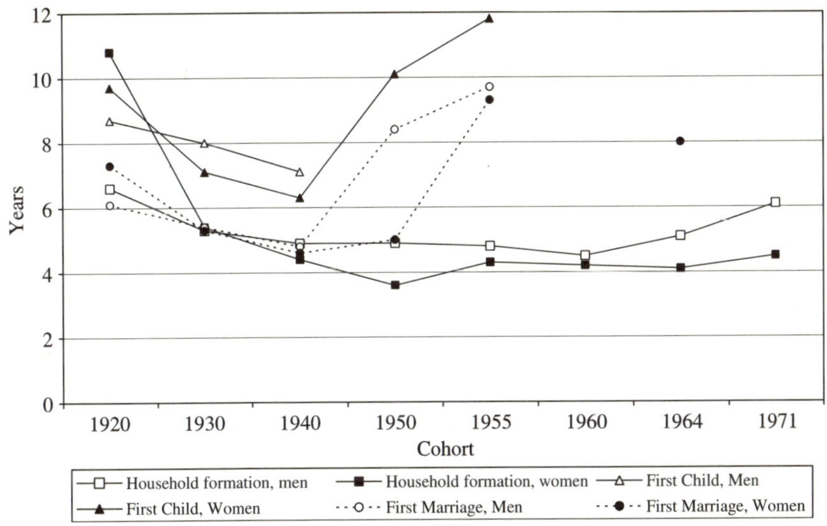

Fig. 3. Interquartile Ranges for Age at Selected Life Course Transitions.

residences without a clear cohort trend. Men of the 1955 and 1960 cohorts are also more likely than others to move two or three times (data not shown). For women, the distribution is quite similar across cohorts with the exception of the 1920 cohort.

By U.S. standards, West Germans marry late and have their first child even later. First marriage occurred late for men in 1920, at almost 28 (see Table 1). Later cohorts married somewhat earlier, but by the 1955 cohort, age at first marriage increased again, extending to 30. The youngest cohort was interviewed at age 27 and thus median marriage age cannot be observed. We can, however, compare the proportion married by this age. Only 17% of the men of this cohort were married at the time of the interview (see Table 3). In contrast, men of the 1940 cohort were almost four times more likely to be married at this age (64%), and even those born in 1960 were twice as likely to be married at this age (36%). The proportion of women who had married by age 27 follows a similar pattern albeit at a higher level. The proportion married increased between the 1920 and the 1940 cohort from 67% to 85%, and then began to fall again. Women born in 1971 were half as likely to be married by age 27 than women born around 1950.

A delay in marriage age does not necessarily imply a de-standardization of family formation patterns – people may marry later, but still experience this event at pretty much the same time than their age peers, albeit later than their parents and older siblings. Fig. 3 shows an early decline and later increase in the inter-quartile range in age at marriage for those cohorts for whom the process was far enough along to calculate the 75th percentile. The same holds for women. Women marry a few years earlier than men but show the same pattern of inter-cohort change, perhaps somewhat attenuated. Variability increased for women, too. Unfortunately, we cannot observe the age at the 75th percentile for most of the younger cohorts and thus follow our earlier strategy of calculating the proportion of people marrying by a certain age for each cohort. The result for women is shown in Fig. 4; for men, in Fig. 5. If there is such a thing as a standardized age for marriage, we should see a steep curve with short tails. The peak of the curve, if any, should contain most cohort members within a narrow age range. When these age standards degrade, we should see curves that flatten out, have less pronounced peaks, and long tails. For the marriage behavior of West German women, this idea is illustrated in Fig. 4.

Women born around 1940 and 1950 indeed tend to marry at the same ages, and the marriage process is essentially completed by their late 20s. For those born around 1920 and 1930, the peaks are somewhat lower, the spread is greater, and there is a fatter tail. Beginning with the 1955 cohort, the

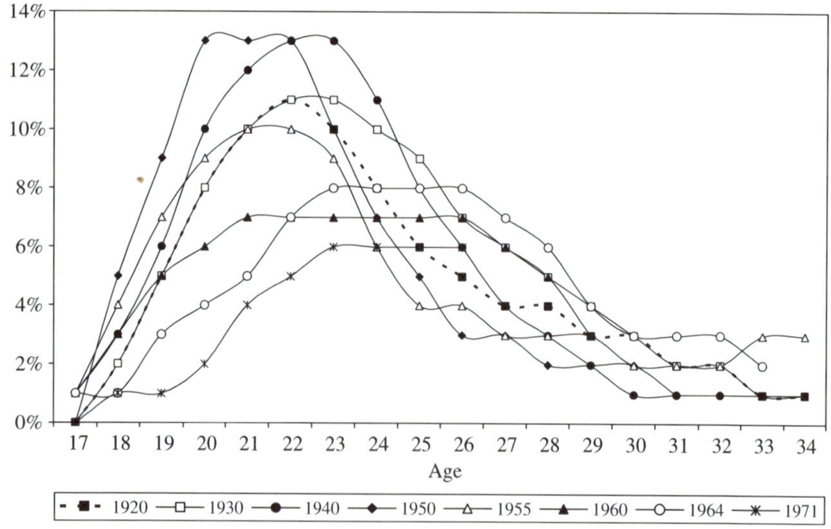

Fig. 4. Marriage Timing by Cohort and Age, Women.

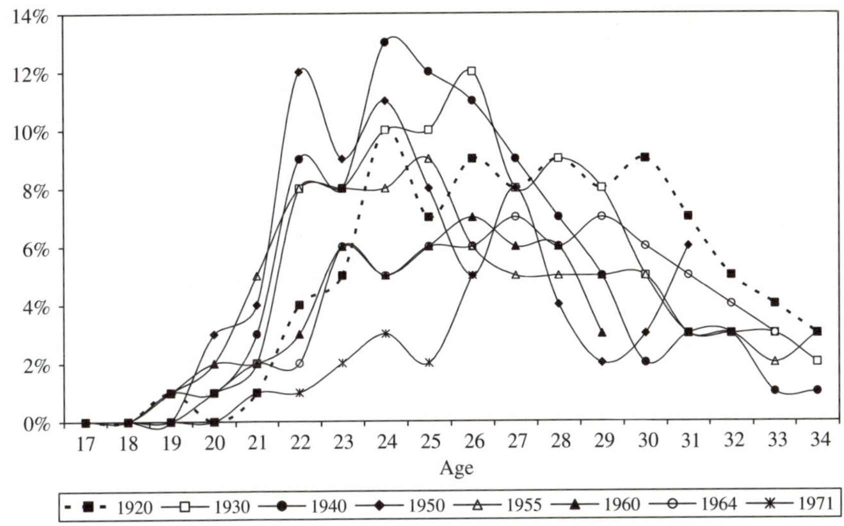

Fig. 5. Marriage Timing by Cohort Age, Men.

curves begin to flatten out. There is still a peak for the 1955 cohort in the early 20s. In contrast, the 1960 cohort has more of a plateau during the late teens through the late 20s and wide tails in both directions. Thus, experiences for these younger cohorts have certainly become more varied. For men, the place of marriage in the life course is much less neat (see Fig. 5). Men born around 1930 and 1940 have peaks in the early and mid-20s, but a substantial proportion marries later. For the 1920 cohort, the curve is much flatter and extends into the early 30s. The younger cohorts, similar to women, have even flatter curves. It is noteworthy that for men born around 1950, there are several peaks, in the early 20s, mid-20s, and early 30s. This is consistent with the thesis that marriage behavior for this cohort is structured by educational participation (Huinink & Mayer, 1995). For the younger cohorts, no such pattern is discernible.

Finally, we turn to age at first childbirth. Median age for this life course event was late for men born around 1920 (29.7) and declined for the 1940 cohort (see Table 1). These are the only cohorts for whom we can observe the inter-quartile range (see Fig. 3), because childbirth occurred even later for the younger cohorts. Comparison of prevalence again illuminates changes over time. By age 27, 43% of men born around 1940 had had a first child, but only 30% of those born around 1950. The proportion declined

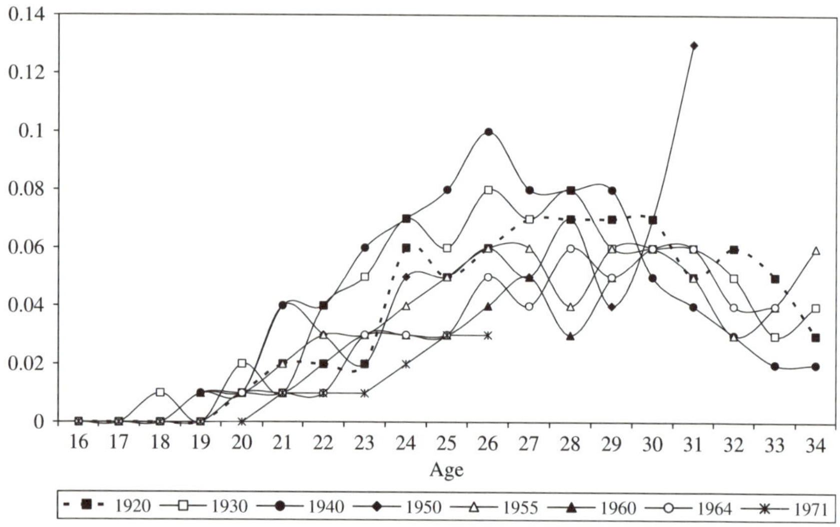

Fig. 6. Timing of First Childbirth by Cohort and Age, Men.

further, reaching its low at 11% for men born in 1971. Although it is therefore difficult to make generalizations about changes in the nature of this transition from our data, Fig. 6 shows the process as far as we can follow the cohort members. Even more so than for marriage, the age at first childbirth varies within cohorts. The line that most resembles a curve with one peak is that for the 1940 cohort. Men born around 1950 have a peak at age 30, but the curve comprises only 55% of the cohort members. The remaining 45% had not yet had children by the time they were interviewed. For the other cohorts, the curves are spread out and rather flat.

Women have their first children earlier than men, but the pattern of inter-cohort change is similar to that of men. The proportion of women who had a child by age 27 reached a peak for women born around 1940 at 68% and then steadily declined to a low of 27% for those born in 1971. Fig. 7 shows that similar to marriage, first childbirth is more age structured for women than for men. The peak is most pronounced for women born around 1940. For women born around 1950, there is a pattern of differentiation with an early peak between age 19 and 24 and a later, flatter peak in the late 20s. Especially for the cohorts born around 1955 and 1960, the curves spread out, and a substantial proportion of women born around 1955 have their first child in their 30s (the 1960 cohort was interviewed when they were around 28 and therefore we cannot observe what happens later on).

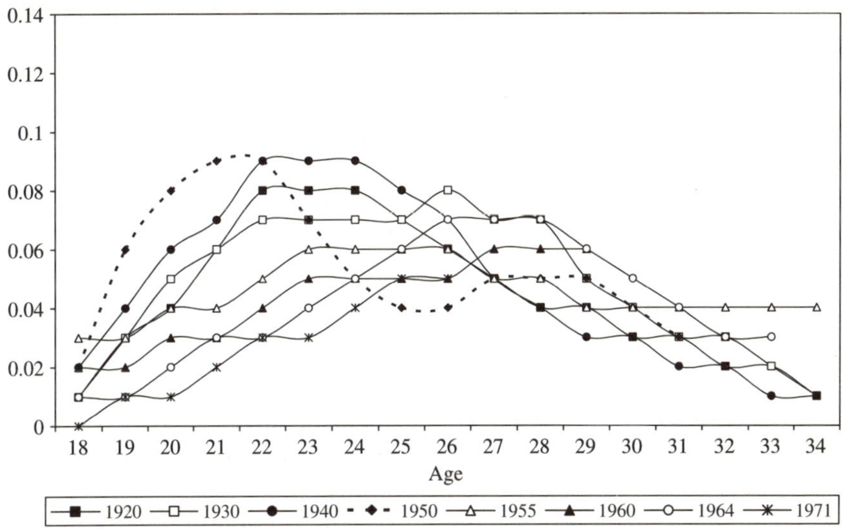

Fig. 7. Timing of First Childbirth by Cohort and Age, Women.

CONCLUDING REMARKS

In regard to the nexus between the end of schooling and the entry into the labor market, we find two clear time trends in these data for West Germany. First, transitions occur later and later. Second, the life courses of men and women become more similar in the school–training–work nexus. With respect to the issue of *de-standardization*, we note that the sequencing of training and work becomes fuzzier as participation in the educational system increases. This is especially so for the cohorts born around 1955 and 1960. Especially for women, one consequence of increased participation in training is that the variation of age at completion of education increases as well, converging on that for men. Otherwise, the expected increase of age variation under the de-standardization hypothesis turns out to be non-existent or rather small – at least for the transitions in the school–training–work nexus reported here. There are, however, strong period effects that go against the assumption of linear or generalized transformations of the life course. Of particular note, people born around 1920 experienced quite heterogeneous life courses due to the war, and the cohorts born around 1955 and 1960 took longer to complete their training than any other cohort.

With respect to family formation, we observe a pattern of age consolidation for the cohorts 1920–1940, but this is followed by *differentiation* that may reflect the educational system for the 1950 cohort. There is even more variability for cohorts born after 1950. We also note that this pattern is stronger for women than for men. Variability is pronounced for all female cohorts. Nevertheless, we are still left with the question of whether the trend towards greater variability in the timing of life course transitions can truly be described as *de-standardization*. Imagine a kindergarten in the mid-1960s and assume for simplicity that it is frequented only by first-borns. The age distribution of parents in this kindergarten would certainly be more uniform than that of a similar kindergarten in the mid-1980s or mid-1990s. Conversely, children born to our younger cohort members will be caring for their elderly parents during variable stages of their own life course and there probably will be greater variability in the availability of grandparents to help with childcare when these children form their own families. Lasting friendships formed in the mid-1950s by young adults will be characterized by a more synchronous experience of important life course transitions than those formed by younger cohorts. This may have implications for the usefulness, the formation and the maintenance of support networks that reach beyond one's kin. With respect to the relationship between these various life course transitions, we note the increasing *de-coupling* of events in the connections between the school–training–work nexus and family formation. For example, forming an independent household is no longer coupled with marriage or having a stable job (Konietzka & Huinink, 2003). Nevertheless, the institutional environment continues to structure the school–training–work nexus and not much change was seen in the way in which cohort members undergo these transitions. On the contrary, by the standard employed in this paper there is actually a homogenization of life course as women's and men's life courses converge in terms of education and labor force participation. It is the family formation nexus that shows the most pronounced changes, and is also the realm in which gender differences persist across cohorts.

In sum, our observations from West Germany across half a century show considerable evidence supporting the de-standardization thesis in the area of private lives. Combined with the extant evidence on the rapid spread of non-marital unions before marriage and the rise of divorce, this de-standardization in the family sphere was coupled with some degree of de-institutionalization and a pluralization of family forms. This development seems to have been triggered by three long-term developments: educational expansion with its delaying effects on family formation, the pronounced value changes starting in the early 1970s and – related to both of these processes – the women's movement. In contrast, we find little support for the de-standardization thesis in the spheres of education, training and work. These transitions have been prolonged, have become more differentiated in regard to multiple training spells and as a result have become less standardized in the degree of orderliness of the sequences of leaving school, training and employment. But with the possible exception of an increasing variability of the age at completing training for women we find a high degree of long-term stability. Rather than a trend or massive changes in recent cohorts we find exceptional experiences of some earlier cohorts, especially the "war" and "post-war" cohorts born around 1920 and 1930 and the "labor market crunch" cohort born around 1955, as well as an increase in job mobility for the 1971 cohort.

Obviously, we should be cautious in passing a final verdict on the de-standardization thesis even in the non-family sphere. On the one hand, we have only examined here a selective set of indicators which all relate to the early part of life. On the other hand, (West) Germany might be a special case where the specific institutions of training and occupationally segmented labor markets still exert strong influences in shaping life courses. While we are constantly baffled by the contrast between what our data show for the past and how contemporary commentators interpreted the social condition, we would be hesitant to extrapolate this into the future. Indeed, there are indications of a profound change in the most recent cohort that may require time to see their full significance. We have only scratched the surface in understanding the mechanisms that produce the life course patterns we observe. But these facts are a better starting point for this task than the speculations that tend to dominate this field both in social science and the public debate.

NOTES

1. The West German educational system has three main tracks: the *Hauptschule* is the main avenue into vocational training in the manual occupations. Graduation took place after normally 8 years until a ninth grade was added in the late 1970s. Lower secondary school is the main avenue into vocational training in clerical occupations and takes 10 years to complete. The *Gymnasium* is the academic track and takes 13 years to complete (Cortina, Baumert, Leschinsky, Mayer, & Trommer, 2003).

2. The duration of labor force participation until age 27 is very similar for the cohorts born after 1950 and therefore the increase of job shifts for the 1971 cohort cannot be contributed to search behavior at the beginning of the work life – if that

were the case, we should see the same numbers of job shifts for the 1960 and 1964 cohorts.

ACKNOWLEDGMENTS

The surveys of the GLHS have been funded by the Deutsche Forschungsgemeinschaft (Sonderforschungsbereich 3 "Mikroanalytische Grundlagen der Gesellschaftspolitik"), the Max Planck Society for the Advancement of Science, the Institut für Arbeitsmarkt- und Berufsforschung, and the European Social Fund.

We gratefully acknowledge the support of the Max Planck Institute for Human Development in Berlin, especially the research assistance of Karola Rockmann and the help of Helena Maravilla in preparing the manuscript. We thank Ross Macmillan for his helpful comments. The paper is part of a research project supported by the Center for Research on Inequalities and the Life Course (CIQLE) at Yale University.

REFERENCES

Anderson, M. (1985). The emergence of the modern life cycle in Britain. *Social History, 10*, 69–87.

Baltes, P. B., & Mayer, K. U. (Eds) (1999). *The Berlin aging study: Aging from 70 to 100*. Cambridge and New York: Cambridge University Press.

Beck, U. (1986). *Risikogesellschaft. Auf dem Weg in eine andere Moderne*. Frankfurt am Main: Suhrkamp Verlag [English translation: *Risk society. On the road to another modernity*].

Beck, U. (1999). *Schöne neue Arbeitswelt: Vision: Weltbürgergesellschaft* (2nd ed.). Frankfurt am Main et al.: Campus.

Becker, H. A., & Hermkens, P. L. J. (Eds) (1993). *Solidarity of generations: Demographic, economic and social change, and its consequences*. Amsterdam: Thesis Publishers.

Boudon, R. (1974). *Education, opportunity and social inequality: Changing prospects in western society*. New York, NY: Wiley.

Brückner, E. (1993). *Lebensverläufe und gesellschaftlicher Wandel. Konzeption, Design und Methodik der Erhebung von Lebensverläufen der Geburtsjahrgänge 1919–1921 (Teil I–V)* (*Materialien* aus der Bildungsforschung 44). Berlin: Max-Planck-Institut für Bildungsforschung.

Brückner, E., & Mayer, K. U. (1998). Collecting life history data. Experiences from the German Life History Study. In: J. Z. Giele & J. G. H. Elder (Eds), *Methods of life course research: qualitative and quantitative approaches* (pp. 152–181). Thousand Oaks: Sage Publications.

Brückner, H., & Mayer, K.U. (1995). *Lebensverläufe und gesellschaftlicher Wandel. Konzeption, Design und Methodik der Erhebung von Lebensverläufen der Geburtsjahrgänge 1954–1956, 1959–1961 (Teil I–III)* (*Materialien* aus der Bildungsforschung 48). Berlin: Max-Planck-Institut für Bildungsforschung.

Büchel, F., Grip, A. de, & Mertens, A. (Eds) (2003). *Overeducation in Europe: Current issues in theory and policy*. Cheltenham, UK: Edward Elgar.

Buchmann, M. (1989). *The script of life in modern society. Entry into adulthood in a changing world.* Chicago/London: University of Chicago Press.

Bumpass, L. L., & Lu, H.-H. (2000). Trends in cohabitation and implications for children's family contexts in the United States. *Population Studies*, *54*, 29–41.

Cortina, K. S., Baumert, J., Leschinsky, A., Mayer, K. U., & Trommer, L. (Eds) (2003). *Das Bildungswesen in der Bundesrepublik Deutschland. Strukturen und Entwicklungen im Überblick*. Reinbek: Rowohlt.

DiPrete, T. A. (2002). Life course risks, mobility regimes, and mobility consequences: A comparison of Sweden, Germany, and the United States. *American Journal of Sociology*, *108*, 267–309.

Ebbinghaus, B. (2002). *Exit from labor. Reforming early retirement and social partnership in Europe, Japan, and the USA*. Unpublished Habilitation thesis, University of Cologne, Cologne, Germany.

Espenshade, T. J. (1985). Marriage trends in America: Estimates, implications, and underlying causes. *Population and Development Review*, *11*, 193–245.

Esping-Andersen, G. (1999). *Social foundations of postindustrial economies*. Oxford: Oxford University Press.

Freeman, R. B. (1976). *The overeducated American*. New York: Academic Press.

Hareven, T. (1981). *Family time and industrial time*. Cambridge: Cambridge University Press.

Hareven, T. (1986). Historical changes in the social construction of the life course. *Human Development*, *29*, 171–180.

Heinz, W. R. (2003). From work trajectories to negotiated careers: The contingent life course. In: J. T. Mortimer & M. J. Shanahan (Eds), *Handbook of the Life Course* (pp. 185–204). New York: Kluwer Academic.

Held, T. (1986). Institutionalization and deinstitutionalization of the life course. *Human Development*, *29*, 157–162.

Hillmert, S. (2002). Labour market integration and institutions: an Anglo-German comparison. *Work, Employment and Society*, *16*, 675–701.

Hillmert, S. (2003). Ausbildungs- und Berufsverläufe der Geburtskohorten 1964 und 1971 in Westdeutschland: die jüngeste Teilstudie der Lebensverlaufsstudie als Scientific Use File. *ZA-Information*, *53*, 167–177.

Hillmert, S., & Mayer, K. U. (2004). *Geboren 1964 und 1971. Ausbildungs- und Berufschancen in den achtziger und neunziger Jahren*. Wiesbaden: VS-Verlag für Sozialwissenschaften.

Hoffmann-Nowotny, H.-J. (1980). *Auf dem Wege zur autistischen Gesellschaft? [Dokumentation der Jahrestagung 1980 der Deutschen Gesellschaft für Bevölkerungswissenschaft e.V.]*. Bielefeld: Deutsche Gesellschaft für Bevölkerungswissenschaft e.V.

Huinink, J., & Mayer, K. U. (1995). Gender, social inequality, and family formation, in West Germany. In: K. O. Mason & A.-M. Jensen (Eds), *Gender and family change in industrialized countries* (pp. 168–199). Oxford: Clarendon Press.

Inglehart, R. (1977). *The silent revolution. Changing values and political styles among western publics*. Princeton, NJ: Princeton University Press.

Jacob, M. (2003). *Ausmaß, Struktur und Ursachen von Mehrfachausbildungen. Eine Analyse von Ausbildungsverläufen in den achtziger und neunziger Jahren in Westdeutschland*. Unpublished Dissertation, Freie Universität, Berlin.

Junge, M. (Ed.) (2002). *Individualisierung*. Frankfurt/New York: Campus [*Individualization*] (In German).

Kohli, M. (1985). Die Institutionalisierung des Lebenslaufs. Historische Befunde und theoretische Argumente. *Kölner Zeitschrift für Soziologie und Sozialpsychologie*, *37*(1), 1–29 [*The Institutionalization of the Life-Course. Historical Evidence and Theoretical Arguments*.] (In German).

Kohli, M. (1986). Gesellschaftszeit und Lebenszeit - Der Lebenslauf im Strukturwandel der Moderne. In: J. Berger (Ed.), *Die Moderne - Kontinuitäten und Zäsuren (Soziale Welt, Sonderband 4)* (pp. 183–208). Göttingen: Schwartz.

Kohli, M. (1989). Institutionalisierung und Individualisierung der Erwerbsbiographie: Aktuelle Veränderungstendenzen und ihre Folgen. In: D. Brock, H. R. Leu, C. Preiss & H.-R. Vetter (Eds), *Subjektivität im technischen und sozialen Wandel* (pp. 249–278). München: DJI Verlag.

Konietzka, D., & Huinink, J. (2003). Die De-Standardisierung einer Statuspassage? Zum Wandel des Auszugs aus dem Elternhaus und des Übergangs in das Erwachsenenalter in Westdeutschland. *Soziale Welt*, *54*, 285–312 [*The de-standardization of a status passage? On the changing process of leaving the parental home and the status passage to adulthood in West Germany*.] (In German).

Kurz, K., Hillmert, S., & Grunow, D. (2002). *Increasing instability in employment careers? Men's job mobility and unemployment in West Germany: A comparison of the birth cohorts 1940, 1955, and 1964*. University of Bielefeld: Global Life Working Paper no. 34.

Mayer, K. U. (1991). Life courses in the welfare state. In: W. Heinz (Ed.), *Theoretical advances in life course research* (pp. 171–186). Weinheim: Deutscher Studien Verlag.

Mayer, K. U. (1994). The postponed generation. Economic, political, social, and cultural determinants of changes in life course regimes. In: H. A. Becker & P. L. J. Hermkens (Eds), *Solidarity of generations. Demographic, economic, and social change, and its consequences* (pp. 47–69). Amsterdam: Thesis Publishers.

Mayer, K. U. (1995). Gesellschaftlicher Wandel, Kohortenungleichheit und Lebensverläufe. In: P. A. Berger & P. Sopp (Eds), *Sozialstruktur und Lebenslauf* (pp. 27–47). Opladen: Leske + Budrich [*Social Change, Cohort Inequality and Life Courses*.]. (In German).

Mayer, K. U. (2001). The paradox of global social change and national path dependencies: Life course patterns in advanced societies. In: A. E. Woodward & M. Kohli (Eds), *Inclusions and exclusions in European societies* (pp. 89–110). London: Routledge.

Mayer, K. U. (2004). Whose lives? How history, societies and institutions define and shape life courses. *Research in Human Development*, *1*, 161–187.

Mayer, K.U., & Brückner, E. (1989). *Lebensverläufe und Wohlfahrtsentwicklung. Konzeption, Design und Methodik der Erhebung von Lebensverläufen der Geburtsjahrgänge 1929–1931, 1939–1941, 1949–1951 (Teil I-III)* (Materialien aus der Bildungsforschung 35). Berlin: Max-Planck-Institut für Bildungsforschung.

Mayer, K., & Müller, W. (1986). The state and the structure of the life course. In: A. B. Sorensen, F. E. Weinert & L. R. Sherrod (Eds), *Human development and the life course: Multidisciplinary perspectives* (pp. 217–245). Hillsdale, NJ: Lawrence Erlbaum Associates Publishers.

Mayer, K., & Schoepflin, U. (1989). The state and the life course. *Annual Review of Sociology*, *15*, 187–209.

Mertens, A. (1998). *Labor mobility and wage dynamics: An empirical study for Germany in comparison with the United States*. Aachen: Shaker Verlag.

Mertens, A., & Mayer, K.U. (2004). *The disappearance of life-long jobs – A myth? Job stability of labour market entrants in West Germany from the 1950's to the 1990's.* Max Planck Institute for Human Development. Unpublished Manuscript.

Meyer, S., & Schulze, E. (1983). Nichteheliche Lebensgemeinschaften – Alternativen zur Ehe? Eine internationale Datenübersicht. *Kölner Zeitschrift für Soziologie und Sozialpsychologie, 35*, 735–754.

Mills, M., & Blossfeld, H.-P. (2003). Globalization, uncertainty and changes in early life courses. *Zeitschrift für Erziehungswissenschaft, 6*, 188–219.

Modell, J., Furstenberg, F. F., Jr., & Hershberg, T. (1976). Social change and transitions to adulthood in historical perspective. *Journal of Family History, 1*, 7–32.

Myles, J. (1993). Is there a Post-Fordist life course? In: W. R. Heinz (Ed.), *Institutions and gatekeeping in the life course* (pp. 171–185). Weinheim: Deutscher Studien-Verlag.

Newman, K. (1999). *Falling from grace: Downward mobility in the age of affluence.* Los Angeles/ Berkeley: University of California Press.

Oppenheimer, V. K., & Kalmijn, M. (1995). Life cycle jobs. *Research in Social Stratification and Mobility, 14*, 1–38.

Riley, M. W., Kahn, R. L., & Foner, A. (1994). *Age and structural lag. Society's failure to provide meaningful opportunities in work, family, and leisure.* New York: Wiley.

Rhodes, M., & Apeldoorn, B. V. (1997). Capitalism versus capitalism in Western Europe. In: M. Rhodes, P. Heywood & V. Wright (Eds), *Developments in West European politics* (pp. 171–189). London: Macmillan.

Rowntree, B. S. (1914). *Poverty. A study in town life.* London: Nelson.

Schelkle, W., Krauth, W. -H., Kohli, M., & Elwert, G. (Eds) (2000). *Paradigms of social change: Modernization, development, transformation, evolution.* Frankfurt/New York: Campus/ St. Martin's Press.

Schimank, U. (2002). *Das zwiespältige Individuum. Zum Person-Gesellschaft-Arrangement der Moderne.* Opladen: Leske + Budrich [*The Ambivalent Individual. The Modern Person-Society Arrangement.*] (In German).

Sennett, R. (2000). *The corrosion of character.* New York: W. W. Norton and Company.

Shanahan, M. J. (2000). Pathways to adulthood in changing societies: Variability and mechanisms in life course perspective. *Annual Review of Sociology, 26*, 667–692.

Wagner, M. (1996). Lebensverläufe und gesellschaftlicher Wandel: Die westdeutschen Teilstudien. *ZA-Information, 38*, 20–27.

Wohlrab-Sahr, M. (1992). Institutionalisierung oder Individualisierung des Lebenslaufs?—Anmerkungen zu einer festgefahrenen Debatte, 1-19. *BIOS Zeitschrift für Biographieforschung und Oral History, 5.*

Zapf, W. (1987). *Individualisierung und Sicherheit: Untersuchungen zur Lebensqualität in der Bundesrepublik Deutschland.* München: Beck.

Zapf, W. (1991). The role of innovations in modernization theory. *International Review of Sociology New Series, 3*, 83–94.

THE STRUCTURE OF THE LIFE COURSE: GENDER AND RACIOETHNIC VARIATION IN THE OCCURRENCE AND SEQUENCING OF ROLE TRANSITIONS

Pamela Braboy Jackson and Alexandra Berkowitz

ABSTRACT

There is an on-going debate among life course scholars regarding the extent to which the life course is becoming more individualized. Before speaking on this issue, however, more information is needed regarding the actual structure of the life course. We define that structure in terms of the occurrence of role transitions and the order in which these transitions are experienced. Using current and retrospective life history information reported in the National Survey of Families and Households (NSFH), we focus on the distribution of work and family role transitions and then examine the order (or sequencing) in which these social roles are both acquired and lost across the life course, paying particular attention to gender and racioethnic variation. Findings suggest that most adults are situated in three role positions: married, working parent. Yet, men are more likely than women to be working and African-American and Latino women are more likely to be in parenting roles. In an examination of role

The Structure of the Life Course: Standardized? Individualized? Differentiated?
Advances in Life Course Research, Volume 9, 55–90

ISSN: 1040-2608/doi:10.1016/S1040-2608(04)09003-3

sequencing, evidence was found for a "normative" life course where most adults first find work, then marry, and later have children. Despite this prevalent pattern, more variation in sequencing behavior is evident among women and ethnic minorities. We conclude with a discussion of the implications of these findings for the individualization hypothesis.

What is the structure of the life course? The answer to this question depends upon how one defines the life course. For some, it is best characterized by entry into adulthood (Buchmann, 1989; Hagan & Wheaton, 1993). For others, the life course goes beyond the type or timing of the first role transition to encompass a series of transitions that occur in some order or sequence over a specified period of time (Jackson, 2004; Hogan, 1978; Marini, 1984; Rindfuss, Swicegood, & Rosenfeld, 1987). Other scholars define the life course in lieu of an end point, such as retirement (Kohli, Rein, Guillemard, & van Gunsteren, 1991; Han & Moen, 1999; Moen, Kim, & Hofmeister, 2001; O'Rand, 1988).

We combine these different perspectives by documenting initial entry into adulthood and the ordering of roles across the life course among adults. We include those who are still engaged in social roles, as well as those who may be permanently disengaged from certain roles, notably marriage and employment. Specifically, we examine the sequencing of transitions within particular configurations of primary social roles. Primary roles are defined as highly valued social positions acquired through some degree of personal effort, including those of worker, spouse, and parent. With this approach, we extend contemporary discussions by addressing two oft-ignored issues: (1) the extent of gender and racioethnic[1] diversity in life course patterns; and (2) conceptual and measurement limitations associated with sole focus on initial entry into, rather than entry into and exit from roles.

DIVERSITY IN THE LIFE COURSE

The life course has been defined as the ordered sequence of events or role transitions that individuals experience as they age from birth to death (Elder, 1985, 1997; Pallas, 1993). Sequencing refers to which social roles precede or follow other social roles (Brim & Ryff, 1980). This definition implies that it is as important to understand *what* people are doing (i.e., are they working) as it is to know *how* they did it (were they married before they started working). Thus, role sequencing is a useful concept for studying the structure of the life course (Modell, 1989).

In examining the transition to adulthood among a sample of men born between 1907 and 1952, Hogan (1978) reports on the sequencing of three role transitions: school exit, entry into first marriage, and entry into first job. He found that 46% of the sample first completed their schooling, then started working full-time, and were later married. A further 8% were on track to do so, as they had completed their schooling and later found a job, but had yet to marry. Hogan typified these patterns as "normative" in U.S. society. A variety of other sequences were reported but the percentages of men following these were considerably smaller and quite dispersed. Hogan's (1981) later work characterizes 75% of the men born before 1910 as having a normative order of role transitions.

Hogan's central focus was cohort variation. He did, nonetheless, examine the impact of racioethnicity on the likelihood of following the normative order or falling into two nonnormative categories. Here, white men were more likely than African-American or Spanish men to complete school, find a job, and then get married. Spanish men were more likely than other men to begin work prior to finishing school or to complete school, marry, and then find a job (labeled as intermediate nonnormative). African-American men were more likely than other men to marry while in school (labeled as extreme nonnormative). Despite these differences, it remains unclear whether the modal sequence differs by racioethnicity as these findings could simply reflect wider variation in life course patterns among minority men. The current study explores racioethnic differences in the degree of variation in sequencing behavior. In particular, we examine sequencing patterns among a group of African-American and Latino men (of varying cohorts) and compare them with those found for white men.

Women's life course patterns may be more complicated than men's since women's worklives are often affected by decisions regarding family formation (Hogan, 1985; Sweet, 1973). Marini (1984), utilizing information gathered in a 15-year follow-up study of a sample of Illinois high school students initially interviewed in 1957 and 1958, finds that the modal sequence among the resulting sample of adults (aged 29–33) was school completion, beginning first job, getting married, and then having a first child. Forty-three percent of men and women were classified as experiencing these four role changes in this order.[2]

Marini's (1984) study raises a further issue of considerable importance. While the majority of women followed the normative order, white women were more than twice as likely as white men to enter adulthood through marriage (26.7% of women compared with 12.8% of men). This is consistent with Hogan's earlier work, showing that marriage before work was much less prevalent than work before marriage (14% versus 64%). This may indicate that the normative prescription in American society to work before starting a family is much stronger for men than women. Other research suggests even more diversity, noting that employed, low-income African-American men are more likely to have their first child before getting married (Testa & Krogh, 1995). Thus, both research and general conclusions on the centrality of the work to marriage transition in the structure of the modern life course may, more generally, represent a white male perspective (Giele, 1982; Hogan, 1978; Rossi, 1980; Ryff, 1985). The current study will allow us to reconstruct life course patterns by using retrospective life history information provided by a national sample of men and women of different ethnic backgrounds. As such, we address the following set of unanswered questions regarding the structure of the life course: (1) is the work-to-marriage transition more prevalent among men than women, regardless of racioethnicity; (2) are there sequencing patterns that seem to define the life course experiences of women moreso than men; and (3) if so, are these patterns consistent across racioethnicity?

THE SCHOOL-TO-WORK TRANSITION

Much of the work in the area of role sequencing focuses on the transition from school to work using reconstructed life histories. The emphasis on the school-to-work transition reflects the influence of the status attainment and human capital approaches (see Featherman & Carter, 1976; Sewell & Hauser, 1975). Two important consequences of this influence, however, are a general perception of extreme disorderliness in the life course and a focus on the timing of the *first* occurrence of the role transition in question. This latter issue is especially relevant because ignoring role exits or repeat role entry can generate overly simple and misleading images of life course patterns.

One of the most dramatic changes in life course patterns has been the reordering of school, work and family events in recent decades. In particular, the timing of school exit has impinged upon the typical ordering of other role transitions. Individuals are remaining in school longer, marrying while in school, and returning to school after getting married or having children (Featherman, Hogan, & Sorensen, 1984; Rindfuss et al., 1987). The inclusion of school exit in the identification of a typical sequence, therefore, has led a number of researchers to conclude that the life course is increasingly more flexible than rigid and that the transition from school to work is more fluid than implied by typical status attainment and life course models (Rindfuss et al., 1987).

Pallas (1993) most explicitly links the role of schooling to the more general concern of differentiation in life patterns. He argues that "…as young people's participation in schooling has increased, sharp differentiation among schooling, work, and adult family formation has given way to increasing overlap and reversibility in the roles individuals hold in their early careers" (p. 439). An important issue raised in much of the life course literature, then, is the intimate interplay between the role played by the state in structuring the life course (e.g., determining when individuals must attend school or when young people can work a full-time job) and the subsequent autonomy provided to individuals (via the loosening of controls by family and peers) that may then lead to a higher prevalence of differentiated or individualized life patterns (Buchmann, 1989; Mortimer & Shanahan, 2003; Shanahan, 2000). Our goal is not to fully disentangle these issues but instead to examine whether a certain level of stability and disorder in the life course can be expected.

THE PERCEPTION OF DISORDER IN THE LIFE COURSE

We contend that because people move in and out of academic and training programs (after high school) before, during, and after other adult role transitions, the inclusion of school transitions creates a strong and perhaps misleading impression of extreme disorder in the life course. For example, Marini's (1984) research revealed 16 patterns effectively characterized adults in terms of the order in which they experienced school exit, entry into marriage, entry into first job, and/or entry into parenthood. However, if one were to bracket variation due to the order of school exit, only five sequencing patterns emerge. The most dominant pattern was to enter the paid labor force, marry, and then have children, similar to findings for Hogan's (1978, 1981) sample of men born earlier in the century.[3]

While schooling may be a reversible transition, the resources that characterize one's level of education are not as reversible (Voydanoff & Kelly, 1984). Schooling, in fact, is often viewed as a means to an end rather than an end, in and of itself (Cameron & Heckman, 1998; Murnane, Willett, & Boudett, 1995; Pallas, 1993, 1994). Thus, some view schooling as a resource rather than a role transition. We adopt this view to capture the fact that the educational role can be used to grant access to other types of resources, especially cultural, material, and social capital (Callero, 1994). This is not always the case with primary social roles (such as spouse and worker) whose benefits depend heavily upon the context in which they are acquired (Jackson, 2004). Unlike the student role, primary social roles are both transition events when they happen and then become personal contexts after they have occurred. It is much more useful for our purposes here, then, to think of this particular role as carrying benefits regardless of the order in which the maximum level is achieved.

MIS-REPRESENTATION OF THE LIFE COURSE

Rindfuss et al. (1987) have noted how focusing on role entry (at the expense of role exit) can jeopardize understandings of the school-to-work transition. These investigators explored the relationship between making a smooth transition from school to another activity state (e.g., work, homemaking, military) and the timing of parenthood among a sample of white men and women drawn from the National Longitudinal Survey of the High School Class of 1972 (NLS) and studied through age 24. They found that only 51% of the men and 40% of the women actually followed a consistent sequence (measured in 13 ways with the underlying requirement being no interruptions between the defined states) over the 8-year period. The findings from the study were so complex that the investigators concluded that both "orderly" and "disorderly" sequences affect when people become parents. This conclusion was attributed to the fact that the category labeled disorderly included individuals who were engaged in activities not typically considered in life course studies, such as being out of the paid labor force.

By bringing attention to the importance of work exits, even brief ones, that may characterize school-to-work transitions, Rindfuss et al. (1987) raise a critical issue in the methodology of role sequencing: mis-representation. This problem arises in cases where certain decision-making rules are not clearly specified. Although we do not believe that investigators purposefully ignore information about the ways in which life course patterns are determined, an unintended consequence of brevity in a methods section results in the possibility that the percentage of any particular pattern is inflated. This can occur in two ways.

First, inflated percentages can occur when individuals who have acquired particular social roles are included in role configurations with those who have not or at least not yet. For example, the men in Hogan's (1978) study who only experienced two of the three role transitions (e.g., E = school exit and J = entry into first job) were classified according to the sequencing of the two events they had experienced, assuming then that the missing transition (e.g., M = marriage) would be the third transition experienced across the life course. An example of such a group is unmarried men who had completed their educations and then entered the job market. A similar decision-making model was applied to those who had yet to enter the paid labor force (EMs were combined with EMJs). Other investigators adopt similar approaches (see Daniels & Weingarten, 1982; Marini, 1984).

A second way in which inflation occurs is when information on role exits is ignored. A life course pattern cannot be fully evaluated without information on role exits. Imagine the case where the sequencing of entry into first job (J) and first marriage (M) are examined. Individuals who experienced a job exit (U) but who worked prior to marriage (JUM) will be included in the sequencing category with those who have not experienced a job exit, but worked prior to marriage (JM) if information on role exits is not utilized. This issue, in fact, may help explain why Marini (1984) found so few gender differences in the proportion of adults following the normative sequencing of role transitions. Women's lives often involve interruptions in their employment careers, whereas men tend to be in the workforce continuously from the time they leave school until retirement (Ryff, 1985). Relatedly, prior research does little to denote whether social roles are occupied simultaneously. This omission in life course research is especially problematic given the growing (and often fluctuating) divorce and unemployment rates. The mounting evidence of role overlap and the simultaneous movement in and out of multiple roles cast doubt on these practices (see discussions in Featherman et al., 1984; Macmillan & Eliason, 2003; Shanahan, 2000).

In general, there is little research describing life course patterns that include role exits. Abeles, Steel, and Wise (1980) use panel data initially collected in 1960 from a national sample of high school students (9th–11th graders born between 1942 and 1945) who were re-interviewed 1, 5, and 11 years after each class's high school graduation. This study actually examined three role characteristics (timing, order, and discontinuity) across four social roles (school, marriage, job, and parenthood). However, order was conceptualized for pairs of transitions only (e.g., school to work, school to marriage, work to marriage). In addition, role discontinuity was measured as a separate phenomenon not clearly linked to the ordering of social roles. This is a third form of mis-representation. As such, it is not clear where in the life course the interruption occurred relative to other roles held at the time. Ultimately, the approach adopted in the current study will address these sources of misrepresentation and allow for a more complete picture of the sequencing of social roles.

THE COMPLEXITY OF ROLE SEQUENCING

In general, some criteria are necessary to adopt in this type of study of the life course since the number of possible sequencing categories increases exponentially (see Rindfuss et al., 1987). The mathematical formula that represents this phenomenon for the occurrence of each transition event with no role reversals (or exits) is a simple *n factorial* (Agresti & Finlay, 1986). For example, the number of possible combinations of two irreversible role transitions is

$$2! = 1 \times 2 = 2 \tag{1}$$

for three transitions,

$$3! = 1 \times 2 \times 3 = 6 \tag{2}$$

for four transitions,

$$4! = 1 \times 2 \times 3 \times 4 = 24 \tag{3}$$

As a concrete example, we consider the process of becoming a married, working parent. There are, in fact, six ways to enter into this role configuration: (1) first find work, then get married, and later have a child; (2) first find work, then have a child, and later marry; (3) first marry, then find work, and later have a child; (4) first marry, then have a child, and later work; (5) first have a child, then get married, and later work; and (6) first have a child, find work, and marry later.

In cases where the investigator is interested in those who may still be at risk for experiencing role transitions *and* those who have experienced the total number of events possible (as defined by the investigator), the formula is revised to include information on the number of possible combinations for the reduced order of transitions. This situation requires consideration of binomial probabilities where n choose k (Pulskamp, 1990). For example, in the case of three events, we would have

$$\begin{aligned}&(3/3 \times 3!) + (3/2 \times 2!) + (3/1 \times 1!)\\&\quad = (1 \times 3 \times 2 \times 1) + (1.5 \times 2 \times 1) + (3 \times 1)\\&\quad = 6 + 3 + 3 = 9\end{aligned} \tag{4}$$

The number of possible combinations increases when we consider five transitions that include exits from primary roles (i.e., entry into labor force, first marriage, first parenthood, job exit, marital exit)

$$5!/4 = (5 \times 4 \times 3 \times 2 \times 1)/4 = 30 \qquad (5)$$

And, of course, there are an infinite number of possibilities with multiple exits. Using a priori categorization, then, is a necessary and efficient strategy.

In this study, we handle the complexity of the life course by first describing the type of role configurations occupied by an ethnically diverse national sample of adults. From there, we present information on the actual life course patterns (using retrospective life history information) reported by adults situated in five of these role configurations: (1) childless, married workers; (2) never-married, working parents; (3) married, working parents; (4) married, unemployed parents; and (5) divorced, working parents.

Research conducted by Hogan (1978) and Marini (1984) bring important attention to the structure of the life course (also see Modell, Furstenberg, & Hershberg, 1976). Recent work by Rindfuss (1991), however, describes the disproportionate concentration of certain role exits (especially divorce and job loss) during the young adult years. It becomes important, then, to extend the life course beyond role gains. We focus on the sequencing of marriage (gains and losses); employment (gains and losses), and parenthood (viewed as irreversible). In essence, we contribute to the life course literature as well as general research on social roles by: (1) having a diverse sample of adults (gender, racioethnicity, and birth cohorts); and (2) recreating the life course that includes role entries and role exits. This is largely a descriptive task but extends research in this area by depicting the life course "...as it is actually lived" (Rindfuss et al., 1987, p. 799).

DATA AND METHODS

This study uses data from the National Survey of Families and Households (NSFH), conducted in 1987 and 1988 (Sweet, Bumpass, & Call, 1988). Respondents were selected from a multistage probability sample of households in the United States. The main sample included 9,643 households. Minorities, single-parent families, families with step-children, cohabiting couples, and recently married persons were oversampled to yield a total sample of 13,017 respondents. The sample can be viewed as representative of noninstitutionalized persons in the United States. The response rate for the study was 75%. This study focuses on the sample of whites, African

Americans, and Hispanics (98% of the total sample). The category "Hispanic" includes respondents who report their racioethnicity as Mexican American, Puerto Rican, Cuban, or other Latino. Sixty-three percent of Hispanics, however, were Mexican. The groups were combined to increase the number of respondents in each cell, for purposes of comparison. We acknowledge that family structure can vary widely between these groups (see Bean & Tienda, 1987) but regard this as the subject of future research.

The NSFH is particularly useful for examining diversity in life course patterns because of the extensive life history information collected throughout the survey and the wide age range of the sample (18–65).[4] Respondents were asked the month and year of their marriage and could list up to five marital unions. The formerly married were also asked the dates of dissolution and how their marriage ended (divorce or widowhood). Work history tables were compiled for 10 job changes and fertility tables allowed for the listing of 12 birth dates.

Spouses are defined here as individuals who are currently married and living in the same household as their marital partner (formal marital unions). Persons who report being currently married but who are living away from their spouses for work-related reasons (not separated due to marital problems) also are included in the married categories. Separate categories are designated for respondents who are in transition from the spousal role (i.e., divorced, separated, widowed). *Workers* include individuals who have been employed (full- or part-time) for at least 6 months (see Marini, 1984; Menaghan, 1989 for a similar approach). The unemployed are those who have worked in the past but are currently out of the paid labor force. Those who are currently looking for work are not distinguished from those who are not seeking work. *Parents* have given birth to or fathered children. This restriction was included because the date on which respondents became step-, foster, and adoptive parents is not included in the fertility history table. Eight percent of the study sample are or have been nonbiological parents.

The ordering, or sequencing, of roles is a major focus of this research. Order is operationalized using information from the work, marital, and fertility history tables provided by respondents. If two role changes occurred in the same month, they were assumed to occur in the most common sequence (see Hogan, 1978; Marini, 1984 for a similar approach). Entry into (and exit from) these adult social roles is represented throughout this chapter by symbols. Entry into the workforce is represented by [W], marriage is [M], and parenthood is symbolized with a [P]. Exit from the marital arena through divorce, separation, or widowhood, are represented by a [D], [Sp], and [Wd], respectively. Unemployment is symbolized by a [U].

It is necessary to keep the possible ordering combinations manageable. Analyses that focus on role sequencing handle the number of possible combinations by limiting the number of times an individual can experience a role transition (see Rindfuss et al., 1987). Those eligible for this portion of the study includes respondents who have *experienced one marriage*, who have *held one or two full-time jobs* (30 + hr/week), and who have *no more than three children.* While this may seem restrictive, it captures the majority of the respondents in the selected sample. For example, among those who ever-married, 76% have only been married once. Similarly, among those who have ever been employed, 84% report having held only one or two jobs, and most parents (81%) have three or fewer children. In general, this selected sample represents nearly 70% of the respondents in the NSFH who belong to various role categories.[5]

In some instances, all of the information necessary to recreate role sequencing patterns was not available (missing dates). This issue was most pronounced among adults who are (or were) in their second job and these individuals are captured in the category of nonclassifiables. The largest group included in the nonclassifiables, however, include those whose life course patterns were, somehow, disrupted by divorce or job loss. While the actual sequencing patterns are known, the tables become too extensive to report every possible permutation. As such, only the primary role sequencing patterns are reported; individuals who actually followed the noted pattern. Tables that include parents have sequences represented by a single letter [P], regardless of the number of children. Here, however, it is important to note that all parenting occurred in succession. In order to provide the reader with some information on the structure of the life course for individuals included in the nonclassifiable categories, we created a variable that assessed the initial role transition (or what happened first). We report this information in the appropriate sections for each role configuration.

RESULTS

The Occurrence of Role Transitions/Role Configurations

Our analyses begin with role transitions and role configurations. These serve as the basis for our later analyses of role sequences in the transition to adulthood. Role configurations of the selected sample from the NSFH are shown in Table 1. The table is divided according to the type of role configuration currently occupied by the respondent. Role configurations include roles gained or lost. For example, the first row (entitled no roles) includes the frequencies for respondents who have never held the roles of spouse, worker, or parent. Information on those who have held only one role and who still occupy that role is presented in rows 2–4 (spouses, workers, parents), whereas those single role occupants who have lost their single role are represented in rows 5–8 (divorced, separated, widowed, and unemployed). This table is quite comprehensive of the selected samples. Between 83% and 95% of respondents are included in the table and most of the eligible sample is represented in this table (between 93% and 98%). As noted in the bottom panel of the table, in order to be classified in the role configuration table, respondents had to report no prior parenting experience (e.g., step-, foster, adoptive children) and at least one of the children ever born must still be living.

As shown in Table 1, very few adults are completely disengaged from society. While African-American men are approximately twice as likely as non-Hispanic white men to occupy no primary social role (row 1), the relative frequency of adults falling in this category is low for both groups (2.9% vs. 1.3%). In moving down the table to single role occupancy (rows 2–8), it is clear that the most traversed pathway into adulthood is through employment, regardless of gender or racioethnicity. On the other hand, there is a higher ratio of parents to workers among minority women compared to all other groups. For example, among African-American women approximately 4% are mothering while 6% are working. Similarly, among Latino women approximately 3.4% are mothering and 6.8% are working. Among non-Hispanic white women, we find that only 0.4% are mothering and 7.6% are working. In other words, there is diversity in the life course pathway even in the initial entry into adulthood that is evidenced most pronouncedly among minority women.

Although it may seem then that minority women have a common experience in regards to role diversity, a closer look at dual role occupants adds yet another twist to the story (rows 9–11). The majority of white men and women occupy the roles of married workers (9.2% and 5.8%, respectively). This is also the highest percentage of dual role occupants for Latino men (6.1%). In contrast, African-American men and women are situated in the dual role category of working parent (row 11). Here, approximately 6.6% of African-American men and 10% of African-American women occupy this configuration. Latino women similarly stand out from the other gender and racioethnic groups. The highest percentage of dual role occupants are married parents (row 10) at 8.8%. As shown in rows 12–22, very few adults situated in two roles have experienced role losses, with the exception of African-American women who are currently unemployed parents (7.2%).

Table 1. Frequency Distribution Showing the Current Role Configurations of Selected Sample of Adults in the NSFH (1987–1988) by Racioethnicity and Gender.

Role Configuration[a]	Racioethnicity											
	White				African-American				Latino			
	Men		Women		Men		Women		Men		Women	
	N	%	*N*	%	*N*	%	*N*	%	*N*	%	*N*	%
No roles												
1 Never-married, Never-employed childless	46	1.3	51	1.0	22	2.9	21	1.5	8	2.2	4	0.7
Single roles												
Gained												
2 [M] Spouses/married	3	0.1	16	0.3	1	0.1	3	0.2	0	0.0	0	0.0
3 [W] Workers/employed	543	15.6	391	7.6	126	16.3	85	6.0	64	17.7	40	6.8
4 [P] Parents	1	0.0	22	0.4	3	0.4	57	4.0	0	0.0	20	3.4
Lost												
5 [D] Divorced	1	0.0	0	0.0	0	0.0	0	0.0	0	0.0	0	0.0
6 [Sp] Separated	0	0.0	0	0.0	0	0.0	0	0.0	0	0.0	0	0.0
7 [Wd] Widowed	0	0.0	5	0.1	0	0.0	8	0.6	0	0.0	1	0.2
8 [U] Unemployed	85	2.4	81	1.6	33	4.3	31	2.2	12	3.3	9	1.5
Dual roles												
Gained												
9 [MW] Married workers	320	9.2	297	5.8	24	3.1	16	1.1	22	6.1	23	3.9
10 [MP] Married parents	6	0.2	149	2.9	0	0.0	25	1.8	1	0.3	52	8.8
11 [WP] Working parents	26	0.7	61	1.2	51	6.6	143	10.0	6	1.7	18	3.1
Lost 1												
12 [DW] Divorced workers	75	2.2	92	1.8	6	0.8	10	0.7	3	0.8	4	0.7
13 [SpW] Separated workers	9	0.3	9	0.2	6	0.8	13	0.9	1	0.3	1	0.2
14 [WdW] Widowed workers	3	0.1	13	0.3	1	0.1	7	0.5	1	0.3	0	0.0

Table 1. *(Continued)*

Role Configuration[a]	Racioethnicity											
	White				African-American				Latino			
	Men		Women		Men		Women		Men		Women	
	N	%	*N*	%	*N*	%	*N*	%	*N*	%	*N*	%
15 [Dp] Divorced parents	0	0.0	32	0.6	1	0.1	9	0.6	0	0.0	7	1.2
16 [SpP] Separated parents	0	0.0	13	0.3	1	0.1	14	1.0	0	0.0	19	3.2
17 [WdP] Widowed parents	3	0.1	89	1.7	0	0.0	21	1.5	0	0.0	9	1.5
18 [UM] Unemployed spouses	50	1.4	86	1.7	7	0.9	7	0.5	1	0.3	1	0.2
19 [UP] Unemployed parents	3	0.1	41	0.8	12	1.6	103	7.2	1	0.3	16	2.7
Lost 2												
20 [UD] Unemployed divorcees	20	0.6	21	0.4	1	0.1	3	0.2	1	0.3	0	0.0
21 [Usp] Unemployed separatees	5	0.1	7	0.1	4	0.5	4	0.3	1	0.3	1	0.2
22 [Uwd] Unemployed widows	27	0.8	81	1.6	11	1.4	32	2.2	1	0.3	2	0.3
Multiple roles												
Gained												
23 [MWP] Married working parents	1,223	35.2	1,167	22.7	200	25.9	207	14.5	142	39.3	118	20.0
Lost 1												
24 [DWP] Divorced working parents	212	6.1	505	9.8	50	6.5	135	9.5	15	4.2	41	6.9
25 [SpWP] Separated working parents	44	1.3	100	1.9	31	4.0	75	5.3	10	2.8	24	4.1
26 [WdWP] Widowed working parents	33	1.0	118	2.3	9	1.2	45	3.2	3	0.8	10	1.7
27 [MUP] Married unemployed parents	406	11.7	855	16.6	66	8.6	89	6.2	40	11.1	80	13.6
Lost 2												
28 [DUP] Divorced unemployed parents	32	0.9	183	3.6	24	3.1	47	3.3	4	1.1	34	5.8
29 [SpUP] Separated unemployed parents	8	0.2	58	1.1	18	2.3	70	4.9	3	0.8	23	3.9
30 [WdUP] Widowed unemployed parents	75	2.2	436	8.5	17	2.2	94	6.6	7	1.9	19	3.2

No. of respondents in table	3,259	4,979	725	1,374	347	576
No. of respondents in NSFH	3,874	5,545	872	1,519	397	608
% of NSFH sample captured	84.1	89.8	83.1	90.5	87.4	94.7
No. of eligible respondents	3,472	5,141	7,71	1,426	361	590
% of eligible sample captured	93.9	96.8	94.0	96.4	96.1	97.6
Criteria for ineligibility						
Prior parenting experience	376	341	92	78	33	15
(ex: step-, foster, adoptive children only)						
Child death	26	63	9	15	3	3
TOTAL Ineligible	402	404	101	93	36	18

[a]Role configuration categories are mutually exclusive.

The largest percentage of adults is situated in the role configuration of married, working parent (row 23). Furthermore, regardless of racioethnicity, a larger percentage of men is represented in this multiple role category (see Menaghan, 1989 for similar findings). However, a higher percentage of non-Hispanic white (22.7%) and Latino (20%) women are married, working parents compared to African-American women (14.5%). A higher percentage of women than men can be found among the divorced (rows 24 and 28) and widowed (rows 26 and 30). Interestingly, the pattern for unemployment among multiple role occupants varies by gender and racioethnicity. Among non-Hispanic whites and Latinos, women are more likely to be married, unemployed parents (16.6% vs. 11.7%, for whites and 13.6% vs. 11.1%, respectively), whereas African men are more likely than African-American women to fall into this role configuration (8.6% vs. 6.2%, respectively).

In essence, there is much diversity in the initial pathway into adulthood as well as the subsequent unfolding of the life course. At the same time, there is enough commonality across some groups to possibly speak of an "American" experience. Employment, in fact, seems to be the role transition that links the experiences of men and women, whites and nonwhites. The diversity in role configurations becomes most evident when family roles are appended to the work role, with African Americans being much more likely to append parenting to the work role, whereas non-Hispanic whites and Latinos append marriage to the work role. By the time all groups experience multiple role transitions (marriage, work, and parenthood), however, the distribution in regards to other configurations appear equivalent. That is, the married, working parent configuration is the most prevalent across all groups.

This consideration of the structure of the life course as evidenced by role configurations has extended past research by more clearly demonstrating both the similarities and differences in role acquisition across three ethnic groups. Furthermore, we have demonstrated that societal images of men as workers and women as mothers fail to capture the dynamic nature of life course processes. Social roles are seldom held singly (one at a time) and movement into another transition creates a new configuration that can alter the meaning and consequences of role behaviors (Jackson, 2004).

Sequencing Patterns

Tables 2–6 describe life course patterns and thus incorporate the sequencing of roles as a theoretical and methodological issue. Only role configurations that would allow for comparison across at least two subgroups are included. For example, all groups can be compared in terms of the sequencing of role transitions among married, working parents. Yet, only a small percentage of Latino women occupy the category of married parents, therefore the sequences within this role category are not reported. Overall, sequencing patterns within five role categories are presented: (1) married workers (childless); (2) working parents (never-married); (3) married, working parents; (4) married, unemployed parents; and (5) divorced, working parents. Again, these individuals represent a subset of the selected role configurations. Information regarding the degree to which individuals in these configurations represent the total configuration (in the NSFH) is available at the bottom of each table.

Married Workers without Children

The first role configuration examined is childless, married workers. As shown in section A of Table 2, the vast majority of employed spouses entered the paid labor force and later married (WM – row 1). A higher percentage of women than men, however, married and later found work (MW – row 2). This gender difference is not as evident among non-Hispanic white adults who held two jobs.

The trend for African-American men and women is also quite striking, although the sample sizes of these groups becomes very small. As demonstrated in section A, a higher percentage of African-American men compared to their female counterparts are entering adulthood through paid employment. Latino men are also more likely than Latino women to work first (rather than marry first). Similar to the findings for whites, however, there is more variability among those who have held two jobs. Unfortunately, the small number of cases in this role configuration for African-American and Latino adults precludes any definitive statements about these patterns.

Working Parents, Never-married

Table 3 shows the sequencing behavior of non-Hispanic whites and African-American men and women who are never-married working parents. The sequencing of parenthood introduces a great deal of complexity in the life course. The more children people have the more distant work events can be from the initial experience of parenting. Nonetheless, there are striking race and gender differences among this group. Categories with one or fewer

Table 2. Percentage Distribution of Sequencing Patterns for White, Black, and Latino Married Workers (Childless) by Number of Jobs.

	White		African-American		Latino	
	Men %	Women %	Men %	Women %	Men %	Women %
Section A: 1 Job						
1 **WM**	93.5	83.7	100.0	71.4	92.3	75.0
2 **MW**	6.1	16.3	0.0	21.4	7.7	25.0
Nonclassifiables	0.4	0.0	0.0	7.2	0.0	0.0
	100.0	100.0	100.0	100.0	100.0	100.0
N	231	184	13	14	13	20
Section B: 2 Jobs						
3 **WMUW**	28.9	36.0	0.0	0.0	14.3	0.0
4 WUW**M**	47.4	46.0	100.0	100.0	28.6	0.0
5 **MW**UW	2.6	6.0	0.0	0.0	14.3	50.0
6 WU**MW**	13.2	10.0	0.0	0.0	28.6	50.0
Nonclassifiables	7.9	2.0	0.0	0.0	14.3	0.0
	100.0	100.0	100.0	100.0	100.0	100.0
N	38	50	7	1	7	2
Representation (%)	84	79	83	94	91	96

Note: W = work; M = marriage; U = unemployment/job loss.

respondents \are excluded from the table but included in the final tabulations in section C of this table.

First, we see in section A that the majority of white men and women and African-American men were working prior to the birth of their first child. Although this is the dominant pattern for white women, the variation in life course patterns is greater compared to their male counterparts: while 92.9% of white men found work prior to having a child, only 53.8% of white women followed this role sequence. Furthermore, among white women with two children, the majority had a child prior to working (71.4%). There were too few white men in the other categories to explore patterns. Among African-American men who have one or two children, the majority were gainfully employed prior to the birth of their children. This pattern is reversed, however, among those with three children. Like non-Hispanic white women, there is more variation in sequencing behavior among Black men compared to white men.

Second, African-American women are most likely to have children prior to full-time employment although white women are not much different.

Table 3. Percentage Distribution of Sequencing Patterns for White and Black Working Parents (Never-Married) by Number of Jobs and Number of Children.

	White				African-American			
	Men		Women		Men		Women	
	N	%	*N*	%	*N*	%	*N*	%
Section A: 1 job								
1 kid	14		26		22		41	
1 **WP**		92.9		53.8		72.7		39.0
2 **PW**		7.1		46.2		22.7		61.0
Nonclassifiables		0.0		0.0		4.6		0.0
		100.0		100.0		100.0		100.0
2 kids			7		4		20	
3 **WPP**				28.6		75.0		55.0
4 **PWP**				57.1		25.0		30.0
5 **PPW**				14.3		0.0		15.0
				100.0		100.0		100.0
3 kids					7		8	
6 **WPPP**						28.6		37.5
7 **PWPP**						14.3		25.0
8 **PPWP**						14.3		25.0
9 **PPPW**						14.3		12.5
Nonclassifiables						28.6		0.0
						100.0		100.0
Section B: 2 jobs								
1 kid			12		4		18	
10 **WP**UW				33.4		50.0		33.3
11 WU**WP**				8.3		25.0		0.0
12 **PW**UW				8.3		0.0		38.9
Nonclassifiables				50.0		25.0		27.8
				100.0		100.0		100.0
2 kids			3		2		9	
13 **WPP**UW				33.3		0.0		11.1
14 **WP**UWP				0.0		50.0		11.1
15 **PPW**UW				0.0		0.0		11.1
16 **PW**UWP				33.3		0.0		11.1
17 **PW**UPW				0.0		50.0		22.2
18 WU**PPW**				33.3		0.0		0.0
19 WU**PW**P				0.0		0.0		11.1
Nonclassifiables				0.0		0.0		22.2
				100.0		100.0		100.0

Table 3. *(Continued)*

	White				African-American			
	Men		Women		Men		Women	
	N	%	*N*	%	*N*	%	*N*	%
3 kids							6	
20 **WPUWPP**								33.3
21 PWUWPP								16.7
22 PWPUWP								33.3
23 WUPWPP								16.7
								100.0
Section C: Pooled	19		49		40		102	
24 **WP**		89.5		44.9		62.5		39.2
25 **PW**		10.5		42.8		25.0		49.1
Nonclassifiables		0.0		12.3		12.5		11.7
		100.0		100.0		100.0		100.0
Representation (%)		73		80		78		71

Note: W = work; P = parenthood; U = unemployment/job loss.

Among African-American women with one child, only 39% were working prior to parenthood. Among those with two children, 55% were working prior to parenthood. Among those with three children, 37% were working prior to parenthood. Although the numbers become more dispersed when those who are currently in a second job are considered (see section B), overall the pattern remains the same: African-American women are more likely than non-Hispanic white and African-American men to have children prior to entering the paid labor force.

When the groups are combined across number of jobs and number of children (see section C), we see that the majority of men (89.5% of white men and 62.5% of African-American men) report having worked a full-time job prior to becoming fathers. A much lower percentage of women engaged in work before motherhood (44.9% of non-Hispanic white and 39.2% of African-American women).

Married, Working Parents

Table 4 shows the sequencing of three transitions (work, marriage, and parenting) for all racioethnic groups of married, working parents. This table is also divided into several parts. Section A presents the sequencing of these role transitions among parents who have only held one job. Section B describes sequencing for those who are currently engaged in their second job. Panel B1 describes individuals who engaged in the three primary social roles simultaneously (without disruption between any particular role), while panel B2 describes patterns for respondents where unemployment occurred before the acquisition of the third primary role.

Regardless of racioethnicity, most men and women first enter the paid labor force, get married, and later have children (WMP – row 1). The second most frequent pattern for white and Latino men is to marry, work, and then have children (MWP – row 3), while women (regardless of racioethnicity) are more likely to marry, have children, and later enter the paid labor force (MPW – row 4) (see Hogan, 1978; Marini, 1984 for similar findings). The second most prevalent sequence for African-American men is to work first, have children, and later get married (WPM – row 2). This particular sequence is also indicative of the pathway taken by a substantial proportion of Latino women (20%).

It is instructive to note three themes illustrated in Table 4 (especially section A): (1) the dominance of the "normative" sequence among white and Latino men, relative to the second most frequently followed life course pattern; (2) the greater diversity of the life course among African-American and Latino women as evidenced by the three other life course sequences where at least 10% of the sample reported the pattern; and (3) the diversity within racioethnic groups that clearly falls along gender lines where men are much more likely to work first while women are more likely to enter family roles prior to paid employment.

When we look at those multiple role occupants who are situated in their second jobs, a similar pattern emerges (see section B). Among men, the modal sequence continues to be employment, followed by marriage, and then children (WMP – rows 7 and 8), suggesting that little information was lost in Hogan's (1978) original study of men's life course patterns. A fifth of the group of non-Hispanic white males, however, could not be classified within any of the sequencing patterns presented in the table. Among those in this group, however, we found that 32% married first, 3.6% parented first, and 64.3% worked first (28 of the 33 had enough information available to determine which transition occurred first). In other words, even among those whose life course patterns were disrupted by loss, we find a substantial proportion entering the paid labor force prior to starting a family. Approximately 40% of African-American men follow life course patterns that are not represented in the table. Among this group (a sample size of 11), we

Table 4. Percentage Distribution of Sequencing Patterns for White, African-American, and Latino Samples of Married Working Parents by Number of Jobs.

	White		African-American		Latino	
	Men %	Women %	Men %	Women %	Men %	Women %
Section A: 1 job						
1 **WMP**	78.3	56.4	58.7	34.7	74.2	32.1
2 **WPM**	2.5	3.7	15.0	10.1	8.1	20.0
3 **MWP**	11.9	8.1	8.7	7.2	12.9	14.3
4 **MPW**	5.9	25.8	6.2	20.3	1.6	25.0
5 **PWM**	0.8	3.7	7.5	18.8	0.0	7.1
6 **PMW**	0.0	0.7	1.3	4.3	0.0	3.5
Nonclassifiable	0.6	1.5	2.5	4.3	3.2	0.0
	100.0	100.0	100.0	100.0	100.0	100.0
N	589	271	80	69	62	28
Section B: 2 jobs						
B1						
7 **WMP**UW	19.4	10.0	24.1	23.0	23.5	10.5
8 WUWMP	35.6	4.4	20.7	4.1	17.6	5.3
9 **WPM**UW	0.0	0.0	0.0	0.0	0.0	0.0
10 WUWPM	1.3	0.0	0.0	0.0	0.0	0.0
11 **MWP**UW	3.1	2.5	6.9	0.0	0.0	0.0
12 WUMWP	7.5	0.9	0.0	0.0	5.9	5.3
13 **MPW**UW	1.9	9.1	0.0	4.1	5.9	0.0
14 WUMPW	2.5	4.1	6.9	6.3	11.7	5.3
15 **PWM**UW	0.0	0.0	0.0	2.1	0.0	0.0
16 WUPWM	0.0	0.3	0.0	2.1	0.0	5.3
17 **PMW**UW	0.0	0.0	0.0	2.1	0.0	0.0
18 WUPMW	0.6	0.0	0.0	0.0	0.0	0.0
B2						
19 WMUWP	3.1	2.2	0.0	2.1	5.9	0.0
20 WMUPW	1.9	20.4	0.0	12.5	5.9	10.5
21 MWUWP	0.6	0.0	0.0	0.0	0.0	0.0
22 MWUPW	1.3	4.1	0.0	4.1	0.0	5.3
23 PWUWM	0.0	0.3	0.0	0.0	0.0	0.0
24 PWUMW	0.0	0.3	0.0	0.0	0.0	0.0
Nonclassifiable	21.2	41.4	41.4	37.5	23.6	52.5
	100.0	100.0	100.0	100.0	100.0	100.0
N	160	319	29	48	17	19
Representation (%)	61	51	55	57	76	40

Note: W = work; M = Marriage; P = Parenthood; U = unemployment/job loss.

found that 36.4% parented first while 63.6% entered the paid labor force first. Fewer Latino men were lost to complicated role sequencing patterns (23.6% or 3). Nonetheless, we did find that one married first and the other two entered the paid labor force prior to engaging family roles.

Similar to the findings for women with one job, many women who are currently engaged in their second job follow the typical pathway of work followed by marriage and then parenthood. Among these women, however, a new sequence emerges. There are a substantial group of women who exit the paid labor force before having children and who then return after the birth of the child (WMUPW – row 20). This pattern was also found among women who are *currently* unemployed, as will be shown below. It is also interesting to find here that a substantial percentage of women continue to follow the traditional female pathway of marriage followed by parenthood and then employment (MPW – rows 13 and 14).

Among these married working mothers who are currently engaged in their second job, there is also a large percentage whose life course patterns are not represented in the table. Among this group (patterns are known for 95 out of 130), 27.4% married first, 7.4% parented first, and 65.3% worked first. Thus, employment remains a strong characteristic of women's initial entry into adulthood. Latino women demonstrate a similar pattern. Over 50% fall within the nonclassifiable group (6 of the 10 women have discernable sequencing patterns). Of these women, we found that 33.3% parented first while 66.7% worked first. This finding is in stark contrast to African-American women (37.5% fall in the nonclassifiable category). Of these women, 16.7% married first, 50% parented first, and 33.3% worked first (12 of 17 had enough information to determine the sequencing pattern).

Married, Unemployed Parents

Table 5 describes the sequencing patterns followed by married, unemployed parents. This table continues to illustrate the dominance of the WMP pattern among men (rows 1, 12, and 13), although it is also the modal pattern for African-American women. At the same time, there are significant numbers of white and Latino women who either exit jobs before having children (WMUP – rows 7 and 20) or start their families and later enter and exit the paid labor force (MPWU – rows 4 and 18).

Although the numbers become quite dispersed among the small samples of African Americans and Latinos who held two jobs, which makes conclusions only tentative, it appears that African-American women are more similar in some regards to their male counterparts within this role configuration (WMPU sequence). A substantial percentage (from 13% to

20%), however, also follow patterns indicative of those found among white and Latino women (WMUP sequence).

Upon closer examination of those who fall in the "nonclassifiable" categories, the majority of white men report having worked first. Among married, unemployed fathers in their first job (some information is known for 7 out of the 11 men), 28.6% married first, 14.3% parented first, and 57.1% worked first. Those in their second jobs were also more likely to work first. Additional information was available for all 12 of the men in this group. We found that 16.7% married first while 83.3% first entered the paid labor force. A similar pattern was found for the few African-American men who fell into the nonclassifiable categories. For example, among the three men currently engaged in second jobs (only one missing case), 66.7% worked first. Approximately 36% of Latino men in their first jobs could not be classified into the sequencing patterns. Of these men (3 of 4 had sufficient information available to document the first transition), one became a father prior to finding a job and two worked prior to starting a family.

Among white women in their first job, approximately 14% (or 38) followed role sequencing patterns not described in the table. Of these women, half (50%) of them entered adulthood through marriage, 8.3% had children first, and 41.7% entered the paid labor force prior to experiencing a family role transition. African-American and Latino women were the highest percentage of adults in this multiple role configuration who did not fall within the sequencing patterns described in Table 5. As indicated in section A, 16.6% of Black women and 33.3% of Latino women were nonclassifiables. Here, African-American women report having children prior to work or marriage (three of the three women in the nonclassifiables for a single job). Only two (of six) Latino women currently engaged in their first jobs had enough information to discern the actual sequencing pattern. Among these women, one married first and the other became a mother prior to any other transition.

Approximately half of white women in their second jobs fell into role sequencing patterns not designated in the table (see section B). Among this group, information was available for most of them (66 of the 80) but the actual sequencing patterns followed by these women are not easily distinguishable. Of these women, approximately 21.1% (14) married first, 1.5% (1) parented first, and 77.3% (51) entered the paid labor force prior to starting a family. It would appear that having many children "stretches" out the life course to the point where it becomes difficult to determine all of the possible ways in which adults order four events that include two losses and

Table 5. Percentage Distribution of Sequencing Patterns for White, African-American, and Latino Samples of Married, Unemployed Parents by Number of Jobs.

	White		African-American		Latino	
	Men %	Women %	Men %	Women %	Men %	Women %
Section A: 1 job						
1 **WMPU**	80.0	9.4	38.4	37.5	63.6	5.6
2 **WPMU**	0.0	0.3	15.4	0.0	0.0	0.0
3 **MWPU**	7.9	2.6	0.0	8.3	0.0	5.6
4 **MPWU**	2.1	9.4	0.0	0.0	0.0	11.1
5 **PWMU**	0.0	0.3	0.0	0.0	0.0	0.0
6 **PMWU**	0.0	0.0	15.4	12.5	0.0	5.6
7 **WMUP**	0.0	43.0	7.7	12.5	0.0	22.2
8 **WUMP**	0.7	13.2	7.7	4.2	0.0	16.6
9 **WPUM**	0.0	0.0	0.0	4.2	0.0	0.0
10 **WUPM**	0.0	1.1	0.0	4.2	0.0	0.0
11 **MWUP**	0.7	6.4	0.0	0.0	0.0	0.0
Nonclassifiable	8.6	14.3	15.4	16.6	36.4	33.3
	100.0	100.0	100.0	100.0	100.0	100.0
N	140	265	13	24	11	18
Section B: 2 jobs						
B1						
12 **WMPUWU**	19.0	0.6	18.2	0.0	20.0	0.0
13 WUWMPU	34.6	3.1	0.0	0.0	80.0	0.0
14 **WPMUWU**	0.0	0.0	0.0	0.0	0.0	0.0
15 WUWPMU	1.7	0.0	0.0	0.0	0.0	0.0
16 **MWPUWU**	1.7	0.6	18.2	10.0	0.0	0.0
17 WUMWPU	15.5	0.6	9.0	0.0	0.0	0.0
18 **MPWUWU**	0.0	5.6	0.0	0.0	0.0	0.0
19 WUMPWU	3.4	8.1	9.1	0.0	0.0	8.3
B2						
20 WMUPWU	0.0	25.4	9.1	20.0	0.0	8.3
21 MWUPWU	1.7	2.0	0.0	0.0	0.0	8.3
22 WUMWUP	0.0	2.5	0.0	0.0	0.0	0.0
23 WUWUMP	1.7	0.6	0.0	0.0	0.0	0.0
Nonclassifiable	20.7	50.9	36.4	70.0	0.0	75.1
	100.0	100.0	100.0	100.0	100.0	100.0
N	58	161	11	10	5	12
Representation (%)	49	50	36	38	40	38

Note: W = work; M = marriage; P = parenthood; U = unemployment/job loss.

as many as three children. Similar complexity is apparent when attempting to determine role sequencing patterns among the divorced (as shown below).

The concern with indistinguishable life course patterns among the group of minority women is less pressing since there are so few in this configuration. The majority of African-American women in their second jobs entered the paid labor force first (80% or 4 of the 5). A substantial percentage of Latino women in their second jobs (75%) are nonclassifiable. This percentage, however, is tempered by the fact that we began with very few married, unemployed mothers. Here we found, nonetheless, that 28.6% married first, 28.6% parented first, and 42.9% worked first (sequencing patterns were determined for 7 of these 9 women).

Divorced, Working Parents

In Table 6, we present the sequencing patterns for divorced, working parents. Unlike the findings for married parents, this table shows a great deal of similarity across groups. Both divorced, men and women are more likely to experience their divorce in the context of a "normative" life course (rows 1, 11, 12, and 23). The percentages continue to favor men in regards to the likelihood of following the typical life course pattern (even among white men in section B). Women, across racioethnic groups, continue to have significant representation in life courses that involve getting married and having children before entering the paid labor force (rows 4, 17, and 18). Consistent with earlier analyses a significant percentage of women and ethnic minorities are nonclassifiable. In general, dual role losses (job loss and marital loss) in conjunction with the possibility of having three children make several life course patterns undetectable. Nonetheless, we report below the type of role transition that occurred first in the life course among the top three groups represented in this role configuration: white men, white women, and African-American women.

Close examination of the initial entry into adulthood among the nonclassifiables shows that employment is the primary role transition experienced by white adults (51% of men and 56% of women across both sections), followed by marriage with approximately 35% of white men and women getting married first. Although the majority of African-American women also enter the paid labor force prior to starting their families (43%), a substantial percentage have children prior to work and/or marriage (41%), with the remaining 16% reporting marriage as the first transition.

While the information provided in Tables 2–6 illustrates the kind of diversity described by other life course sociologists, it does not provide a parsimonious image of what people are doing. To aid in this endeavor, we

Table 6. Percentage Distribution of Sequencing Patterns for White, African-American, and Latino Samples of Divorced, Working Parents by Number of Jobs.

	White		African-America		Latino	
	Men %	Women %	Men %	Women %	Men %	Women %
Section A: 1 job						
1 **WMPD**	75.3	41.0	66.6	22.0	28.6	33.4
2 **WPMD**	3.2	2.0	5.6	4.0	0.0	0.0
3 **MWPD**	6.4	9.5	5.6	8.0	14.3	13.3
4 **MPWD**	2.2	16.0	0.0	12.0	14.3	26.6
5 **PWMD**	0.0	0.0	0.0	2.0	0.0	0.0
6 **PMWD**	0.0	1.0	0.0	2.0	0.0	0.0
7 **WMDP**	0.0	1.0	0.0	0.0	14.3	0.0
8 **MWDP**	0.0	1.0	0.0	0.0	0.0	6.7
9 **MPDW**	0.0	9.5	0.0	8.0	0.0	6.7
10 **PMDW**	0.0	0.0	0.0	0.0	0.0	0.0
Nonclassifiable	12.9	19.0	22.2	42.0	28.5	13.3
	100.0	100.0	100.0	100.0	100.0	100.0
N	93	106	18	50	7	15
Section B: 2 jobs						
B1						
11 **WMPD**UW	9.5	1.6	0.0	0.0	—	0.0
12 WUWMPD	52.4	1.6	0.0	0.0	—	0.0
13 **WPMD**UW	0.0	0.0	0.0	5.0	—	0.0
14 WUWPMD	0.0	0.0	0.0	0.0	—	0.0
15 **MWPD**UW	4.7	1.0	16.7	0.0	—	14.3
16 WUMWPD	0.0	0.0	0.0	5.0	—	0.0
17 **MPWD**UW	0.0	2.6	0.0	0.0	—	14.3
18 WUMPWD	0.0	1.6	0.0	5.0	—	42.8
19 **PWMD**UW	0.0	0.0	0.0	0.0	—	0.0
20 WUPWMD	0.0	1.0	0.0	0.0	—	0.0
21 **PMWD**UW	0.0	1.0	0.0	0.0	—	0.0
22 WUPMWD	0.0	0.0	0.0	0.0		0.0
B2						
23 WMPUWD	19.1	11.3	16.7	5.0	—	0.0
24 MWPUWD	0.0	0.0	0.0	0.0	—	0.0
25 MPWUWD	0.0	2.6	0.0	0.0	—	0.0
26 WUMPDW	0.0	0.0	0.0	0.0	—	14.3
27 WMUPDW	0.0	3.5	0.0	0.0	—	0.0
Nonclassifiable	14.3	72.2	66.6	80.0	—	14.3
	100.0	100.0	100.0	100.0	—	100.0
N	21	115	6	20	1	7
Representation (%)	54	44	48	51	53	54

Note: W = work; M = marriage; P = parenthood; U = unemployment/job loss.

adopt the strategy utilized by Hogan (1978) and Marini (1984) and simply examine the sequencing of first job, first marriage, and birth of first child among the multiple role occupants. We further restrict this part of the analysis to individuals who have only held one job to make comparisons across studies more feasible. This information is provided in three sections in Table 7 (this table excludes the nonclassifiables).

As shown in Table 7, regardless of role configuration, men can be found among those whose first transition into adulthood is employment, followed by marriage, and then childbirth (WMP). The WMP sequence is also the modal pattern for women who are married, working parents but a significantly smaller percentage of minority women follow this pathway compared to their white peers. When we consider those adults who have experienced a role loss (married, unemployed parents and divorced, working parents), we find an interesting divergence. First, the normative sequence becomes even more pronounced among white men. That is, even loss events (unemployment and divorce) are experienced in the context of a "normative" life course (WMPU and WMPD). This feature of the life course seems to vary for minority men according to the role configuration under consideration. In fact, among the divorced, African-American men have the highest percentage in the WMP category (an even higher percentage than white and Latino men). Second, a higher percentage of African-American women who have experienced job loss follow the normative order compared to their male counterparts (WMPU). Although the sample size is rather small for these groups, the distribution of sequencing patterns is still informative. Third, a higher percentage of women than men report acquiring both family roles (spouse and mother) before entering the paid labor force.

DISCUSSION

According to Runyan (1982), one of the most important questions facing the life course orientation is what kinds of regularity may be found in the sequence of events in individual lives. Before documenting sequencing, however, it is important to know what people are doing. The presentation of role configurations addresses this initial step in describing the structure of the life course.

We found that most adults are (or have) engaged in multiple social roles. The majority of adults in the NSFH, in fact, are situated in the multiple role category of married, working parent regardless of gender and racioethnicity. The type of roles gained and lost, however, differs by status characteristic.

Table 7. Percentage Distribution of Sequencing Patterns for White, African American, and Latino Men and Women in Three Role Configurations.

	White		African-American		Latino	
	Men %	Women %	Men %	Women %	Men %	Women %
Married, Working Parents						
Section A						
1 WMP	78.7	57.4	60.3	36.4	76.7	32.1
2 WPM	2.6	3.7	15.4	10.6	8.3	17.9
3 MWP	11.9	8.2	8.9	7.6	13.3	14.3
4 MPW	5.9	26.2	6.4	12.2	1.7	25.0
5 PWM	0.9	3	7.7	19.7	0.0	7.1
6 PMW	0.0	1.5	1.3	4.5	0.0	3.6
	100.0	100.0	100.0	100.0	100.0	100.0
N	586	267	78	66	60	28
Married, Unemployed Parents						
Section B						
7 WMPU	87.5	16.0	45.4	50.0	100.0	7.7
8 WMUP	0.0	45.3	9.1	16.6	0.0	30.7
9 WUMP	0.8	9.4	9.1	5.6	0.0	23.1
10 WPMU	0.0	0.5	18.2	0.0	0.0	0.0
11 MWPU	8.6	5.2	0.0	11.1	0.0	7.7
12 MWUP	0.8	6.6	0.0	0.0	0.0	0.0
13 MPWU	2.3	16.0	0.0	0.0	0.0	15.4
14 PWMU	0.0	0.5	0.0	0.0	0.0	0.0
15 PMWU	0.0	0.5	18.2	16.7	0.0	15.4
	100.0	100.0	100.0	100.0	100.0	100.0
N	128	212	11	18	7	13
Divorced, Working Parents						
Section C						
16 WMPD	80.9	50.0	92.3	36.1	40.0	33.3
17 WMDP	0.0	1.2	0.0	0.0	20.0	0.0
18 WPMD	3.6	2.3	7.7	5.6	0.0	0.0
19 MWPD	7.1	10.2	0.0	11.1	20.0	13.3
20 MWDP	0.0	0.0	0.0	0.0	0.0	6.7
21 MPWD	7.1	26.1	0.0	19.5	20.0	40.0
22 MPDW	0.0	10.2	0.0	11.1	0.0	6.7
23 PWMD	0.0	0.0	0.0	8.3	0.0	0.0
24 PMWD	1.3	0.0	0.0	8.3	0.0	0.0
	100.0	100.0	100.0	100.0	100.0	100.0
N	84	88	13	36	5	15

Note: W = work; M = marriage; P = parenthood; U = unemployment/job loss; D = divorce.

For example, a higher percentage of men enter adulthood through employment, whereas initial entry into adulthood for women is either through marriage or parenthood. Not surprisingly, then, a higher percentage of men who have lost the only role they occupied tend to be unemployed, compared to women. Similarly, African-American and Latino women are more likely than whites to be parents although Latino women are more likely than others to be among those who have never worked and who are currently married parents. The type of role configurations documented for individuals within the NSFH reveal that Latino men are more similar to white men than they are to African-American men. In general, men are more similar across role categories and sequencing patterns than are women.

Our findings underscore the importance of incorporating information about role entries and role exits in a description of the life course. Demographic evidence clearly shows some variation in the extent to which various groups have experienced certain role transitions. For example, African Americans have lower marriage rates (Tucker & Mitchell-Kernan, 1995), higher unemployment rates (Testa & Krogh, 1995), and a younger age at parenthood (Ventura, Mosher, Abma, & Henshaw, 2001) compared to whites. Similarly, women have higher unemployment rates than men (Browne, 2000) and are more likely to be raising children (Berk, 1985; Hochschild & Machung, 1989). These demographic realities translate into very different social role configurations.[6]

In terms of role sequencing, there is a recognizable life course pattern. The typical pathway into adulthood for men and women is to first work, then marry, and later have children (WMP). This pattern is more consistent for white men than any other group, regardless of subsequent role losses such as divorce or unemployment. Nonetheless, the results from this study suggest that the model described within most life course research is a "male" model rather than a "white male" model (Cooney & Hogan, 1991). Minority men, especially Latinos, also report working before acquiring these family roles.

Gender also surfaced as a defining marker for life course patterns. The sequencing of job loss, in particular, reveals a set of sequencing patterns that were most prevalent among women. Although Marini (1984) does not bring this issue to our attention, there is some discussion of the ways in which women's life course patterns differ from men's life course patterns (Treiman, 1985) including the particular phenomenon of leaving the labor force prior to having children (Cowan, Cowan, Heming, & Miller, 1991; Daniels & Weingarten, 1982).

The consistent emergence of patterns that depict women experiencing job loss immediately before the birth of a child highlight the intimate link between women's labor force participation and childbearing. In fact, this gender variation speaks most clearly to the second form of mis-representation in previous life course research. It is clear, for example, that the WMUP pattern differs from the typical WMP pattern but a single focus on entry into roles and not role exits would lead one to combine those who have experienced job loss (the former group) with those who have not experienced this role loss (the latter group).

As such, this study raises a number of potentially interesting questions regarding the structure of the life course. For example, what factors predict the likelihood of following the typical life course pattern? Do these factors vary by gender and/or racioethnicity? Does the sequencing of role loss matter for overall well-being? These are but a few of the questions that could be addressed in future work in light of the demonstrated consistency and diversity in the structure of the life course.

The findings from this study have important implications for the ongoing debate among some life course sociologists regarding the growing individualization of the life course (e.g., see Buchmann, 1989; Mayer & Schoepflin, 1989; Ramirez, 1989). The individualization hypothesis contends that life course decisions are less constrained by regulations set forth by state agencies and people are, therefore, creating their own life course patterns (Modell, 1989; Pallas, 1993). In the context of the sequencing of primary social roles, individualization can be defined as deviation away from a previously labeled "normative" life course, which is primarily represented by school-to-work-to-marriage-to-parenthood sequence (Elder, 1975; Hogan, 1978; Hogan & Astone, 1986).

We have also found some evidence that certain groups in society are more likely to embrace a more individualized life course. This includes African Americans in general, African-American women in particular, and white women of particular birth cohorts (Jackson, 2004). At this juncture, we must leave it up to the reader to determine whether any degree of variation from the modal pattern is indicative of increasing individualization or if extreme variation is the most clear-cut evidence for a hyper-individualized life course. Some scholars would not be too surprised to find the type of variation in life course paths as described in this study, given the emphasis on individualism and independence in the socialization practices of the U.S. (Arnett, 1995).

In conclusion, we can now answer the three questions initially posed by this chapter. First, the work to marriage transition *is* much more prevalent among men compared to women. This pattern is evident in the subsamples of childless adults as well as those who have children, have experienced a divorce, and/or have been unemployed. For ethnic minorities, Latino men seem to more consistently follow this life course pattern compared to

African-American men. Second, there *is* a particular life course pattern that defines the structure of womens' lives more than it does mens' lives: WMUP(W). This study has thus provided even more empirical evidence for the discussions about job exits that coincide with the birth of the first child (Cowan et al., 1991). To answer our third question, we build on this finding by further demonstrating that Latino women are very similar to their white female counterparts in regards to having followed this sequencing pattern. African-American women, on the other hand, seem to be the most likely group to start their families (either via marriage or parenthood) before entering the paid labor force.

Each of these patterns has important implications for areas of vulnerability across the life course. The importance of having resources to help meet personal and family needs is obvious, yet some scholars argue that certain groups experience a "life-cycle squeeze" in which a disparity occurs between economic rewards (i.e., actual income) and expectations (i.e., appetite for consumer goods) at particular stages in the life cycle (Estes & Wilensky, 1978; Oppenheimer, 1974, 1979). More specifically, when the person becomes married and has children there is often a drain on existing financial resources. The life-cycle squeeze is also associated with role stress since a family's vulnerability to stress and their capacity to adjust to the demands of new transitions is a function of its crisis-meeting resources (see Anderson, 1988; Boss, 1987; Pearlin, 1985). Childbearing families, families with teenagers, and retired elderly adults experience the most acute life-cycle squeeze (Oppenheimer, 1981). More broadly, however, a squeeze occurs whenever a person acquires a financially draining role, thus having children first or having children after a job loss can trigger a "squeeze" since parenting is a resource-depleting transition rather than a resource-enhancing transition (Jackson, 2004). Women, and ethnic minorities, who are already economically disadvantaged in society, may find that their role sequencing serves to further exacerbate their status in society – a process many aging researchers refer to as cumulative disadvantage (Tausig, Michello, & Subedi, 1999). In essence, drawing attention to the processes through which individuals embark onto certain life trajectories and documenting the sequencing patterns that result from such decision-making remains fertile ground for theorizing about the structure of the life course.

NOTES

1. A large national study identified the most preferred racial/ethnic term for each group. Sixty-two percent of whites prefer "white" (17% prefer "Caucasian"), 58% of

Hispanics prefer "Hispanic" (12% prefer "Latino"), 44% of Blacks prefer "Black" (28% prefer "African American"), and 50% of American Indians prefer "American Indian" (37% prefer "Native American") (Tucker et al., 1996). In an effort to recognize individual dignity, we use the most preferred terms for each group interchangeably.

2. The narrow range of birth cohorts may help explain the discrepancy between Hogan's (1978) and Marini's (1984) findings for men. Hogan did find that the percentage of men who conformed to the normative pattern decreased with each successive cohort. Marini's cohorts (born between 1940 and 1943) represent the younger birth cohorts in Hogan's study (born between 1907 and 1952). The sharpest decrease in the percentage of men who followed the normative pattern began with the 1947–1952 birth cohorts (49%).

3. If one were to exclude schooling from the role sequencing histories, one would actually find that 56% of women and 61% of men (those who had experienced work, marriage, and parenthood) report following the "normative" pathway. This figure is much closer to Hogan's (1978) 63% of men in the 1942–1946 birth cohort who found work prior to getting married.

4. Data sets often exclude information on fertility behavior (e.g., OCG-II) or use cohort data which limits the age range of the sample (e.g., NLS). Therefore, descriptions of the life course are either restricted to experiences in the school and work arena or are based on the experiences of young adults. While it is true that most social roles are acquired between the ages of 18 and 30, some role losses are experienced later in life (e.g., widowhood).

5. This restriction may introduce a potential age bias, but the direction of this bias is difficult to assess. On one hand, we expect that older people have had longer exposure to the risk of these transitions than younger people. As such, they may be overrepresented in this part of the study. On the other hand, older people may have more complicated lives (e.g., more job losses) and may therefore, be excluded from this part of the study. These issues cannot be addressed here but are being explored by the authors.

6. Growing up in single-parent families has been linked to school drop-out, juvenile delinquency, and early pregnancies (Loury, 1985; McLanahan, 1985; Shaw, 1983). Thus, the life course decisions made by parents may have generational consequences for their children's life course patterns as well.

REFERENCES

Abeles, R., Steel, L., & Wise, L. (1980). Patterns and implications of life-course organization: Studies from Project TALENT. In: P. Baltes & O. Brim (Eds), *Life-span development and behavior*, (Vol. 3, pp. 307–337). New York: Academic Press.

Agresti, A., & Finlay, B. (1986). *Statistical methods for the social sciences*. San Francisco, CA: Dellen Publishing Co.

Anderson, S. (1988). Parental stress and coping during the leaving home transition. *Family Relations*, *37*, 160–165.

Arnett, J. (1995). Broad and narrow socialization: The family in the context of a cultural theory. *Journal of Marriage and the Family*, *57*, 617–628.

Bean, F., & Tienda, M. (1987). *The Hispanic population of the United States*. New York: Russell Sage Foundation.

Berk, S. (1985). *The gender factory*. New York: Plenum Press.

Boss, P. (1987). Family stress: Perception and context. In: M. Sussman & S. Steinmetz (Eds), *Handbook of marriage and the family* (pp. 695–723). New York: Plenum Press.

Brim, O., & Ryff, C. (1980). On the properties of life events. In: P. Baltes & O. Brim (Eds), *Life-span development and behavior*, (Vol. 3, pp. 367–388). New York: Academic Press.

Browne, I. (2000). Opportunities lost? Race, industrial restructuring, and employment among young women heading households. *Social Forces*, *78*, 907–929.

Buchmann, M. (1989). *The script of life in modern society: Entry into adulthood in a changing world*. Chicago, IL: The University of Chicago Press.

Callero, P. (1994). From role-playing to role-using: Understanding role as resource. *Social Psychology Quarterly*, *57*, 228–243.

Cameron, S., & Heckman, J. (1998). *Life cycle schooling and dynamic selection bias: Models and evidence from 5 cohorts of American males*. Cambridge, MA: National Bureau of Economic Research.

Cooney, T., & Hogan, D. (1991). Marriage in an institutionalized life course: First marriage among American men in the twentieth century. *Journal of Marriage and the Family*, *53*, 178–190.

Cowan, C., Cowan, P., Heming, G., & Miller, N. (1991). Becoming a family: Marriage, parenting, and child development. In: P. Cowan & M. Hetherington (Eds), *Family transitions* (pp. 79–109). Hillsdale, NJ: Lawrence Erlbaum Assoc., Inc.

Daniels, P., & Weingarten, K. (1982). *Sooner or later: The timing of parenthood in adult lives*. New York: W.W. Norton and Co.

Elder, G. (1975). Age differentiation and the life course. *Annual Review of Sociology*, *1*, 165–190.

Elder, G. (1985). Perspectives on the life course. In: G. Elder Jr. (Ed.), *Life course dynamics: Trajectories and transitions, 1968–1980* (pp. 23–49). Ithaca, NY: Cornell University Press.

Elder, G. (1997). The life course and human development. In: R. Lerner (Ed.), *Handbook of child psychology, volume 1: Theoretical model of human development* (pp. 939–991). New York: Wiley.

Estes, R., & Wilensky, H. (1978). Life cycle squeeze and morale curve. *Social Problems*, *25*, 277–292.

Featherman, D., & Carter, T. (1976). Fatalism and familism among Anglos and Mexican Americans in San Antonio. *Sociology and Social Research*, *60*, 393–402.

Featherman, D., Hogan, D., & Sorensen, A. (1984). Entry into adulthood: Profiles of young men in the 1950's. In: P. Baltes & O. Rim Jr. (Eds), *Life-span development and behavior*, (Vol. 6, pp. 159–202). New York: Academic Press.

Giele, J. (1982). *Women in the middle years: Current knowledge and directions for research and policy*. New York: Wiley.

Hagan, J., & Wheaton, B. (1993). The search for adolescent role exits and the transition to adulthood. *Social Forces*, *71*, 955–979.

Han, S., & Moen, P. (1999). Clocking out: Temporal patterning of retirement. *The American Journal of Sociology*, *105*, 191–236.

Hochschild, A., & Machung, A. (1989). *The second shift*. New York: Avon Books.

Hogan, D. (1978). The variable order of events in the life course. *American Sociological Review*, *43*, 573–586.

Hogan, D. (1981). *Transitions and social change*. New York: Academic Press.

Hogan, D. (1985). The demography of life-span transitions: Temporal and gender comparison. In: A. Rossi (Ed.), *Gender and the life course* (pp. 65–78). New York: Aldine.

Hogan, D., & Astone, N. (1986). The transition to adulthood. *Annual Review of Sociology, 12*, 109–130.

Jackson, P. (2004). Role sequencing: Does order matter for mental health? *Journal of Health and Social Behavior, 45*, 132–154.

Kohli, M., Rein, M., Guillemard, A.-M., & van Gunsteren, H. (1991). *Time for retirement: Comparative studies of early exit from the labor force*. New York: Cambridge University Press.

Macmillan, R., & Eliason, S. (2003). Characterizing the life course as role configurations and pathways: A latent structure approach. In: J. Mortimer & M. Shanahan (Eds), *Handbook of the life course*. New York: Plenum/Kluwer Academic Press.

Marini, M. (1984). The order of events in the transition to adulthood. *Sociology of Education, 57*, 63–84.

Mayer, K., & Schoepflin, U. (1989). The state and the life course. *Annual Review of Sociology, 155*, 187–209.

Menaghan, E. (1989). Role changes and psychological well-being: Variations in effects by gender and role repertoire. *Social Forces, 67*, 693–714.

Modell, J. (1989). *Into one's own: From youth to adulthood in the United States, 1920–1975*. Berkeley, CA: University of California Press.

Modell, J., Furstenberg, F., & Hershberg, T. (1976). Social change and transitions to adulthood in historical perspective. *Journal of Family History, 1*, 7–32.

Moen, P., Kim, J. E., & Hofmeister, H. (2001). Couples' work/retirement transitions, gender, and marital quality. *Social Psychology Quarterly, 64*, 55–71.

Mortimer, J. T., & Shanahan, M. J. (Eds). (2003). *Handbook on the life course*. New York: Plenum/Kluwer Academic Press.

Murnane, R., Willett, J., & Boudett, K. (1995). Do high school dropouts benefit from obtaining a GED? *Educational Evaluation and Policy Analysis, 17*, 133–147.

Oppenheimer, V. (1974). Life cycle squeeze: The interaction of men's occupational and family life cycles. *Demography, 11*, 227–246.

Oppenheimer, V. (1979). Structural sources of economic pressure for wives to work: An analytical framework. *Journal of Family History, 4*, 177–197.

Oppenheimer, V. (1981). The changing nature of life-cycle squeezes: Implications for the socioeconomic position of the elderly. In: R. Fogel (Ed.), *Aging: Stability and change in the family* (pp. 47–81). New York: Academic Press.

O'Rand, A. (1988). Convergence, institutionalization, and bifurcation: Gender and the pension acquisition process. *Annual Review of Gerontology and Geriatrics, 8*, 132–155.

Pallas, A. (1993). Schooling in the course of human lives: The social context of education and the transition to adulthood in industrial society. *Review of Educational Research, 63*, 409–447.

Pallas, A. (1994). *Research in sociology of education and socialization*. Greenwich, CT: JAI Press, Inc.

Pearlin, L. (1985). Life strains and psychological distress among adults. In: A. Monat & R. S. Lazarus (Eds), *Stress and coping: An anthology* (pp. 192–207). New York: Columbia University Press.

Pulskamp, R. (1990). A note on the estimation of binomial probabilities (in teacher's corner). *The American Statistician, 44*, 293–295.

Ramirez, F. (1989). Reconstituting children: Extension of personhood and citizenship. In: D. Kertzer & K. Schaie (Eds), *Age structuring in comparative perspective* (pp. 143–165). Hillsdale, NJ: Erlbaum.

Rindfuss, R. (1991). The young adult years: Diversity, structural change, and fertility. *Demography*, *28*, 493–512.

Rindfuss, R., Swicegood, C., & Rosenfeld, R. (1987). Disorder in the life course: How common and does it matter? *American Sociological Review*, *52*, 785–801.

Rossi, A. (1980). Life-span theories and women's lives. *Signs*, *6*, 4–32.

Runyan, W. (1982). *Life histories and psychobiography*. New York: Oxford University Press.

Ryff, C. (1985). The subjective experience of life-span transitions. In: A. Rossi (Ed.), *Gender and the life course* (pp. 97–113). New York: Aldine.

Sewell, W. H., & Hauser, R. M. (1975). *Education, occupation and earnings*. New York: Academic Press.

Shanahan, M. J. (2000). Pathways to adulthood in changing societies: Variability and mechanisms in life course perspective. *Annual Review of Sociology*, *26*, 667–692.

Sweet, J. (1973). *Women in the labor force*. New York: Seminar Press.

Sweet, J., Bumpass, L., & Call, V. (1988). *The design and content of the National Survey of Families and Households*. Working Paper NSFH-1. Center for Demography and Ecology, University of Wisconsin-Madison.

Tausig, M., Michello, J., & Subedi, S. (1999). *A sociology of mental illness*. Upper Saddle River, NJ: Prentice-Hall.

Testa, M., & Krogh, M. (1995). The effect of employment on marriage among black males in inner-city Chicago. In: M. Tucker & C. Mitchell-Kernan (Eds), *The decline in marriage among African Americans: Causes, consequences, and policy implications* (pp. 59–95). New York: Russell Sage Foundation.

Treiman, D. (1985). The work histories of women and men: What we know and what we need to find out. In: A. Rossi (Ed.), *Gender and the life course* (pp. 213–231). Hawthorne, \New York: Aldine.

Tucker, M. B., & Mitchell-Kernan, C. (1995). Marital behavior and expectations: Ethnic comparisons of attitudinal and structural correlates. In: M. Tucker & C. Mitchell-Kernan (Eds), *The decline in marriage among African Americans* (pp. 145–171). New York: Russell Sage Foundation.

Tucker, H., McKay, R., Kojetin, B., Harrison, R., de la Puente, M., Stinson, L., & Robinson, E. (1996). Testing methods of collecting racial and ethnic information: Results of the current population survey supplement on race and ethnicity. *Bureau of Labor Statistical Notes*, *40*, 1–149.

Ventura, S., Mosher, W., Abma, J., & Henshaw, S. (2001). Trends in pregnancy rates for the United States, 1976–1997: An update. *National Vital and Health Statistics Reports* (Vol. 49, no. 4). Hyattsville, MD: National Center for Health Statistics.

Voydanoff, P., & Kelly, R. F. (1984). Determinants of work-related family problems among employed parents. *Journal of Marriage and the Family*, *46*, 881–892.

MEASURING THE EARLY ADULT LIFE COURSE IN MEXICO: AN APPLICATION OF THE ENTROPY INDEX

Elizabeth Fussell

ABSTRACT

This research uses a new method of analyzing the demographic statuses of a synthetic cohort of Mexicans to examine how the life course has changed in Mexico during the past 30 years. I examine young people's statuses as students, workers, parents, spouses, and their position in the household using Mexican census data for 1970 and 2000. In doing so, I describe the stages of the early life course and the social institutions that structure each of these life stages. During this period there has been relatively little change in the timing of life stages and limited change in the statuses that structure each stage. The most significant changes have been the increase to nearly universal participation in education throughout childhood and early adolescence and the increase in women's employment during adulthood. Surprisingly, the age of first marriage and childbearing has not been significantly delayed during this period despite the substantial social change. Consequently, the structure of the life course for the majority of young people in 2000 is not much different from that of their

The Structure of the Life Course: Standardized? Individualized? Differentiated?
Advances in Life Course Research, Volume 9, 91–122

ISSN: 1040-2608/doi:10.1016/S1040-2608(04)09004-5

parent's generation who came of age 30 years earlier despite dramatic economic and demographic changes that have occurred.

There are few summary statistical tools for describing the early life course in a holistic manner. Measures of timing, spread, and the integration of status transitions are useful, but difficult to pull together into a cohesive picture (Modell, Furstenberg, & Hershberg, 1976; Fussell & Furstenberg, 2005). Such measures are desirable, however, for describing how the life course has changed over time and for exploring the impact of social structures on the life course. I propose a new summary measure – the entropy index of age-specific status combinations – that accomplishes several tasks of interest to life course scholars. The entropy index measures the degree to which individuals of a given age are similar in their combination of demographic statuses, thus describing the age-graded stages of the life course. By comparing the entropy index at different points in time it shows whether and how the age-gradedness of the life course has changed over time. Furthermore, decomposition of the entropy index shows the extent to which particular demographic statuses structure particular stages of the life course. In accomplishing these tasks, the entropy index contributes a new descriptive tool to the life course tool box. In this paper I demonstrate the use of this tool by applying it to the 1970 and 2000 Mexican census.

YOUTH IN MEXICO

Life course analysis is fundamentally concerned with the interaction between individual biographies and social structures (Elder, 1999). A central debate in research on the life course focuses on how social structures have institutionalized the life course into a series of age-graded life stages and whether that process of institutionalization has now halted or even reversed (Fussell, 2004a; Held, 1986; Kohli & Meyer, 1986; Shanahan, 2000). This debate has centered on research in advanced industrial societies in Western Europe and North America. In contrast, life course research in Mexico is typically focused on analyzing labor and family formation processes (Blanco, 2002; Escobar Latapí & Roberts, 1991; Gomes, 2001; González de la Rocha, 1994; LeVine, 1993; Pries, 1996; Quilodrán, 1996; Tuirán, 2001, 2002).

Research on the life course in Mexico has focused on how economic transformations, particularly the prolonged period of economic crisis and restructuring, has impacted labor and family trajectories of men, women, and families. However, economic transformations are not the only changes that might be expected to have an impact on early life course transitions. Demographic changes, such as prolonged life expectancy and declining fertility, and trends in marriage and divorce and educational expansion also significantly shape the context as well as the timing and level of life course transitions (Tuirán, 2002). This set of social structures differs qualitatively from that posited to influence the institutionalization and individualization of the life course in Western European and North American contexts. In Mexico and many non-Western nations, families play a stronger role than public institutions in shaping the life course. In the following paragraphs I review demographic, social and economic trends, and how they might affect the timing and context of early life course transitions.

Mexico is far along in the demographic transition from high to low fertility and mortality, though there is still positive population growth. Life expectancy at birth in Mexico has increased from 60 to 74 years between 1970 and 2000, while fertility has declined from an average of 6.5 to 2.4 births per woman in that period (Instituto Nacional de Estadística, Geografia e Informática, 2004a; United States Census Bureau, 2004a). The government campaign that began in the mid-1970s to promote the use of family planning methods and lower ideal family sizes has contributed to the rapid and widespread fertility decline (Cabrera, 1994).

The decline in fertility, for the most part, came about from lower fertility within marriage rather than a delay in family formation. Women's average age at first marriage only increased from 21.2 to 22.7 between 1970 and 2000 (Fussell & Palloni, 2004; Moreno & Singh, 1996). However, adolescent fertility declined substantially as it was increasingly defined as a social problem (United States Census Bureau, 2004b). In other words, demographic changes primarily affected the context in which youth became adult. Demographic changes lowered the average number of siblings and increased the likelihood that children would reside in a household with both parents throughout their youth. However, these changes had relatively small effects on the timing of the transition to marriage and childbearing.

In addition to the family planning initiative, the Mexican government also expanded education at all levels (Mier y Terán & Rabell, 2003). In 1970, 65.7% of 6–12 year olds and 52.6% of 13–15 year olds were attending school. By 2000, enrolment had grown to 93.8% and 76.6%, respectively, for these children (Instituto Nacional de Estadística, Geografia e Informática, 2004b). Consequently, the percentage of the adult population (age 15 and older) that had no instruction declined from 31.6% in 1970 to 10.3% in 2000 and the adult literacy rate increased from 65.5% to 90.5%

(Instituto Nacional de Estadística, Geografia e Informática, 2004c). The expansion of education marked a significant intervention by the state in children's lives.

While the Mexican government has long-valued education as an institution for preparing citizens for civic and economic participation for much of its history, it only mandated education to the primary level (grades 1–6). In 1993, it changed the national constitution to make secondary education compulsory. At the same time, however, the rhetoric and public opinion about the value of education shifted. While government emphasized the role of education in preparing workers, the public saw very little evidence of returns to their investments in human capital (Levinson, 1999; Muñoz García & Suárez Zozaya, 1990). Therefore, there was little incentive for most Mexicans to prolong their education. Still, the expansion of education certainly represents a change in the context of childhood and adolescence as more children spend this period of their lives in school. However, few stay in school through their late-teens and early 20s so there it causes no significant delay in the transitions to family and household formation (Lindstrom & Brambila, 2001).

These social and demographic changes have occurred during a period of intense economic change in Mexico. In the 1970s, Mexico was experiencing the end of the economic miracle – a period of economic growth and early industrialization. This period closed with the first economic crisis in 1982 when Mexico defaulted on its foreign debt, resulting in economic contraction and widespread unemployment. The recovery from this economic crisis was very slow and painful, based on the implementation of austerity measures that reduced public services and the opening of the economy to foreign trade. Consequently, the 1980s are often referred to as the lost decade (Lustig, 1990; Weller, 2000). A second crisis occurred in 1994 when government devalued the peso by half. The recovery was quicker this time, but Mexico continues to suffer from high levels of underemployment, poverty, and economic inequality (Alarcón, 1994; Lustig, 1997; Portes & Hoffman, 2003). The consequences of economic crises and restructuring for families have been amply documented in the literature about employment and family survival strategies (Benería, 1992; González de la Rocha, 1994; Martin, 1996). The bottom line is that the crisis has minimized both the Mexican government and Mexican families' investments in the large cohort of youth coming of age. This means that few welfare state institutions have emerged that strongly structure the life course beyond childhood and families may be more likely to rely on adolescents and young adults as earners.

Research on family survival strategies, the set of adaptive strategies families employ to cope with the economic crisis, has demonstrated the ways in which families have operated as a social safety net during this period. A primary adaptive strategy is to send more household members into the labor force. Thus, during the past 30 years women from all socio-economic strata have entered the labor force and children and adolescents in many poor and working class households have also engaged in productive activities (Fussell 2004a, b; Fussell and Palloui, 2005; Garcia & Olivera, 1994; Robles-Vásquez & Abler, 2000). This income-generating strategy is often complemented by consumption-based strategies, where families find ways of stretching the budget and substituting lower- for higher-cost goods (Benería, 1992; González de la Rocha, 1994; Martin, 1996). Women are also likely to stop childbearing within marriage earlier in order to enter the labor force, another means to ensure adequate income (Mier y Terán Rocha, 1996). Clearly, the economic crisis has affected the context of youth by depriving families of resources to invest in youth. However, reliance on family and social networks for economic support is part of a long cultural tradition of consolidating family networks through extended family households and early marriage and childbearing. Therefore it is unlikely to strongly influence the timing of home-leaving, marriage, or childbearing for the youthful generation (Fussell & Palloni, 2004).

This brief review of social and economic trends in Mexico suggests that there is little reason to expect the timing of events in the youthful life course to have changed though the context is most certainly different in 2000 than in 1970. In particular, young adults have fewer siblings but their families may be struggling hard to make ends meet economically. The observed stability in the timing of life course transitions into family roles suggests that early family formation represents a resource for families, even more so during a period of national instability. Evidently the social institutions that have guided and shaped the life course in Western Europe and North America do not operate the same way as the social institutions in Mexico and presumably other regions of Latin America. In the following section I review life course methods that have been applied to examine changes in the structure of the life course and introduce a new method which I apply to Mexican census data.

LIFE COURSE ANALYSIS OF SYNTHETIC COHORTS

Most research on the life course is based on the individual behavior of a cohort of individuals. By following a group who were born in a particular period, a birth cohort, or who experienced a particular transition at a point in time, such as a graduating class of high school seniors, life course scholars have examined how the experiences of a generation shape their later life outcomes (e.g. Elder, 1999 [1974]; Michael, 2001). However, longitudinal data are relatively scarce, particularly for historical populations or in countries with few well-funded universities and social science research centers. To answer questions regarding changes in the life course in historical populations or developing countries scholars must use available sources. Typically, these are cross-sectional census data.

Demographic research has a long tradition of using synthetic cohort measures based on census data (Shryock & Siegel, 1976). For example, the total fertility rate (TFR) is a synthetic cohort measure of fertility that estimates the number of children that would be born to a hypothetical woman who gave birth over her lifetime at prevailing age-specific fertility rates. It assumes that women who are aged 20 to 24 today will give birth when they are aged 30 to 34 at the same rate as 30 to 34-year-old women today. One can only make this assumption if one assumes that prevailing conditions that shape fertility will not change radically in the next 10 years, a somewhat risky assumption, but one that has widespread acceptance. Thus, measures of synthetic cohorts are useful tools for describing individual life course transitions.

Life course analysis is concerned with studying not just single transitions, but the set of transitions that make up the life course, particularly those that are observable as demographic transitions. For example, Modell et al. (1976) examined a set of observable transitions – leaving school, entering the workforce, leaving household of origin, marriage, and establishing one's own household – for two synthetic cohorts, the 1880 Philadelphia cohort taken from census records and the 1970 U.S. census. They examined the prevalence, timing, and spread of individual transitions, as well as the congruity and integration of pairs of transitions. While their study is widely cited, their methods have rarely been duplicated (for an exception see Stevens, 1990).

The expansion of computer technologies and statistical packages that allow for stochastic approaches to modeling life course transitions has improved on descriptive approaches by providing causal models of life course transitions (Mayer & Tuma, 1990). However, these stochastic approaches depend on individual life histories from longitudinal or retrospective data and thus their historical-comparative scope is limited. Furthermore, they are limited to examining one or at most two transitions at a time, leaving life course scholars who desire a more holistic or contextual view of the life course unsatisfied. Thus there is a need for methods that treat the life course more holistically and move beyond simple description.

The method I present here is an entropy analysis of age-specific status combinations of synthetic cohorts. This method makes use of basic census data to broaden the range of times and places in which life course scholars do research. Analytically, it demarcates the ages in the life course when there are relatively more status combinations, a condition that is interpreted as a period of change in the life course, such as the transition to adulthood, or a period of stability when the life course is more structured and there are fewer status combinations, such as childhood or full adulthood. Furthermore, by comparing the life course in different periods it demonstrates change in the construction of the life course over time. In this way it addresses the question of whether the life course has become more standardized or individualized in recent decades (see discussion in Shanahan, 2000). In this chapter I apply an entropy analysis of age-specific status combinations in a synthetic cohort to Mexican census data from 1970 and 2000 to analyze how the early life course is structured and whether it has changed.

DATA AND METHODS

The Mexican census data used for this analysis come from the IPUMS International Database website managed by the Minnesota Population Center at the University of Minnesota (Sobek, Ruggles, McCaa, King, & Levinson, 2002). I use one-percent samples of the 1970 and 2000 censuses to compare two synthetic birth cohorts. I analyze school attendance, employment status, relationship to household head, marital and parental statuses, while differentiating by gender and urban or rural residence. Individuals in these two synthetic cohorts were born nearly a generation apart from one another, allowing analysis of differences between the life course of the contemporary generation of youth and that of its parents 30 years before. Over this period of time, there is evidence of population aging and urbanization, with the average age increasing by a little more than 2 years and the percentage living in urban areas increasing from 59 to 76% (see Table 1). These demographic shifts are controlled in the sense that analyses are distinguished by age and residence, as well as gender.

The analysis of the data proceeds in two steps. In the first section I examine individual transitions drawing on the analysis by Modell et al. (1976) to show the prevalence, timing and spread of individual statuses, as well as the congruity of pairs of transitions. I examine individuals of ages 6 to 45 in the two census years to capture the full range of the early life course. This analysis demonstrates the transition patterns for single transitions, but

Table 1. Frequency Distributions of 1% Census Samples of those Aged 6–45, Mexico.

	1970	2000
Mean age of men (s.d.)	20.3 (11.1)	22.4 (11.1)
Mean age of women (s.d.)	20.6 (11.0)	22.9 (11.1)
% female	50.4	51.5
% urban (more than 2,500 inhabitants)	58.7	75.6
% head or spouse		
Rural men	27.9	31.4
Urban men	28.3	36.4
Rural women	32.8	36.9
Urban women	35.5	40.5
% child of head		
Rural men	66.8	60.8
Urban men	68.3	53.9
Rural women	59.9	52.6
Urban women	61.2	48.4
% Other relative/non-rel.		
Rural men	5.4	7.7
Urban men	3.3	9.8
Rural women	7.2	10.4
Urban women	3.4	11.1
% attending school		
Rural men	37.9	39.4
Urban men	27.0	40.0
Rural women	32.0	36.2
Urban women	23.7	36.8
% In the labor force		
Rural men	46.1	50.4
Urban men	48.9	57.6
Rural women	15.5	14.9
Urban women	8.2	30.9
% Ever-married		
Rural men	33.2	35.7
Urban men	34.3	40.7
Rural women	40.4	44.4
Urban women	43.6	46.8
% parents		
Rural men	—	—
Urban men	—	—
Rural women	37.0	41.7
Urban women	38.7	45.4
N	74,013	323,859

Note: Data are weighted.

does not show easily how the transitions combine. In the second step of the analysis I introduce the entropy analysis of age-specific status combinations. This measure demonstrates the extent to which specific status combinations are held by groups at each age. The entropy index is constructed by coding each individual in the sample according to their current statuses (see the Appendix A for coding scheme). Each respondent receives a seven-digit code that describes their marital and parental status, their status as student and worker, and their relationship to the head of household (head/spouse, child, or other).[1] Therefore, the statuses held at any given age indicate whether a person is currently attending school, currently employed, currently living in their parents household, their own household (as head or spouse of head), or living in the household of someone else, whether they are ever-married, and have ever had children (in the case of women). For each age, the percentage distribution in all of the status combinations is calculated according to the gender and residence of each individual thus making four synthetic cohorts (rural and urban men and women) for each year. This measure is summarized with an entropy index to describe the heterogeneity of age-specific status combinations telling us how much differentiation by statuses there is at each age.

The entropy index is a measure of heterogeneity ranging from 0 when there is perfect homogeneity (everyone in a single status combination) to a maximum heterogeneity defined when there is an equal distribution of cases in all status combinations. I based my entropy index on Theil's (1972) general entropy index. It is calculated as

$$E = \sum_{s=1}^{S} p_s \log(1/p_s)$$

where S is the number of states and p_s is the proportion of the population in state s. This measure has been used by Billari (2001) to show the heterogeneity of state distributions by age for longitudinal life course data. Here it is applied to synthetic cohort data to show the heterogeneity of status combinations at given ages. To make the entropy index more intuitively comprehensible I present the age-specific entropy as a percentage of the maximum entropy so it can be understood as the extent to which the status combinations are structured (closer to 0%) or not (closer to 100%) for a particular age group. Women have 128 possible status combinations, while men have only 64 since parental status is not measured for men. Therefore, women's maximum entropy is calculated as $E = 128 * (1/128) * (\log(1/(1/128))) = 2.10$ and men's maximum entropy is calculated as $E =$

$64 * (1/64) * (\log(1/(1/64))) = 1.81$. The observed entropy is presented as a percentage of the maximum entropy, for example, the entropy value for urban women aged 22 in 2000 was 58.4% of the maximum entropy showing that there was a relatively high degree of heterogeneity of status combinations at that age. In contrast, the entropy value for urban women aged 12 in 2000 was only 14.1% of the maximum entropy showing that there was relative homogeneity of status combinations at this age.

To evaluate how much of the entropy index is due to a single status I calculate the percentage difference between total entropy (E_f) and entropy as measured when one of the statuses is not included in the calculation (E_r). When the percentage change is greater than would be expected if the statuses in both the full and reduced indices were equally divided (max E_f and max E_r), I interpret this as meaning that the omitted status contributes more than expected to the full entropy index. The formula for calculating these is generalized as follows:

$$P = \left(\frac{\max E_f - \max E_r}{\max E_f}\right)100$$

$$O = \left(\frac{E_f - E_r}{E_f}\right)100$$

When the observed heterogeneity, O, is greater than the predicted heterogeneity based on maximum entropy distributions, P, the omitted status contributes relatively more to the entropy index. When O is less than P, the omitted status contributes relatively little to the entropy index. For example, if the reduced entropy index is calculated for women without including school in the model the maximum entropy is 1.81, while the full entropy index maximum is 2.10; thus, the predicted percentage decline in the maximum entropy index is −17.9%. If the observed percentage change between the full and reduced entropy indices is greater than −17.9%, I interpret this as implying that leaving out schooling decreases heterogeneity in status combinations more than expected and therefore schooling contributes relatively more than other statuses to the heterogeneity of age-specific status combinations in the full index. If the observed decline is less than −17.9%, I interpret this as meaning that schooling contributes relatively less to the heterogeneity of age-specific status combinations in the index with all possible status combinations. This decomposition of the entropy index is useful for determining which social statuses and their associated social structures shape the life course at particular life stages.

The entropy index of age-specific status combinations as applied to life course research generally demonstrates the ages at which transitions from one life stage to another occur and provides a lens on the degree to which social institutions and social norms shape the life course. The degree of heterogeneity of status combinations is expected to be smaller when social institutions strongly influence statuses held at a given age and larger during periods in the life course when status changes are concentrated or when social norms guiding the life course are relaxed and a greater variety of status combinations can be held. In sum, the entropy index of age-specific status combinations captures the timing of change in the life course and encapsulates the degree to which young people depart from what may be considered the normative pathway from childhood to adulthood. By comparing the age-specific entropy measure between the 1970 and 2000 censuses I show the direction of that change during this period and assess whether there is increasing or decreasing variability in the combinations of statuses held by young people.

TIMING, SPREAD AND PREVALENCE OF LIFE COURSE TRANSITIONS

The basic questions life course scholars ask about the transition to adulthood are, at what age does the process begin, how long does it take for everyone who is going to experience the transition to experience it, and what proportion of the population actually experiences it? In other words, what is the timing, spread, and prevalence of role acquisition and exit in the transition to adulthood. Adapting the Modell et al. (1976) approach, I measure the age at which the greatest percentage of the population ever held the status at a given age (prevalence) and the ages at which the first, fifth, and ninth deciles had made the transition to estimate the mid-point and the ages marking the timing of the transition for the central 80% (timing and spread). The deciles are based on the prevalence of the status.[2] These figures are presented graphically for urban and rural men and women (Figs. 1a–d). The bars represent the prevalence and are measured on the left-hand *y*-axis. The lines represent the range of ages at which the central 80% make the transition with the central dot marking the age at which the fifth decile made the transition (metric defined by the right-hand *y*-axis).

Looking first at men's transitions (Figs. 1a and b), apart from the greater prevalence of school attendance, it is notable how little change has occurred between the two censuses in the prevalence and timing of the transitions. Greater rates of school attendance have not slowed men's entry into the labor force because most men end their secondary education in their early teens at around the same time they enter the labor force. Indeed, leaving school and starting employment occurred at very similar ages and proceeded at very similar paces in both 1970 and in 2000, though there was undoubtedly a great degree of overlap in which boys combined school and work that cannot be observed with these measures. In 2000, the process of leaving school and entering the labor force was well underway for half of rural men by age 15, with the central 80% accomplishing the transition by age 21. Their urban counterparts make the transition somewhat later as only half of the urban men had left school and joined the labor force by age 18 and the central 80% have made the transition by their mid-20s. For rural men, and to a lesser extent urban men, this transition occurs well before the transition into adult family roles.

Men's transitions from dependent to independent family statuses show relatively little change between the two periods. The process of home-leaving and family formation for young men occurred at slightly later ages in 2000, but were closely timed and nearly universal in both periods. In Mexico, leaving the parental home is not necessarily timed prior to marriage and household formation, as it is in the U.S. and other Western European countries. Here, the timing of home-leaving, becoming a household head, and marrying occur at similar ages. On average, marriage occurs slightly earlier than the transition out of the parental home and into an independent household. The lack of change in this pattern over time suggests that family formation is guided by institutional processes that have been unchanged by the economic, social and demographic changes characterizing Mexican society in the last quarter of the 20th century.

Women's life course has undergone slightly more change than men's (Figs. 1c and d). Between 1970 and 2000 much of the gender differential in educational participation was eliminated in rural and urban areas alike. By 2000 more than 95% of women attended school. Half of rural women at this time have left school by age 15 with the central 80% having made the transition by age 19. In urban areas schooling is more prolonged, so that half of women have left school by age 17 and the central 80% have finished with school by age 24. However, on average, schooling does not last long enough to bring about a significant delay in women's average age at entry into marriage and parenthood. Women's ages at these transitions remain very similar to those in 1970.

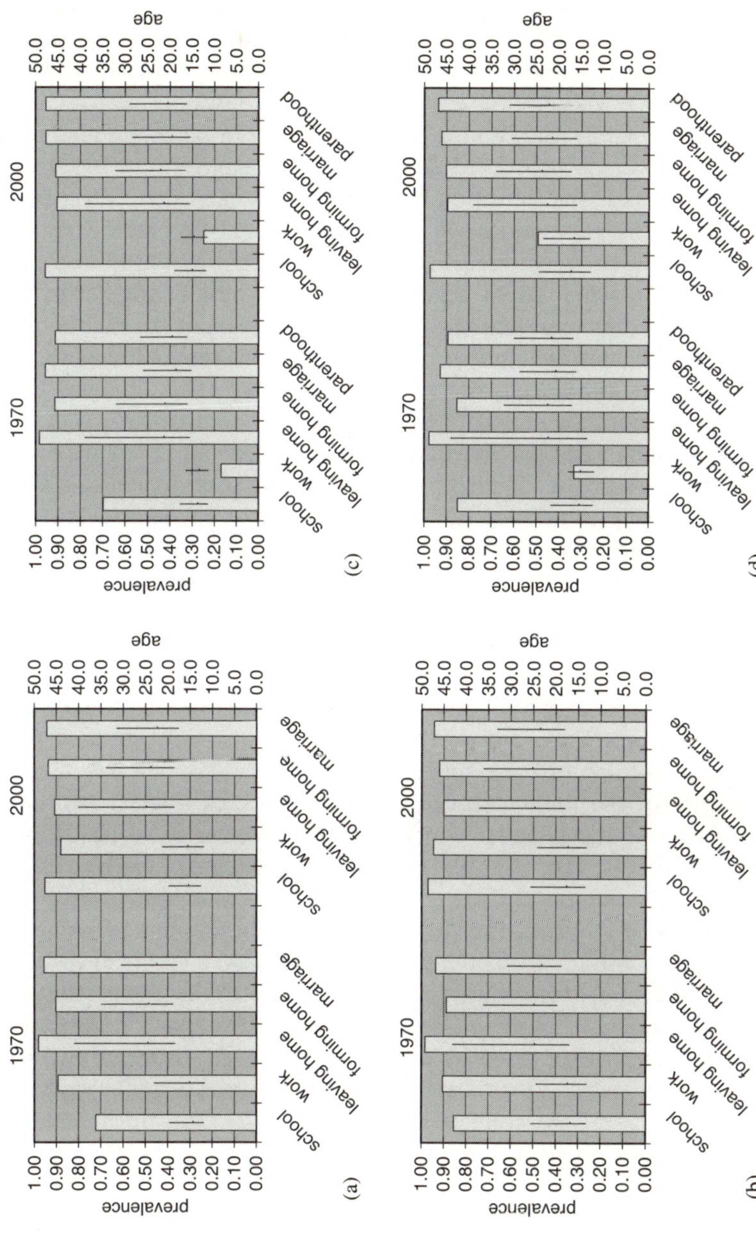

Fig. 1. Prevalence and Timing of Transitions for (a) Rural and (b) Urban Men. Prevalence and Timing of Transitions for (c) Rural and (d) Urban Women.

For women in Mexico, family formation is expected to be the focus of women's lives. The processes of leaving the parental home, marrying, having children, and becoming a household head or spouse of a household head occurred earlier than those of men, were closely timed in both periods, and were nearly universal in rural and urban areas alike. In both periods more than 90% of all women proceeded through the marriage and childbearing with half having made the transition by their early 20s and the central 80% completing the transitions before age 30 in rural areas and by their early 30s in urban areas. The process of leaving home and becoming the spouse of a household head or a household head is more prolonged because it is not uncommon for young couples to reside with their parents or parents-in-law. In general, the transition to family formation for both males and females has proven resistant to the social, economic, and demographic changes that have occurred during this period.

One way in which early marriage and household formation has persisted is through women's greater labor force participation. Census indicators of transition prevalence and timing however do not capture women's employment patterns well, since women's employment is often a reversible and temporary status. Thus, the low prevalence of women's labor force participation fails to capture the fact that many women cycle in and out of the labor force and their participation at older ages is as high or even higher than it is at younger ages. Often women combine employment with family statuses as their earning ability becomes more important to maintaining the household. The failure to capture these kinds of reversible statuses is a significant shortcoming of these single transition indicators.

This analysis of the timing, spread and prevalence of single transitions also does not reveal the extent to which each of these individual transitions is linked to the others. Modell et al. (1976) investigated the inter-relatedness of transitions by looking at the age-congruity of pairs of cohort transitions, the degree to which any two transitions overlapped in timing and duration for the cohort as a whole. Age congruity is calculated as the years of overlap between the two cohort transitions multiplied by two in the numerator and the sum of the spread of the two transitions in the denominator. Age congruity can be seen in Figs. 1a–d by examining the extent to which the age ranges overlap. Substantial overlap in the timing of two transitions suggests that many members of the cohort experience the two transitions simultaneously, while relatively little overlap suggests that the two transitions are not tightly related.

The age congruity of transition pairs are presented in Table 2 and grouped as the non-family-status transition pairs, the family-status transition pairs, and the non-family/family-status transition pairs. As already noted, transitions out of school and into work are tightly integrated since most young men and women who leave school do so around the time they enter into employment.[3] Among rural men, age congruity in the school-to-work transition increased between 1970 and 2000 as a greater percentage of boys stayed in school into their early teens, about the time they were eligible to begin working. Among urban men, age-congruity in the school-to-work transition decreased slightly, though levels are still very high. This is likely to be a result of some urban men staying in school into their early 20s when the majority of men have already entered the labor force. Age congruity in the

Table 2. Age Congruity of Transition Pairs, by Sex, Residence, and Year.

	Men				Women			
	1970		2000		1970		2000	
	Rural	Urban	Rural	Urban	Rural	Urban	Rural	Urban
Non-family transitions								
Leaving school-starting work	0.83	0.97	0.91	0.90	0.83	0.78	0.76	0.83
Family transitions								
Leaving home-first marriage	0.95	0.94	0.70	0.86	0.91	0.87	0.78	0.82
Leaving home-household formation	0.84	0.72	0.96	0.92	0.75	0.73	0.95	0.91
Leaving home-parenthood	—	—	—	—	0.92	0.85	0.81	0.84
First marriage-household formation	0.78	0.76	0.74	0.85	0.65	0.74	0.75	0.76
First marriage-parenthood	—	—	—	—	0.81	0.89	0.92	0.89
Household formation-parenthood	—	—	—	—	0.81	0.85	0.81	0.87
Mixed transitions								
Leaving school-leaving home	0.00	0.35	0.00	0.31	0.00	0.37	0.04	0.44
Leaving school-first marriage	0.00	0.25	0.00	0.37	0.00	0.18	0.14	0.53
Leaving school-household formation	0.00	0.07	0.00	0.24	0.00	0.00	0.00	0.32
Leaving school-parenthood	—	—	—	—	0.00	0.07	0.00	0.43
Starting work-leaving home	0.00	0.31	0.00	0.18	0.00	0.15	0.00	0.26
Starting work-first marriage	0.00	0.03	0.00	0.10	0.00	0.00	0.00	0.12
Starting work-household formation	0.00	0.02	0.00	0.09	0.00	0.00	0.00	0.11
Starting work-parenthood	—	—	—	—	0.00	0.00	0.00	0.21

Formula: Congruity = 2 × (years of overlap of central 80% between two transitions)/spread transition a + spread transition b.

school-to-work transition is lower among women than men, decreasing between the two periods for rural women and increasing for urban women. In other words, for women who are employed, their entry into the labor force overlaps significantly with the timing of their exit from school. The age congruities of family related transitions, leaving the parental household, marrying, having children, and forming a new household, are highly age congruent in both periods. Though there are slight shifts in patterns of age-congruity over time and between urban and rural contexts, these changes are unlikely to be meaningful given the small magnitude of change.

There is very little overlap in the age congruities of mixed-transition pairs, that is, non-family and family transition pairs. In rural areas, the transition out of school and into work occurs well before the transition into family and household formation for men and women alike, resulting in age-congruity values of 0 in most cases. Where the age-congruity measures are higher, it is most likely the case that it was not the same individuals who stayed in school and had families, but rather a small, select group that followed a new pattern of staying in school and delaying family formation, while the majority continued to leave school in their mid-teens and formed families several years later. The age-congruity of pairs involving work and family transitions overlap to a lesser degree, though they often increase over time. This suggests that for men and women alike the transition into employment occurs well before the transitions into family and household formation, though there is a slight, and perhaps increasing, degree of overlap, particularly among women.

AGE-SPECIFIC STATUS COMBINATIONS OF A SYNTHETIC COHORT: AN APPLICATION OF AN ENTROPY INDEX

While age-congruity measures describe patterns of cohort behavior, they offer no information about the extent to which an individual's transitions occur simultaneously, the integration of status transitions. Issues of integration motivate further analysis of the life course because they address the question of how contingent one transition is on others. There is a great deal of packaging of statuses that is not captured with the examination of single transitions or even pairs of transitions. The previously described entropy index can answer several questions: At which ages do status combinations become more or less complex within a cohort? How much does each status account for that complexity? Has the complexity in the life course increased or decreased over time and for whom?

Several observations can be made from the graphs of the age-specific entropy indices for men and women (see Figs. 2a and b). First, the entropy index ranges between 0% and 5% of the maximum entropy, emphasizing the degree of fluctuation in the heterogeneity of status combinations. Second, the pattern of the entropy index is quite regular, with heterogeneity increasing after age 11, peaking in the mid-20s, and declining thereafter, a pattern consistent with the concentration of status changes in adolescence and young adulthood. Third, there are few differences in the age patterns between 1970 and 2000, suggesting that there has not been a major structural reorganization of the early life course, with the possible exception of children who were even more likely to attend school in the later period. Fourth, patterns for men and women differ after they peak in the early 20s. Women experience an increase in heterogeneity in 2000 relative to 1970, while there is little difference for men. Fifth, urban–rural differences are small but important. Urban men experienced more heterogeneity between the ages of 16 and 25 than rural men and urban women experience more heterogeneity than rural women from age 18 through 45. Overall, the transition to adulthood in Mexico is relatively short and concentrated in the years between ages 12 and 22.

The regularity in the pattern of heterogeneity and the relative lack of change in the timing of transitions suggests that the early life course is structured quite rigidly. But which social institutions play the most important role in structuring the life course at various ages? Based on Figs. 2a and b, I have broken the life course into four periods; childhood, between ages 6 and 11, when heterogeneity is low; adolescence between ages 12 and 16, when heterogeneity increases rapidly; young adulthood, between ages 17 and 22, when heterogeneity peaks; and adulthood, after age 23, when heterogeneity decreases. I estimate the amount of heterogeneity due to a given status by evaluating the average percentage decrease in heterogeneity for an age range resulting from the removal of that variable from the calculation of the entropy index (see Table 3). As explained earlier, I compare the percentage decrease in observed heterogeneity to that expected by the elimination of one status from the predicted heterogeneity. An observed decrease that is greater than expected is considered to be a significant indication of change in the structuring of the life course. These are marked in bold in Table 3. To add more information of the relative importance of these statuses, I also provide the percentage distributions of the four most common status combinations of all the possible status combinations that

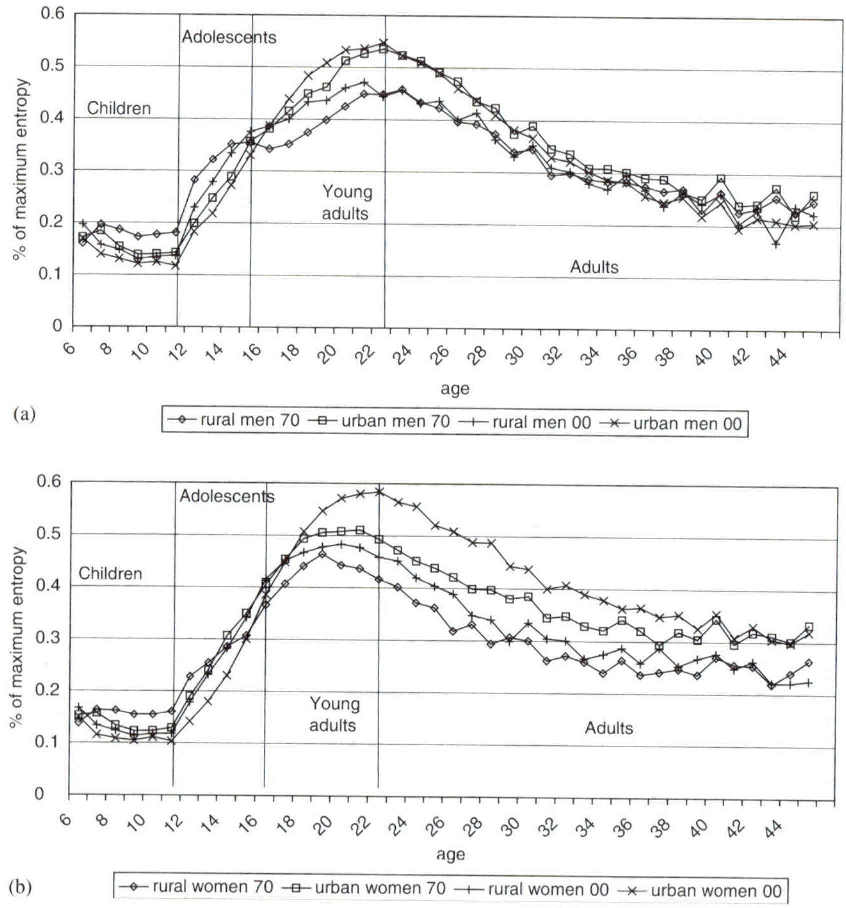

Fig. 2. (a) and (b). Men's and Women's Entropy Indices for Rural and Urban Mexico, 1970 and 2000.

were used to make the entropy index for rural and urban men and women in each period in Appendix B.

Children aged 6–11 have the most structured lives of all the ages considered here, as is evident in Figs. 2a and b. For the most part this is because children were unlikely to be employed, household heads, married or have children. These statuses were not counted either, so they do not contribute to the entropy index for these ages. Rather, most children resided in their parent's homes in both 1970 and 2000, but the percentage attending school

Table 3. Percentage Change in Entropy Index between Periods and Due to Leaving Out Each Status Group.

	1970				2000			
Age ranges	6–11	12–16	17–22	23–45	6–11	12–16	17–22	23–45
Rural men								
% change in total entropy between 1970 and 2000					−21.9	2.8	4.9	−2.6
Without school	**−83.3**	**−39.9**	−15.5	−6.6	**−38.0**	**−32.8**	−14.9	−4.0
Without employment	—	**−36.5**	**−32.0**	**−30.1**	—	**−31.2**	**−26.4**	**−31.0**
Without marital status	—	−3.9	−18.7	−16.5	—	−0.5	−17.9	−15.9
Without household position	−16.7	−13.3	−22.3	−33.0	−60.6	−25.4	−25.7	−29.4
Urban men								
% change in total entropy between 1970 and 2000					−14.6	−3.6	3.0	−9.0
Without school	**−79.4**	**−42.8**	**−24.0**	−10.8	−28.7	**−30.1**	**−22.2**	−11.7
Without employment	—	**−26.7**	**−27.8**	**−26.3**	—	**−28.4**	**−22.7**	−19.2
Without marital status	—	−3.1	−12.7	−17.0	—	−0.7	−13.3	−18.6
Without household position	−20.5	−19.7	−25.0	−33.5	**−70.4**	−31.5	−27.9	**−34.7**
Rural women								
% change in total entropy between 1970 and 2000					−23.5	5.2	9.3	6.5
Without school	**−84.0**	**−41.8**	−9.6	−5.7	**−38.6**	**−39.3**	−10.4	−3.5
Without employment	—	**−21.8**	**−19.0**	**−23.1**	—	**−19.6**	**−20.2**	**−32.9**
Without parental status	—	−5.0	−15.4	**−19.7**	—	−1.4	−11.2	−7.7
Without marital status	—	−7.7	−11.1	−7.0	—	−2.1	−6.6	−2.4
Without household position	−15.0	−16.9	**−24.3**	**−29.6**	−61.5	−26.9	−23.5	−22.8
Urban women								
% change in total entropy between 1970 and 2000					−19.3	−14.1	14.5	11.2
Without school	**−76.9**	**−42.0**	−15.3	−6.7	−27.7	**−32.0**	−17.3	−6.8
Without employment	—	**−18.4**	**−21.3**	**−25.5**	—	**−19.0**	**−20.6**	**−33.8**
Without parental status	—	−3.9	−11.2	−15.1	—	−0.7	−10.3	−11.2
Without marital status	—	−3.6	−8.3	−7.6	—	−1.6	−6.9	−8.6
Without household position	−22.9	**−23.9**	**−25.1**	**−28.9**	**−72.1**	**−35.0**	**−25.1**	**−23.3**

Note: For men aged 12 and above, the percentage change is greater than expected if delta >−21.8; for women aged 12 and above, delta> −17.9; for men and women aged 5–11, delta >−38.7 for "without school" and delta >−61.3 for "without household position"; for men aged 12 and above, delta >−34.6 for "without household position" and for women aged 12 and above, delta >−35.8 for "without household position."

was far greater in 2000 than in 1970. Therefore, in 1970 for rural and urban boys and girls alike, school attendance was a greater source of heterogeneity in children's status combinations than was the child's relationship to the household head since basic education was far from universal. In 2000, when school attendance at these ages was nearly universal, it became a less important source of heterogeneity for children, especially in urban areas where nearly all children attended school. Concurrently, the declining significance of education as a source of heterogeneity leads to a greater significance of children's relationship to the household head, particularly for urban boys and girls. This may also result from the greater likelihood that urban families lived in extended family households (so that the children are recorded as "other" household members). Overall, however, children's heterogeneity decreased between 15% and 24% as the lives of children became more structured between 1970 and 2000.

During adolescence (ages 12–16) the age-specific status combinations of cohorts quickly became much more heterogeneous as youth began to make the transitions associated with adulthood. If the life course were becoming more standardized we would expect to see a decrease in heterogeneity between 1970 and 2000. If the life course were becoming more individualized, it would increase. Instead, we see that this life stage in both periods was similarly heterogeneous. There is very little change in observed heterogeneity between the two periods. The greatest change is for urban girls who experienced a 14% decrease in heterogeneity between the two periods. Observed decreases, also seen for urban boys, are mostly due to less differentiation by school attendance in 2000 than in 1970, while differentiation by household position increased. In other words, more urban teenagers, especially girls, were attending school making household position a relatively more important source of heterogeneity and contributing to an overall decrease in heterogeneity among this group.

During young adulthood (ages 17–22), heterogeneity peaks for men and women alike in both cohorts. This is consistent with the concentration of life course changes occurring at these ages, particularly home leaving, household formation, marriage, and childbearing. There was very little change in the degree of heterogeneity for men, demonstrating that the transition to adulthood for young men is fairly scripted and involves a well-known set of status combinations that did not change over time (4.9% increase for rural men and 3.0% increase for urban men). The status that contributes most to rural men's heterogeneity was employment, as was the case for urban men in 1970. In other words, men who were not employed at these ages were likely to be quite different from those who were employed with respect to school attendance, marital status, and household position. For urban men in 2000, employment and school attendance contribute more than expected to heterogeneity. This demonstrates that in 2000 more urban men were attending school, leading to a greater differentiation among men by both school attendance and employment.

In contrast, heterogeneity increased for young adult women age (ages 17–22) during this period, growing by 9% and 14% for rural and urban women, respectively. Women's heterogeneity at these ages is mainly due to their household position and employment status which have similar contributions to heterogeneity in both 1970 and 2000. For instance, women of these ages who were employed were most often single, childless, children of household heads who were not in school, while women who were employed were most often household heads or their spouses, married with children and not attending school. The increase in heterogeneity between the two periods resulted from a growth in status combinations, most likely in terms of employment and household statuses.

In later adulthood (ages 23–45) most people have finished school, left their parent's home, begun employment (among men), married and formed families. Thus, heterogeneity of age-specific status combinations generally decreases. Between the two periods, the level of men's heterogeneity at all ages was very similar for rural men, but decreased slightly for urban men. As was the case in young adulthood, men's adult roles were tightly prescribed. For rural men, most of the heterogeneity was due to employment status. Rural men who were unemployed differed from their employed counterparts with respect to the other statuses considered. For urban men between 1970 and 2000, the importance of employment as a source of heterogeneity diminished. In 2000 few urban men could afford to be unemployed, reducing this source of heterogeneity.

In contrast, rural and urban women experienced increases in total heterogeneity during adulthood. Among rural adult women heterogeneity was mainly due to their employment status, while for urban adult women both employment status and household position differentiate groups of women. The relative importance of employment in differentiating women increased between 1970 and 2000. This is not surprising given women's large-scale entry into the labor force during this period. The increase in overall heterogeneity of rural and urban adult women is due to the growth of women who combine marriage and parenthood with employment (analyses not shown).

This decomposition analysis of the entropy index reveals much about the content of these stages of the early life course in Mexico. Childhood has become slightly more standardized in the past 30 years as more boys and girls stay in school. Adolescence is a period of rapid change in status combinations in both 1970 and 2000, but heterogeneity has decreased somewhat between time periods. Urban adolescent girls, and to a lesser extent urban adolescent boys, have decreased their heterogeneity as more of them attend school. For adolescents the main sources of heterogeneity are schooling and employment statuses. In young adulthood, heterogeneity of status combinations peaks for all groups. During this period heterogeneity increased marginally for urban and rural women, while it has remained at similar levels for males. The increase for women is mainly due to their increased labor force participation which created a new group of women who combine family and paid work rather than withdrawing from the labor force after marriage and childbearing. Adults continue in the paths established in young adulthood, as more of them enter into the prescribed roles of male breadwinners and female homemakers, consequently heterogeneity diminishes with each additional year of age.

In sum, relatively little has changed during the early life course. Only children's greater participation in education and adult women's greater labor force participation lowered or raised heterogeneity substantively. Furthermore, young adulthood is not significantly prolonged during this period, since there is little change in the timing of these transitions. Instead, the transition to adulthood continues to be highly structured in Mexico.

DISCUSSION AND CONCLUSION

This article makes both a methodological and a substantive contribution to the study of the life course. In it I have introduced a new method for analyzing age-specific status combinations for a synthetic cohort as a way of making better use of historical or cross-national census data to analyze the life course. The entropy index provides a tool for examining periods of relatively more or less differentiation among age cohorts, allowing us to demarcate stages of the life course. This is particularly useful for analyzing the age boundaries for participation in social institutions such as schools, labor markets, and family formation. It also reveals the extent to which participation in these social institutions are part of a package of statuses. In this way, the entropy index is a tool for describing a rather unwieldy amount of information about the life course, contributing to a growing toolbox of algorithmic tools based on descriptive rather than stochastic methods (Billari, 2003).

The entropy index of age-specific status combinations has shortcomings, however, which are important to discuss. In the example presented here I differentiated within the population according to gender and residence in two periods. However, this only partially captures the stratification within the population that may be expected to structure the life course. Ideally, the population would be differentiated by socioeconomic status to see how youth from more or less privileged backgrounds proceed in the transition. However, using the characteristics of the household as an indicator of socioeconomic status is problematic since household departure and formation are part of the process under investigation. Consequently, the entropy index of age-specific status combinations provides a crude description of central tendencies, but offers little leverage for understanding variation between sub-groups in the transition to adulthood. One possibility is to break up the sample according to invariant characteristics of the population, such as race, ethnicity, religion, or region of residence.

An additional problem stems from changes in the composition of the populations at two points in time. In the case of Mexico, the population urbanized quite rapidly during the period in question, which may make the urban and rural samples less comparable over time since the 2000 urban sub-sample includes more rural-to-urban migrants than the 1970 urban sub-sample. Likewise, although the definition of urban remained the same over the two points in time (municipalities with 2,500 people or more), the geographic boundaries of urban places grew, again making the urban and rural sub-samples different in each period. This problem cannot be addressed without more information than the census is able to provide.

The second contribution of this article is substantive, that is, to describe changes and provide a partial explanation of those changes in the early life course in Mexico during the past 30 years. The process of the transition to adulthood in Mexico can be seen as guided by a set of social structures or norms that continue to support a relatively early transition into adult statuses, even as social structures supporting greater educational participation and increased female labor force participation have emerged. This is remarkable since education and female employment have long been associated with later marriage ages for women in Western countries. The puzzle is to explain this relative lack of change in the transition to adulthood during a period of intense economic and social change in Mexico.

Mexican social scientists have argued repeatedly that the economic crisis has stimulated a great deal of change in Mexican society and family life. It has led to changes in the gender division of household labor and greater employment among women. But these changes in women's roles did not necessarily reflect a social revolution in gender norms. Instead, they were driven by economic necessity. The economic crisis also led to a widespread cynicism about an individual's prospects for improving their economic conditions and hence produced "survival strategies" rather than upward mobility strategies. Consequently, although Mexicans value education, education alone may not be perceived as sufficient for improving one's economic status. Thus, the economic roller coaster ride that Mexicans have experienced may lead them to grip more tightly to that which is familiar in other words, to families.

The lack of change in the prevalence and timing of marriage and childbearing, and to a lesser extent home-leaving and household formation, suggests that families provide the stability that is otherwise lacking in Mexican society. Therefore, delaying family formation is not the preferred strategy for coping with unemployment, underemployment, or economic uncertainty generally. Instead, other adaptive strategies are invoked. One widespread strategy is for married women to stop childbearing earlier and enter the labor force, thus reducing the dependency ratio in the household. Alternately, adolescent children may enter the labor force to generate additional income. The migration of some family members to urban areas or the United States is another means of managing the risk of economic shortages. Other household-based strategies that stretch the budget or increase income may also be used, though they are more difficult to detect with census data. While it is difficult to demonstrate empirically, it is possible to argue that the economic crisis has reinforced early family formation patterns.

The study of Mexico provides an interesting contrast to studies of the transition to adulthood undertaken in advanced industrial societies. Mexico has few of the social institutions that contribute to the delay in family formation. In advanced industrial societies, widespread tertiary education, extended occupational apprenticeships, and social sanctions for those who marry before they are economically self-sufficient are all seen to delay family formation. Furthermore, attitudinal and value changes with respect to gender and religion are identified as part of the second demographic transition that reinforce these structural supports for later marriage and childbearing. Mexico contrasts sharply with these societies, in particular highlighting how the early life course is structured differently according to the presence or absence of a state that is able to invest in youth.

NOTES

1. Since men are not asked how many children they have ever given birth to, they do not receive a code for parental status. Although men's parental status could be

inferred from the household composition, this would involve making the untenable assumption that men co-reside with their children. Also, individuals employment, marital, and parental statuses are not recorded at ages 5–11, so the entropy index at these ages is based solely on the relationship to the household head and school attendance. This is consistent with the statuses that children actually hold, with the exception perhaps of employment.

2. Modell et al. (1976) measure the spread or range (the number of years between the age at which the first 10% and the last 10% of the group experiences the transition), and the ages at which the 1st and 5th deciles of the population make the transition, as well as the median age. However, with regard to the median age they state that "the notion of half a population leaving a status which not all of them have ever occupied is self-contradictory." Therefore I do not include the median age in my analysis.

3. Recall that this measure is based on the timing of those who actually make the transition, not the prevalence of those who make the transition. For this reason, the school-to-work transition is as closely integrated for women as for men even though fewer women enter the labor force.

REFERENCES

Alarcón, G. D. (1994). *Changes in the distribution of income in Mexico and trade liberalization.* Tijuana, Baja California: El Colegio de la Frontera Norte.

Benería, L. (1992). The Mexican debt crisis: Restructuring the economy and the household. In: L. Benería & S. Feldman (Eds), *Unequal burden: Economic crisis, persistent poverty, and women's work* (pp. 83–104). Boulder, CO: Westview Press.

Billari, F. C. (2001). The analysis of early life courses: complex descriptions of the transition to adulthood. *Journal of Population Research, 18*, 119–142.

Billari, F. C. (2003). Life course analysis: two (complementary) cultures? Reflections on how to analyze the transition to adulthood. *Presentation at the IUSSP Study Group on Transition to Adulthood in Developed Countries*, Calgary, Canada, August 4–6.

Blanco, M. (2002). Trabajo y familia: entrelazamiento de trayectorias vitales. *Estudios Demográficos y Urbanos, 17*, 447–483.

Cabrera, G. (1994). Demographic dynamics and development: The role of population policy in Mexico. *Population and Development Review, 20s*, 105–120.

Elder, G. (1999) [1974]. *Children of the great depression: social change in life experience.* Westview Press: Boulder, CO.

Escobar Latapí, A., & Roberts, B. R. (1991). Urban stratification, the middle classes, and economic change in Mexico. In: M. González de la Rocha & A. Escobar Latapí (Eds), *Social responses to Mexico's crisis of the 1980s* (pp. 91–113). San Diego, CA: University of California, San Diego.

Fussell, E. (2004a). What's driving the increase in Mexican women's employment? Family status, education, economic crisis and regional restructuring. Manuscript.

Fussell, E. (2004b). Convergence and divergence in the transition to adulthood during the 20th century: Race, nativity, and gender comparisons. Paper presented at the Society for Research on Adolescence Biennial Meeting, March 11–14, Baltimore, MD.

Fussell, E., & Furstenberg, F. F. (2005). The transition to adulthood during the 20th century: Differences by race, nativity and gender. In: R. Settersten Jr., F. Furstenberg Jr. & R. Rumbaut (Eds), *On the frontier of adulthood: Theory, research, and public policy* (pp. 29–75). Chicago: University of Chicago Press.

Fussell, E., & Palloni, A. (2004). Persistent marriage regimes in changing times. *Journal of Marriage and Family*, *66*, 1198–1210.

Garcia, B., & de Oliveira, O. (1994). *Trabajo feminino y vida familiar en México*. México, D.F.: El Colegio de México.

Gomes, C. (2001). *Procesos sociales, población y familia: Alternativas teóricas y empíricas en las investigaciones sobre vida domestica*. Facultad Latinoamericana de Ciencias Sociales y Miguel Ángel Porrúa, México, D.F.

González de la Rocha, M. (1994). *The resources of poverty: women and survival in a Mexican city*. Cambridge, MA: Blackwell.

Held, T. (1986). Institutionalization and deinstitutionalization of the life course. *Human Development*, *29*, 157–162.

Instituto Nacional de Estadística, Geografia e Informática. (2004a). *Sistema de Indicadores para el Seguimiento de la Situatción de la Mujer en México*. Tasa global de fecundidad: coberatura nacional. File accessed on June 3, 2004. http://dgcnesyp.inegi.gob.mx/sisesim/

Instituto Nacional de Estadística, Geografia e Informática. (2004b). Sistemas Nacíonales Estadístico y de Información Geográfica. Porcentaje de la población de 5 años y más que asiste a la escuela por grupos de edad y sexo, 1970–2000. File accessed on June 3, 2004. http://www.inegi.gob.mx/

Instituto Nacional de Estadística, Geografia e Informática. (2004c). Sistemas Nacíonales Estadístico y de Información Geográfica. Indicadores seleccionados sobre nivel de instrucción, promedio de escolaridad, aptitude para leer y escribir y alfabetismo, 1960–2000. File accessed on June 3, 2004. http://www.inegi.gob.mx/

Kohli, M., & Meyer, J. (1986). Social structure and social construction of life stages. *Human Development*, *29*, 145–149.

LeVine, S. (1993). *Dolor y Alegría: Women and social change in urban Mexico*. Madison, WI: The University of Wisconsin Press.

Levinson, B. (1999). Una etapa siempre difícil: Concepts of adolescence and secondary education in Mexico. *Comparative Education Review*, *43*, 129–161.

Lustig, N. (1990). Economic crisis, adjustment, and living standards in Mexico 1982–85. *World Development*, *18*, 1325–1342.

Lustig, N. (1997). The United States to the rescue: Financial assistance to Mexico in 1982 and 1995. *CEPAL Review*, *61*, 41–62.

Martin, C. (1996). Economic strategies and moral principles in the survival of poor households in Mexico: An urban and rural comparison. *Bulletin of Latin American Research*, *15*, 193–210.

Mayer, K., & Tuma, N. (1990). *Event history analysis in life course research*. Madison, WI: The University of Wisconsin Press.

Michael, R. (2001). *Social awakening: Adolescent behavior as adulthood approaches*. New York: Russell Sage Foundation.

Mier y Terán Rocha, M. (1996). The implications of Mexico's fertility decline for women's participation in the labor force. In: J. M. Guzmán, S. Singh, G. Rodríguez & E. A. Pantelides (Eds), *The fertility transition in Latin America* (pp. 323–342). Oxford: Clarendon Press.

Mier y Terán Rocha, M., & Rabell, C. (2003). Inequalities in Mexican children's schooling. *Journal of Comparative Family Studies*, *34*, 435–454.

Modell, J., Furstenberg, F., & Hershberg, T. (1976). Social change and transitions to adulthood in historical perspective. *Journal of Family History*, *1*, 7–32.

Moreno, L., & Singh, S. (1996). Fertility decline and changes in proximate determinants in the Latin American and Caribbean Regions. In: J. M. Guzmán, S. Singh, G. Rodríguez & E. A. Pantelides (Eds), *The fertility transition in Latin America* (pp. 113–134). Oxford: Clarendon Press.

Muñoz García, H., & Suárez Zozaya, M. H. (1990). Educación superior y empleo: Decrece la influencia de la educacion sobre el empleo. *Demos*, *3*, 30–31.

Portes, A., & Hoffman, K. (2003). Latin American class structures: Their composition and change during the Neoliberal era. *Latin American Research Review*, *38*, 41–82.

Pries, L. (1996). Institucionalización o desinstitucionalización del curso de vida? Biografía y sociedad como un enfoque integrativo e interdisciplinario. *Estudios Demográficos y Urbanos*, *11*, 395–417.

Quilodrán, J. (1996). Trayectorias de vida: un apoyo para la interpretación de los fenómenos demográficos. *Estudios Sociológicos*, *14*, 393–416.

Robles-Vásquez, H., & Abler, D. G. (2000). Education and Labor Force Participation by Mexican Children During Structural Adjustment: A Micro-economic Analysis. Paper presented at the Latin American Studies Association Annual Meeting, Miami, FL, March 16–18.

Shanahan, M. J. (2000). Pathways to adulthood in changing societies: Variability and mechanisms in life course perspective. *Annual Review of Sociology*, *26*, 667–692.

Shryock, H.S., & Siegel, J.S. (1976). *The methods and materials of demography. Condensed Edition* by Edward G. Academic Press/Stockwell. San Diego, CA: Harcourt, Brace, Jovanovich.

Sobek, M., Ruggles, S., McCaa, R., King, M., & Levinson, D. (2002). *Integrated public use microdata series-international: Preliminary version 0.1.* Minneapolis: Minnesota Population Center University of Minnesota.

Stevens, D. (1990). New evidence on the timing of early life course transitions: The United States 1900 to 1980. *Journal of Family History*, *15*, 163–178.

Theil, H. (1972). *Statistical decomposition analysis.* Amsterdam: North-Holland Publishing Company.

Tuirán, R. A. (2001). Estructura familiar y trayectorias de vida en México. In: C. Gomes (Ed.), *Procesos socials, población y familia: Alternativas teóricas y empíricas en las investigaciones sobre vida domestica* (pp. 23–60). Facultad Latinoamericana de Ciencias Sociales y Miguel Ángel Porrúa, México, D.F.

Tuirán, R. A. (2002). Transición demográfica, trayectorias de vida y disigualdad social en México: lecciones y opciones. *Papeles de población*, *8*, 25–66.

United States Census Bureau. (2004a). Table 010. Infant mortality rates and life expectancy at birth, by sex. Mexico 1970–2000. Downloaded from International Data Base. Accessed June 3, 2004. http://www.census.gov

United States Census Bureau (2004b). Table 028. Age-specific fertility rates and selected derived measures. Mexico 1980–2000. Downloaded from International Data Base. Accessed June 3, 2004. http://www.census.gov

Weller, J. (2000). Employment trends in Latin America and the Caribbean during the 1990s. *CEPAL Review*, *72*, 31–51.

APPENDIX A. STATUS COMBINATION VARIABLE

Status	Code
Relationship to household head: Not head or spouse	0- - - - - -
Relationship to household head: Head or spouse	1- - - - - -
Relationship to household head: Not child	-0- - - - -
Relationship to household head: Child	-1- - - - -
Relationship to household head: Not other	- -0- - - -
Relationship to household head: Other	- -1- - - -
Marital status: Never-married	- - -0- - -
Marital status: Ever-married	- - -1- - -
Parental status: Has no child	- - - -0- -
Parental status: Has own child	- - - -1- -
Labor force status: Not in labor force	- - - - -0-
Labor force status: In labor force	- - - - -1-
School status: Not attending school	- - - - - -0
School status: Attending school	- - - - - -1

APPENDIX B

Entropy index for rural and urban men and women (see Tables B1 and B2)

Table B1. Men's Status Combinations During Childhood, Adolescence, Young Adulthood, and Adulthood, Mexico 1970 and 2000.

Rural men 1970				Rural men 2000			
6–11	12–16	17–24	25–45	6–11	12–16	17–24	25–45
Child, in school 57.4	Child, in school 36.3	Child, employed 41.1	Head, married, employed 67.6	Child, in school 81.9	Child, in school 53.7	Child, employed 39.7	Head, married, employed 67.5
Child, not in school 40.0	Child, employed 27.4	Head, married, employed 16.4	Head, married, inactive 8.3	Other, in school 10.5	Child, employed 18.7	Head, married, employed 18.0	Head, married, inactive 9.7
Other, in school 1.6	Child, inactive 27.1	Child, inactive 14.9	Child, married, employed 8.2	Child, not in school 6.2	Child, inactive 11.2	Child, inactive 11.7	Child, employed 8.0
Other, not in school 1.0	Child, in school, employed 3.6	Child, married, employed 10.7	Child, employed 6.7	Other, not in school 1.0	Child, in school, employed 6.1	Child, married, employed 7.7	Child, married, employed 4.8

Table B1. *(Continued)*

Urban men 1970				Urban men 2000			
6–11	12–16	17–24	25–45	6–11	12–16	17–24	25–45
Child, in school 69.3	Child, in school 60.4	Child, employed 31.7	Head, married, employed 65.6	Child, in school 84.0	Child, in school 66.9	Child, employed 28.9	Head, married, employed 69.4
Child, inactive 27.9	Child, inactive 17.1	Child, in school 13.8	Child, married, employed 8.4	Other, in school 11.7	Child, employed 9.8	Child, in school 18.5	Child, employed 8.7
Other, in school 2.1	Child, employed 12.2	Child, inactive 13.6	Child, employed 7.3	Child, inactive 3.3	Child, inactive 6.8	Head, married, employed 14.5	Child, married, employed 4.0
Other, inactive 0.6	Other, in school 3.6	Head, married, employed 12.2	Head, married, inactive 6.4	Other, inactive 0.8	Other, in school 6.7	Child, inactive 7.4	Other, married, employed 3.6

Table B2. Women's Status Combinations During Childhood, Adolescence, Young Adulthood, and Adulthood, Mexico 1970 and 2000.

Rural women 1970				Rural women 2000			
6–11	12–16	17–24	25–45	6–11	12–16	17–24	25–45
Child, in school 55.8	Child, inactive 49.9	Spouse, married, parent 31.2	Spouse, married, parent 69.9	Child, in school 81.9	Child, in school 53.3	Spouse, married, parent 26.1	Spouse, married, parent 63.3
Child, inactive 41.6	Child, in school 29.7	Child, inactive 25.0	Child, married, parent 6.6	Other, in school 10.8	Child, inactive 23.6	Child, inactive 21.6	Spouse, married, parent, employed 14.9
Other, in school 1.6	Child, employed 7.5	Child, married, parent 11.8	Spouse, married, parent, employed 5.8	Child, inactive 6.0	Child, employed 6.9	Child, employed 12.7	Other, inactive 4.1
Other, inactive 1.0	Other, inactive 2.5	Child, employed 7.0	Spouse, married 4.6	Other, inactive 1.0	Other, in school 5.4	Other, married, parent 7.4	Other, married, parent 3.6

Table B2. *(Continued)*

Urban women 1970				Urban women 2000			
6–11	12–16	17–24	25–45	6–11	12–16	17–24	25–45
Child, in school 69.7	Child, in school 51.8	Spouse, married, parent 22.1	Spouse, married, parent 60.9	Child, in school 83.8	Child, in school 68.4	Child, in school 17.6	Spouse, married, parent 43.6
Child, inactive 27.0	Child, inactive 26.1	Child, inactive 17.8	Spouse, married, parent, employed 8.6	Other, in school 12.1	Child, inactive 10.1	Spouse, married, parent 15.6	Spouse, married, parent, employed 27.2
Other, in school 2.3	Child, employed 5.9	Child, employed 16.2	Child, married, parent 6.7	Child, inactive 3.2	Other, in school 7.5	Child, employed 15.6	Child, employed 5.1
Other, inactive 1.0	Other, in school 3.9	Child, married, parent 8.8	Child, employed 3.5	Other, inactive 0.7	Child, employed 4.6	Child, inactive 9.3	Child, married, parent, employed 2.7

PART III:
CONTEXTS AND CONTINGENCIES

COLLEGES, CAREERS, AND THE INSTITUTIONAL STRUCTURING OF THE TRANSITION TO ADULTHOOD

Ann E. Person, James E. Rosenbaum and Regina Deil-Amen

ABSTRACT

Life course theorists have variously focused on the role of culture, history, and social structures in shaping the life course, but rarely have they examined concrete social organizations. This chapter looks at how an important group of institutions – 2-year colleges – structures the transition to adulthood in the U.S. Using data from different types of 2-year colleges, we analyze their assumptions about and their impact upon the transition from school to work. With increasing numbers of young adults seeking workforce preparation from 2-year colleges, consideration of these institutions can offer valuable insights into the structuring of the transition from school to work specifically, and the transition to adulthood more generally.

The Structure of the Life Course: Standardized? Individualized? Differentiated?
Advances in Life Course Research, Volume 9, 125–149

ISSN: 1040-2608/doi:10.1016/S1040-2608(04)09005-7

In their consideration of the structure of the life course, theorists have often focused on "contexts," addressing these explicitly in terms of culture (Baltes, 1997), history (Elder, 1998), and social structures (Dannefer, 1984). Yet relatively few life course scholars have studied what Dannefer (1984, p. 107) calls the "concrete forms of social organizations," such as schools and occupations as potential influences on the structure of the life course. Such organizations certainly play a dynamic role in culture, history, and social structures, at once shaping and being shaped by them. So it is unfortunate that organizations have received so little attention in life course research. As Dannefer notes, access to and experiences with such organizations are critical channels of developmental opportunity. Moreover, "concrete" organizations are more readily observed than more abstract phenomena like culture, history, and social structures. Similarly, if many individuals experience problems in a life stage associated with a certain institution, it is easier for reformers to change institutional procedures than to change culture, history, or social structures.

This chapter focuses on a particular group of organizations – 2-year colleges – and their impact upon the transition to adulthood, and especially the transition from school to work. As the lowest tier of higher education, 2-year colleges might be regarded as unimportant. Such a view would be mistaken. They are the fastest growing segment of higher education and now enroll nearly half of all students entering postsecondary education in the United States (Bailey, 2003). Moreover, many of the radical changes in our society over the last 40 years have largely focused here. Great increases in labor market skill demands have encouraged many more students to enter or re-enter college. In addition, new kinds of students have entered, with most of the increased access for minority and disadvantaged students occurring in 2-year colleges. The practice of combining college and work is more feasible in 2-year colleges, which are often located near workplaces and residences and offer occupational curricula relevant to work. Most dramatically, unlike traditional 4-year colleges, virtually all 2-year colleges opened their doors to admit all interested high school graduates, regardless of students' prior academic achievement (Dougherty, 1994).

Two-year colleges also tend to focus more explicitly than high schools or universities on the student's transition into the labor market (Grubb, 1996). The transition from school to work is a key process in the life course, as the individual's experiences moving through education and into a career are commonly understood to have important and lasting influence on subsequent pathways through adulthood (Kerckhoff, 2003). Because 2-year colleges have an explicit focus on work entry, they represent an ideal institution for examining the ways schools can help prepare students for jobs.

At the same time, individuals are judged and judge themselves in terms of their progress in attaining certain markers of life stage accomplishments, including the attainment of college degrees. While individuals' progress toward a college degree is often considered an indicator of their personal accomplishment, institutions differ in the way they facilitate progress. Even the time it takes to attain the same "2-year" degree may systematically differ by institution. What structures and practices underlie these differences? How does this play out in individuals' experiences of progress? Do these differences have implications for individual's developmental trajectories?

We seek to answer these questions by examining a combination of quantitative and qualitative data. Broadly speaking, we find that institutions' distinct missions reflect different assumptions about the life course and result in very different organizational structures. These structures then impact individuals' life course pathways through the way they facilitate progress toward developmental markers. At the same time, these varying organizational structures support different psychological processes associated with the transition to adulthood. Specifically, we find that 2-year colleges that encourage exploration without structuring efficient education and career pathways are associated with delayed progress and decreasing confidence as students continue at the institution. In contrast, colleges that are structured to encourage efficient progress and to reduce mistakes are associated with fewer delays and increased confidence after institutional entry.

SOCIAL AND PSYCHOLOGICAL DEMANDS OF THE TRANSITION TO ADULTHOOD

In a recent review of research on the transition to adulthood within a life course perspective, Shanahan (2000, p. 667) applauds the "heightened sensitivity to transition behaviors as developmental processes" evinced in the literature. Attention to transition behaviors is critical to any discussion of the structure of the life course not only because transitions link one developmental phase to the next, but also because they can determine which pathways will be opened up and which closed off, giving the life course its shape.

Like most developmental phases, the transition to adulthood can be understood in terms of the primary tasks of the phase as well as the psychological processes that support successful movement toward achieving these tasks. Exploring career options and making timely progress preparing for a career are central tasks of the transition to adulthood and are the focus of this chapter. In the transition to adulthood, as well as the transition from school to work, two related psychological processes emerge as especially important. These include development of one's self-concept and a focus on preparation for appropriate roles.

Exploration is usually stressed in the development of self-concept in the transition to adulthood. Through identity exploration, adolescents and young adults test the waters in personal areas (like love and relationships) and in public arenas (like schooling and work). Arnett (2000), however, expands discussion of identity development to include the individual's subjective readiness for adult roles, which dovetails with the ability to choose appropriate roles and avoid inappropriate or unrealistic ones. This psychological process that supports the transition to adulthood is often termed "planful competence" (Clausen, 1991). This refers to the adolescent's or young adult's ability to think through choices in both personal and public spheres, as well as the discernment to avoid unwise choices. In a consideration of the transition from school to work in Germany, Heinz (1999) also intertwines the two processes – development of self-concept and planful competence – in the discussion of "biographical orientation." According to Heinz, a biographical orientation toward schooling and career results from the young adult's "self-socialization" into certain roles through reflection on the outcomes that have resulted from their own investment of assets and agency.

One final characteristic of the transition to adulthood that merits consideration is the high degree of heterogeneity in the timing and sequence of the different markers associated with the period (e.g., leaving school and the parental home, marriage, and parenthood). As Shanahan notes (2000), such high levels of differentiation point toward increasing individualization of the life course that could complicate discussion of the life course as being "structured." Later in this chapter, however, we will argue that heterogeneity of developmental pathways is not synonymous with an unstructured life course. In fact, we will show how certain organizations can actually promote heterogeneity by offering unstructured pathways, even as alternative organizational structures can minimize heterogeneity in the timing of a key developmental marker.

Interestingly, the transition to adulthood encompasses potentially contradictory tasks, making it a highly complex life stage. While this life stage involves exploration, instability, and timetable heterogeneity, it also makes conflicting demands for fateful decisions, commitments, and timely progress to attaining markers of adulthood. While observers have noted these contradictory demands on this life stage (Arnett, 2004), there has been little consideration of how major social institutions organize and influence these conflicting processes.

CONCRETE ORGANIZATIONAL ENVIRONMENTS AND THE STRUCTURE OF THE LIFE COURSE

Given the diverse demands, career goals, and interests among young adults in American society, it should come as no surprise that large numbers of young adults choose to enroll in postsecondary educational institutions. Colleges and universities, after all, have developed with the purpose of both broadening the student's mind through intellectual exploration, as well as preparing students for the labor market by enhancing their human capital. These dual purposes correspond to the young adult's need for identity exploration and preparation for adult roles. At the same time, the great variety of postsecondary programs available to interested students may reflect the high variation in both needs and wants of the young adult student population.

While such a view of postsecondary educational institutions is reasonable, it also assumes a great deal of understanding by institutional actors of the developmental needs of their students. Such understanding, while plausible, may be overly optimistic. In addition to responding to the developmental needs of their clients, organizations (in education and other domains) are also likely to shape individuals' needs, even as the institutions are shaped by them. As Dannefer (1992) has observed, social context is "not only a powerful organizer of individual developmental patterns," but also, "consist[s] of processes that are themselves organized" (p. 91). As such, Dannefer has repeatedly called for life course theorists to pay more explicit attention to the organization of social contexts, including through study of organizations themselves.

One major study of the role of organizations in structuring human development comes from the work of Eccles and her colleagues (Eccles et al., 1993). These scholars have looked explicitly at the role of schools – junior high and middle schools, specifically – in shaping the developmental trajectories of adolescents. The crux of their argument is that many of the typical but problematic developmental changes associated with this stage of the life course, such as declining achievement and self-esteem, are not intrinsic to adolescence. Rather, they stem from a "mismatch between the needs of developing adolescents and the opportunities afforded them by their social environments," especially schools (p. 90). According to Eccles and colleagues, the organizational, instructional, and climate variables that characterize junior high and middle schools are precisely the opposite of what developing adolescents really need. For example, middle schools tend to emphasize control, just as youth need more autonomy. They tend to limit opportunities for deeper relationships with teachers, just as contact with non-familial adults becomes important to young people. This disjuncture between the developmental needs of adolescents and the social context of schooling then sets off negative developmental trends for many young people.

The process described by Eccles and her colleagues reflects what Dannefer (1984) calls the "self-fulfilling prophecy" of human development. Typical ontogenetic life course models, he asserts, limit consideration of human development because causal processes are viewed as maturational and, therefore, invariant (with the limited exception of "error variance" in non-normative individual cases). "The *environment* thus is not analyzed in the *production of normality*," Dannefer (1984, p. 105) contends (emphasis added). Eccles and her colleagues show how, because adolescents are expected to be conflictive, organizations (as well as individuals) respond by asserting control over youth. Young people seeking autonomy then rebel against the source of control. Ultimately, this produces the very behavior the organization is mandated to control.

This chapter takes up Dannefer's challenge to examine the role of concrete social organizations in developmental processes and potential implications for the production of normal development during an important transitional period. Specifically, we will examine the way 2-year colleges provide structures and procedures that influence individuals' timely progress to institutional and developmental milestones and the ways these structures are related to changes in individuals' self-concept and planful competence in the transition to adulthood. In doing this, this chapter provides evidence on the role of an increasingly important organizational form in shaping the life course, and a more general discussion of the structuring of the life course.

TWO-YEAR COLLEGES' EMERGENCE AS A MAJOR SOCIETAL INSTITUTION

Two-year colleges, which offer sub-baccalaureate credentials like certificates and associate's degrees, have become an increasingly important environment for young adults in the United States over the past three decades.

Whereas less than half of high school graduates enrolled in college a generation ago, currently almost 70% do so (Rosenbaum, 2001). Community colleges have played a crucial role in facilitating this change as these increased enrollments have disproportionately gone to such colleges. In fact, community colleges now enroll about half of all students entering postsecondary education (Bailey, 2003).

While public community colleges enroll 96% of all students in 2-year colleges, private occupational colleges provide an alternative institutional form. Private 2-year colleges offer a compelling contrast in terms of organizational structures, and therefore, developmental environment. Like community colleges, private 2-year colleges offer accredited associate's degrees in occupational fields like electronics, information technology, business, and health technician fields. Because they emphasize occupational preparation and offer accredited college degrees, they are called "occupational colleges." Occupational colleges are quite comparable to community colleges in terms of their occupational program offerings and mission to educate students for the workforce. On the other hand, occupational colleges are dissimilar to most other business and technical schools. The latter offer no degree above a certificate and are often not accredited.

Both types of institutions play a role in transforming the life stage of young adulthood for a large segment of American society. Through open admissions, community colleges and some occupational colleges have opened up higher education to a much broader segment of society than in the past, including individuals with disadvantaged backgrounds and individuals with low academic achievement. Two-year colleges have also devised geographic and time-scheduling innovations that would have been inconceivable a generation ago. Satellite campuses are located near residential areas and work places making higher education easily accessible to individuals with family or work responsibilities. They have pioneered scheduling innovations: classes are offered in early mornings before work, in evenings after work, on Saturdays, Sundays, and during vacations, and are even available 24 hours a day on the internet.

Such innovations permit higher education to be more easily superimposed upon traditional life stage activities, adding a new dimension to the young adult life stage. Quite simply, they provide a practical way for individuals to fit college into demanding family or work schedules. At the same time, they have allowed postsecondary education to be completed on a part-time basis and combined with other full-time responsibilities.

Together, public community colleges and private occupational colleges offer appropriate settings for the study of the potential role of the organizational environment in shaping the "normative" development and the life course trajectories of young adults. As noted, great numbers of young adults enroll in these institutions and they represent new kinds of college students with respect to demographic characteristics and prior educational and work experiences. At the same time, however, since we focus on students in similar occupational programs at both types of college, we can observe them in the processes that mark the transition to adulthood, particularly their self-assessments of their educational progress in different institutional settings. Finally, these colleges are also useful places to examine the sense of possibility and self-focus that Arnett (2000) describes as characteristic of emerging adulthood. Both types of 2-year colleges promote themselves to potential students as places for self-improvement that will shape their capacities to meet life's challenges.

STUDY DESIGN

The following sections describe these two types of colleges based on a detailed empirical study of seven public community colleges and seven private occupational colleges. Schools were selected because they offer similar occupational programs and were accredited by recognized accrediting agencies. Of the seven private occupational colleges, three are non-profit and four are for-profit. Within these schools, we draw on several different forms of data. First we interviewed 96 administrators and staff members (including presidents, deans, program chairs, and directors of counseling and advising services) at the different colleges to determine their understanding of students' needs and the ways in which their institution is able to meet students' needs. Next, we interviewed 86 young adult students in a variety of programs at the schools about their experiences at college and their thinking on education, work, and the future. Finally, we surveyed nearly 4,400 students at all 14 schools on the same subjects.

All 14 colleges are located in a single metropolitan area in Illinois. While this geographic limitation raises questions about generalizability, geographic proximity allowed the researchers to visit each campus repeatedly over several years. Indeed, we visited each campus at least 15 times, and several were visited over 40 times. Given that 2-year colleges have experienced extraordinary growth, increased complexity, and emergence of radically new programs and structures serving new types of students, this familiarity provides this research with a distinctive strength for shedding light on how these organizations operate and how they may influence the young adult life stage.

How representative is our sample? Our occupational colleges are *not* typical: only 6% of private for-profit colleges are accredited to offer associate's degrees (Apling, 1993). These private colleges are not a random sample. They offer some of the best programs in these fields, and may be considered to represent an *ideal type* that provides a new perspective on how 2-year colleges can operate. On the other hand, we suspect that our community colleges may be typical, or at least representative of a large segment of community colleges. While it is always difficult to demonstrate generalizability in qualitative research, our community colleges are similar to others on at least one key issue: 50% of the students enrolled in our seven community colleges are enrolled in transfer programs, and the average for the entire state is also 50% (Illinois Board of Higher Education, 2002, Table VI-2).[1] Our findings are also compatible with studies of administrators in 18 community colleges nationally (Cross & Fideler, 1989) and a national sample of 1,725 faculty in 92 community colleges (Brewer, 1999), which report that administrators and faculty are nearly evenly split in their ranking of transfer (general education) and workplace preparation as the top institutional priorities. Our community colleges do not seem atypical on this key issue.

It is important to note that both types of schools enroll students from a wide variety of stages in the life course. From teenagers to senior citizens, and every stage in between, the enrollments at both community colleges and occupational colleges are very diverse. Still, the median age of students sampled at the colleges in our study is in the mid-20s and all our student interviews are with young adults. Moreover, staff respondents at both types of college characterize their student bodies as consisting of what they view as two distinct groups. The first is "straight up adults," as one administrator puts it, referring to older individuals who are often understood to be "non-traditional" students. The second is younger students who have entered college within a few years of leaving high school. Since we are interested in the role of the colleges in shaping the transition to adulthood, we limit the present analyses to younger students, those 30 years old and younger ($N = 3,689$).

DISTINCTIVE MISSIONS OF THE TWO TYPES OF COLLEGES

Community colleges and occupational colleges have markedly different institutional missions, according to the administrators we interviewed.

Community college administrators tend to articulate their mission as broad and holistic and staff members tend to interpret it that way. This translates into an organizational philosophy directed at serving an entire community. "That's the function of the community college," one department chair explains, in a typical statement, "to serve everybody's needs." In contrast, occupational college administrators see their school's mission as more focused, preparing their students to meet labor market demands and to obtain suitable skilled jobs.

Community colleges see their role in terms of human development writ large. One Vice President for Academic Affairs actually describes his college's primary mission as "developmental," followed by "baccalaureate, career, and community." This description clearly maps onto traditional developmental perspectives that view the life course in a linear fashion, moving from personal development, then to professional accomplishments, and eventually turning outward to the community. A Dean of Instruction at another community college speaks in similarly sweeping terms: "We have to have a broadly educated, well versed and ... liberally educated worker and citizen." The priority on personal development and personal exploration is widely shared at these institutions, even among the program chairs of occupational programs.

In contrast, occupational college administrators in our sample tend to view their role as specific to the transition to work. As one administrator at a for-profit occupational college notes, the school does not focus on "education for the sake of education." Rather, the college seeks to provide the student with "the skills and knowledge and other attributes ... useful in securing initial employment, and also career growth." As this statement indicates, occupational colleges have chosen to focus on education for careers, even as they recognize that students have broader developmental goals.

Indeed, administrators believe that broader developmental goals will proceed from educational progress and career success. The Dean of Students at a non-profit occupational college makes this thinking explicit. "[I]f the student wants to be a lifelong learner," she says, "we encourage that and, you know, we recognize some will choose to continue their education. But we still hit it very hard in that mission statement ... that we wanted to continue [the college's] identity with career education." Even when a dean in a non-profit college states that the college's goal is to prepare "valuable citizens with fulfilled personal lives," he emphasizes the importance of work preparation and career readiness for having a fulfilled life.

Since both types of college serve many disadvantaged students, the needs of these students influence the institution's mission. These colleges frequently face challenges in educating students who have little understanding of college because they are the first in their families to attend college. At one non-profit occupational college, administrators often explicitly refer to the school's stated mission "to serve an underserved population," by which they mean low-income, first-generation, and minority students. However, it is clear that the private colleges' priority is to provide workforce preparation for these students. As the Dean of General studies at one private college explains, "we are going to give the employer a good," and here he pauses before finally saying, "product." He continues, capturing the tension between the respective approaches, "I hate to use words like that, but we are very conscious of who our external publics are." He then discusses the school's efforts to build contacts with potential employers, contacts which are viewed as essential if these unselective colleges are to succeed at placing their graduates in jobs.

This difference between community colleges and occupational colleges is not really surprising, given the varying demands of the public and private sectors. The two types of colleges also have very different histories. Community colleges have traditionally emphasized transfer to 4-year colleges, and occupational colleges have emphasized workforce training. More important to our purposes are the ways in which these distinct institutional missions reveal different beliefs about students' developmental trajectories and the role of the institution in shaping them. These varying missions result in very different concrete organizational structures at the two types of college and these may have strong implications for students' life course pathways.

In sum, community college administrators speak in general terms of offering support for individual development. "We help build dreams" and "we improve people's lives through learning," says one administrator, using a phrase that appears often among community college respondents. With such broad objectives, it is not surprising that community colleges structure their programs loosely and offer a wide variety of both program and course choices to students. Speaking of program offerings and structures, this woman continues, noting "We'll try for as long as we can to keep their [students'] options open." Similarly, speaking of strategies to meet students' needs, another community college administrator explains, "Students change. They change degrees, they change fields, they change majors, they change their minds ... I see that our vocational programs will need to become more diverse." These sentiments, which are representative of the administrators from all our community colleges, reflect normative characteristics of the transition to adulthood: young adults explore, they change, they have diverse needs. The community college serves as a context for students' development. It does not, however, provide mechanisms specifically directed toward bringing about any particular developmental goals.

DISTINCTIVE INSTITUTIONAL STRUCTURES FOR IMPLEMENTING COLLEGE MISSIONS

Consistent with their different missions, the two types of colleges have different organizational structures (Deil-Amen & Rosenbaum, 2003, 2004). Community colleges provide students with a vast array of program and course choices. Yet, since community colleges provide little structure and relatively few advising resources, students at community colleges may find themselves engaging in a sort of trial and error approach to education and career development. An Assistant Dean at a community college confirms this as she explains the commonly noted difficulty many students face in getting good information about college procedures: "Oftentimes people don't graduate on time because they're missing courses, clusters of courses, transferable courses … that's a problem for us." It is also a problem for the students, which we discuss later.

In contrast with community colleges' emphasis on helping students "build their dreams," occupational colleges focus on the more immediate goal of obtaining employment and they provide structures toward that end. One occupational college has long used the tagline "where dreams find direction." Indeed, occupational colleges focus heavily on directing students through college and into jobs. Their organizational structures are designed to facilitate this. These colleges offer relatively few major programs with little or no flexibility in course choices and devote a great deal of resources to student advising, sometimes making it mandatory. Unlike community colleges that encourage students' propensities for change and exploration, occupational colleges actively endeavor to focus students' efforts and minimize change. As one dean explains, "we try to minimize the opportunities for students to go down the wrong path in terms of the courses they need to take. So there are a few electives … They all take roughly the same courses." Another administrator notes, "we have a lot of personalized [advisory] services, and we have to because otherwise, a lot of the students … would be lost." Rather than accepting instability or exploration as normative, occupational colleges view them as counterproductive and something to be avoided. Thus, they structure their programs accordingly.

The distinct approaches toward role exploration in the two types of college have clear implications for students' self-conceptions. While community colleges provide many potential roles for students and provide great latitude in paths toward such roles, occupational colleges ask students to select a goal at the outset of their studies and then the institution delineates a clear and relatively structured path toward the goal. Clearly, these two orientations also reveal different assumptions about students' ability to choose appropriate roles, as well as indicating different stances toward the importance of planful competence (cf. Clausen, 1991). Community college structures rely on students' ability to plan wisely, allowing them to choose from a vast array of options. In contrast, occupational colleges remove much of the need for individual planning through institutionally planned curricula that provide efficient pathways toward distinct career goals.

Each of these organizational stances has potential benefits, as well as drawbacks. At community colleges, students are truly able to explore, should they feel the need. Yet exploration can drift into confusion. At occupational colleges, students are presented with a very clear path toward a few career goals, from which they can choose; but they are more or less locked into that path as soon as they enroll. If they subsequently discover that they have chosen poorly at the outset and desire to change programs, they will lose time and may even have to start from scratch.

DO DIFFERENT STUDENTS ENTER THE TWO TYPES OF COLLEGE?

Having seen that these two types of colleges have different missions and different structures, we next examine their implications for students. Before examining the potential impact of college type, we must consider the possibility that these different institutions may attract different types of students.

There are many reasons to think these two types of colleges enroll highly similar students. As noted previously, 2-year colleges are the least selective of postsecondary institutions. They attract large numbers of lower achieving students who would likely be unable to gain admission to 4-year colleges (Rosenbaum, 2001). Since both types of colleges are commuter schools, they attract students from similar geographic areas. Both also increase affordability by allowing students to live at home and maintain full-time jobs. In addition, community colleges provide access to low-income students through low tuitions, while occupational colleges make effective use of federal and state financial aid programs that make their higher tuition affordable to low-income students. Consequently, both types of college attract students with similarly low socio-economic backgrounds.

Our survey restricted the sample to students in similar occupational programs at the two types of colleges. Each type of college offered programs in the same fields, including, for example, computer information systems, electronics engineering technology, administrative assistant, and health technology, all of which offer terminal associate's degrees in fields with relatively strong labor market demand. At both types of college, these students are actively engaged in the transition from school to work, as they are pursuing a program of study explicitly directed toward labor market entry (although many students at each type of college plan eventually to seek a bachelor's degree).

Using our survey of 3,689 students 30 years old and under, we examined student attributes in the two types of college (see Table 1). Even small differences (of only 3%) can be statistically significant in these data (because of the large number of cases). Still, students in the two types of college are

Table 1. Social and Demographic Characteristics of the Student Survey Sample, Age ≤ 30.

	Total Sample (%)	Community Colleges (%)	Occupational Colleges (%)
Lives independently of parents	29	26*	30*
Receives no financial support from parents	49	48	50
Helps support parents financially	22	24*	21*
Married	12	12	12
Has child	18	11*	21*
Male	49	48	50
Racial/ethnic minority	66	56*	71*
Parent education is less than HS graduate	20	16*	23*
Respondent HS grades were C or lower	26	24	27
Total *N*	3,689	1,243	2,446

Source: Authors' data; percentages are valid.

* The difference between college types is statistically significant at $p \leq 0.05$ level.

highly similar with respect to most markers of adulthood, including marriage, financial self-sufficiency, providing support to parents, and living independently.[2] There are substantial differences between the groups of students with respect to having a child, being an ethnic minority, and parental education. Yet, surprisingly, despite their higher tuitions, private colleges enroll the more "disadvantaged" students on each of the three criteria. While we cannot rule out the possibility that different types of students enroll in these different colleges, preexisting differences appear to be minimal and may even put occupational college students at some disadvantage for degree progress.

While unmeasured student attributes are potential confounding influences, there are several ways to address potential selection biases. First, the argument for an organizational effect of the two college types is strengthened by examining the *changes over time* in student outcomes at the two types of college. Second, regression analysis allows us to examine the impact of school structures, while holding individual characteristics constant. Third, triangulation of survey data with qualitative findings from our student interview sample will allow us to examine the contention that organizational structures matter, and will also help to shed light on how and why this might be the case.

"DELAYED PROGRESS" IN THE TWO TYPES OF COLLEGE

Do these differences in structure have implications for students' progress toward a degree? While it often takes a long time to assess degree attainment (4–8 years is common for completion of a 2-year degree), prior to that, there is at least one noteworthy indicator of "delayed progress." This indicator is "timetable increases," which involve personal adjustments in perceptions of how long a degree is likely to take. We study timetable increases by comparing students' initial and current expectations for the time it would take to complete the degree (measured 6–18 months after beginning college) and consider increases to occur if students have increased the time they expect it to take to obtain their planned degree by 1 year or more.[3]

Timetable increases are an important aspect of students' progress and may have importance in both affective and practical ways. When students must extend their degree timelines longer than they had initially planned, this is discouraging and likely to decrease motivation to continue in college.

More time in school is also economically costly and may conflict with students' external commitments, which are especially apt to play a role for students at 2-year colleges, who are often working and/or raising a family.

As previously described, private college administrators highlight distinct procedures and structures that have the explicit goal of reducing student delays. Interviews with students at these schools find that they express fewer frustrations with college procedures. The following analyses examine whether students at these private schools do in fact experience fewer delays and whether these differences remain after controls for student attributes.

Results of analysis of variance (ANOVA) reveal considerable differences between the two types of college with respect to students' upward revisions of their college timetables by a year or more since entering college (see Table 2). While 53% of students in public 2-year colleges increased the time they expected to get their planned degree, only 32% of private college students made such upward adjustments. The difference is large and significant ($F = 120.66$, $p<0.001$). While an associate's degree is supposed to take 2 years, many students who initially believed this have realized it will take longer and this is a more common occurrence for students attending community college.

Additional analyses (not reported here) further support this contention. First, timetable increases are strongly and significantly associated with college type, even after controlling for all variables in Table 1 in multivariate logistic analyses. Second, among students who have taken remedial courses, community college students are significantly less likely than occupational college students to realize that their remedial courses confer no credit toward a degree. Such misconceptions are likely to lead to subsequent unanticipated timetable increases and a great deal of frustration after the mistake is discovered. Students at private colleges rarely have similar misconceptions. Different types of colleges are clearly associated with different

Table 2. Analysis of Variance: Student Increased Timetable for Degree Completion by 1 Year or More.

	Community Colleges	Occupational Colleges	Significance (*F*)
Increased timetable for degree completion by 1 year or more	53%	32%	0.000 (120.66)

Source: Authors' data; percentages are valid.

incidence of timetable increases. The looser structure in community colleges is much more likely to lead to delayed progress (and misconceptions about remedial courses) than the more rigid, less flexible structures in occupational colleges.

STUDENTS' CONFIDENCE IN THE TWO TYPES OF COLLEGE

College structure may have further implications for students' confidence, an important component in self-concept development, one of the major psychological processes associated with the transition to adulthood. We asked students in our survey how confident they felt about a number of their own abilities at the time of the survey, compared to when they first entered the college (again, measured about 6–18 months after beginning). Occupational colleges provide highly structured programs, with clear pathways to degrees and jobs and hence present clear portraits of "possible selves." Students at these institutions thus should experience increased confidence over time more often than their counterparts at community colleges. As Table 3 shows, students at occupational colleges were significantly more likely than

Table 3. Analysis of Variance: "Compared to when I first entered this college, I…".

	Community Colleges (%)	Occupational Colleges (%)	Significance (*F*)
"… feel much more confident about my ability to do school work"	23	29	0.020 (5.45)
"… feel much more confident about my ability to complete a college degree"	28	34	0.001 (10.83)
"… feel much more confident about my ability to get a good job"	25	31	0.001 (10.86)

Source: Authors' data; percentages are valid.

their counterparts at community colleges to report "much higher" confidence in their ability to do school work, to earn their degree, and to get a good job. As noted previously, small differences may be significant in these data, given the large sample size; but further analyses (not reported here) show the effect of occupational colleges on student confidence held, even after controlling for all the student attributes reported in Table 1.

We expect that students' confidence in their ability to get a degree might be influenced by timetable increases. Specifically, students who experience delayed progress might have reduced confidence in their ability to get a degree, and delayed progress might explain part of the college-type effect on confidence. This process of "cooling out" students, lowering their aspirations, has been discussed by sociologists (Clark, 1960; Brint & Karabel, 1989), but without reference to its implications for the structure of the life course. Table 4 reports logistic regression coefficients for the relationship between college type, timetable increases, and decreased confidence about graduating. The results indicate that both college type and timetable increases have large statistically significant effects on students' confidence in their ability to complete the college degree and that college type has smaller influence after controlling for timetable increases. Occupational colleges alone reduce the odds ratio of a decrease in confidence by 64% (1−0.36), but after controlling for timetable increases, the occupational college effect is somewhat smaller (57%, i.e., 1−0.43). Timetable increases also have a strong and significant influence. Students experiencing a timetable increase were much more likely to have lower confidence that they could complete a college degree (183% increase in the odds ratio). While delayed progress

Table 4. Logistic Regression of Decreased Confidence on College Type and Timetable Increases.

	Model 1		Model 2	
	b	exp(b)	b	exp(b)
Occupational college	−1.025*** (0.204)	0.359	−0.852*** (0.208)	0.427
Timetable increase			1.040*** (0.250)	2.830
Constant	−2.970		−3.561	

Source: Authors' data.
$N = 3,689$.
***$p < 0.001$.

appears to have a strong negative impact on students' confidence, it does not entirely account for the effect of college structures.

QUALITATIVE IMPACT: STUDENTS' VIEWS

Some indecision seems perfectly natural – perhaps even normal – among young adults, for whom identity and role exploration are especially salient. The developmental question, however, is *how* the organizational features of the college facilitate the students' thinking about their education and career paths once they have enrolled. In order to assess this, we asked students about their experiences at the college and their thoughts about the future.

First, since the normative developmental pattern in the transition to adulthood includes exploration and instability, it would seem that students in this stage might chafe under occupational colleges' structural arrangements. Indeed, our survey shows that occupational college students are somewhat more likely than community college students to disagree with the statement that their college offers "enough course choices" (about 16% vs. 12%). Yet in our interviews, students at these colleges appear to appreciate the specificity of course requirements, which lets them know in advance what courses they will take. This allows them to plan both in terms of academic preparation, as well as scheduling their time. Many occupational college students express a certain peace of mind in having clear-cut course requirements. As one student explains, "it's pretty much a standardized schedule ... you know, it [choosing courses] wasn't something I had to worry about." For young adults with external obligations to work and family – typical among 2-year college students – having one less thing to worry about at school can be a relief.

In addition to reducing the need to plan, occupational college structures also encourage students to identify with their career path. Dress codes are common at these institutions, and students report that this makes them feel "professional," and that it sets them apart from their peers outside the college, whom they often view as less mature. Occupational colleges' close relationships with employers, faculty who work in industry, and industry influence on curricula also serve this purpose. An information technology student reports that she is satisfied in knowing that "by the time you graduate, you'll have very full working knowledge of exactly what the industry's looking for. Because they [faculty] are coming from the industry." She goes on to say that faculty go out of their way to remind students what they are working toward, recounting how one instructor would look up baseline salaries for the students' degrees as a way of encouraging them. Another occupational college student makes an explicit comparison between institutional practices at his current school and a community college, which he had attended previously. "Basically," says this student who is both studying and working in business, "the thing I've noticed [at the OC] is the credibility of the courses, you know, as far as what you've taken and the actual usage of them [in the workplace]." Occupational colleges also enforce strict rules about attendance, punctuality, homework deadlines, and appropriate behavior in interacting with instructors and students. Rather than expressing resentment, students generally accept these rules as helping them develop the kinds of behaviors that employers will expect.

At community colleges, students speak of exploration, and some view it as a positive experience. Others, however, feel exploration is a luxury they cannot afford. A pharmacy student at one of our community colleges explained how a faculty member who was acting as an advisor during registration (a common practice at community colleges) misguided her. She ended up taking a class in the instructor's department, even though it would not help her achieve her degree goal. The student sought advice from another advisor, after she had already completed the course, but the advisor offered little help: "Well, you can look at it in a positive way, you know, that you learned [something]," the student explains, recounting the episode, "but that wasn't my point." This student continues, describing the experience as a "waste of time and money." Our student survey shows that this is not an isolated incident. In fact, community college students are about twice as likely (about 48% vs. 23%) as occupational college students to report taking a course, which they later found would not count toward their degree. It is clear from our interview data that the variety and flexibility of community colleges' offerings require students to be very capable in making choices on their own, or they are apt to make mistakes and "waste time and money" taking courses that may not contribute to their progress toward the degree.

In sum, our student interviews show that occupational colleges structure out the need for students to plan their studies, and at the same time allow them to plan successfully around predictable schedules. Students at occupational colleges report that policies like dress codes, work-related rules, and internships help them to identify with their eventual profession, even as the explicit applicability of their coursework in the labor market makes them feel confident about their ability to transition into the professional world. In contrast, students at community colleges exercise more discretion in their academic path, but do not always experience this as a benefit.

CONCLUSIONS AND IMPLICATIONS

This study may shed some light on current controversies surrounding the structure of the life course, especially with respect to the increasing individual variability in the transition to adulthood that has been observed in most industrialized societies, as well as the broader de-standardization of the life course. Clearly, the increased access to higher education provided by the vast expansion of 2-year colleges has radically transformed expectations. In contrast to a generation ago, when postsecondary education was available to only a small portion of the population, today the vast majority of youth plan to attend college, and college has become a normative expectation. While this might suggest greater standardization, the loose structures of most community colleges (where about half of all college students enroll) allow many students to move in and out of college over extended periods of time, promoting increased dispersion in the timing of school-to-work transitions. Moreover, the relatively low costs (both economic and opportunity costs) of education at community colleges promote a fluid coupling between education and work as well as variety in the sequencing of these markers. With training and retraining readily available, life course transitions may include school-to-work, work-to-school, and various combinations of school and work. Occupational colleges, on the other hand, would seem to yield tighter coupling between school and work, even as they promote compression in the transition from education into careers. Occupational colleges are unlikely to influence the structure of the life course on a population level, however, as they enroll so few students.

As always in non-experimental studies, one must be cautious about inferring causality from correlations. Students may differ before they enter these two types of colleges in ways that affect the observed associations. Those preexisting student differences that we observed, however, are mostly small and may suggest that private college students would have poorer outcomes. Instead, our results find the opposite. This suggests that college effects may be even stronger than the results indicate, as they appear to overcome the additional disadvantages of their students (assuming that minority students, with less educated parents, or having a child would ordinarily have greater timetable increases and lowered self-confidence (cf. Dougherty, 1994; Grubb, 1996).

Moreover, our conclusions do not rest solely on quantitative association. Qualitative analyses indicate that the two types of colleges have markedly different missions and they have implemented very different structures. Community colleges emphasize exploration and they provide an unstructured curriculum consistent with this – similar to 4-year colleges, which were designed to serve upper middle-class students. In contrast, occupational colleges have devised and implemented extensive structures and procedures to minimize instability and mistakes, or to catch mistakes quickly, before they become serious. It is hard to imagine that these radically different organizations would not have major influences of just the sort we have found.

Triangulation of multiple sources of data and attention to changes over time suggests that the organizational features of the two college types actually shape students' developmental pathways. Structures that respectively emphasize diversity (i.e., at community colleges) and focus (at occupational colleges) shape the observed differences in students' delayed progress and self-concept. When a complex life stage like the transition to adulthood poses contradictory tasks, these alternative institution types may facilitate success at one task more than another, even as they may delimit the range of tasks the individual will be required to face. We find that community colleges mostly respond to the normative tendencies toward exploration, instability, and timetable heterogeneity in the transition to adulthood. Occupational colleges, in contrast, primarily respond to the opposite tasks of this developmental transition: fateful decisions resulting in long-term commitments, and timely progress in attaining the markers of adulthood. Community colleges offer a wide variety of options, and they strongly encourage exploration, change, and taking as long as one needs to make a decision; but they provide few mechanisms to prevent or correct mistakes. In contrast, occupational colleges offer a limited number of options, and they provide extensive structures that prevent and catch mistakes and that limit costly delays in progress; but they require that students make decisions at entry if they hope to finish "on time" in two years.

These two institutions promote different aspects of the transition to adulthood. They place differential emphasis on conflicting processes and their differential approaches are associated with large differences in students' progress toward their goals. While alternative interpretations of these results cannot be rejected with certainty, attention is drawn to the obvious and powerful influence of organizational structures and procedures and the ways these different structures likely influence student experiences in these two types of institutions.

In an ideal world, of course, students would be aware of their own developmental needs and competencies and could select the appropriate developmental environment. Although our findings suggest that students are not aware of these different requirements, and indeed that students are very similar at the two types of colleges, our findings also suggest that different types of students will do better at each college type. If we consider students' development of self-concept and planful competence in terms of "more" or "less" advanced, we might conclude that students with more of both would fare well in either institutional environment because they know what they want and can figure out how to get there. Students whose self-concept allows them to commit to a program, but who have low planful competence would probably do well at an occupational college. There the student makes a single choice, choosing an occupational program, and then a highly structured curriculum specifies most courses for the next two years, requiring little further planning. In the reverse case, where a student has more planful competence, but is not ready to make a commitment to a major, he or she would benefit from the community college setting. Here, the student could explore, choose an eventual program, and select courses to complete it. Finally, students who are unready to make a commitment and who have less planful competence might be at risk in either institution type. At an occupational college, they might choose a major which they later decide is not appropriate for them. While at community college, they might be unable to chart their coursework so as to meet program requirements efficiently. In any case, neither type of college could be expected to fit all students.

Our analyses have focused only on degree seekers and do not address the other missions of the community colleges. Still, private colleges exemplify alternative structures that could potentially be useful for degree-seeking students at public colleges. Since occupational colleges only enroll about 4% of all 2-year college students, they will not have a major impact on the structure of the life course at a population level and pose no serious threat to community colleges. While the answer is not to turn community colleges into occupational colleges, community colleges could better help degree-seeking students by borrowing some lessons from occupational colleges. These would include: (1) creating clear curriculum structures; (2) improving counseling; (3) closely monitoring student progress; (4) implementing information and advising systems that would quickly show signs of student difficulties; and (5) alleviating conflicts with external pressures (Deil-Amen & Rosenbaum, 2003).

More generally, instead of assuming that the characteristic problems that students experience in the young adult life stage are inherent properties of that life stage, the present analysis suggests that institutional structures may have an important impact on students' development. By examining institutional structures, we can identify some of these influences, and policy responses may reduce some problems. Our purpose here has been to pursue

Dannefer's exhortation to investigate the role of concrete organizational environments in producing normative development and giving structure to the life course. In the past generation, American society has taken enormous strides in making college accessible to many new groups of students. The revolution is still incomplete, however, as college completion continues to lag, especially among disadvantaged students. This research has taken a life course perspective to identify ways that alternative structures may affect students' progress in an important class of institutions that serve many of these new students.

NOTES

1. While 50% of the students enrolled at our community colleges are in transfer programs, 90% of the students we sampled are enrolled in occupational programs. In this way, we limit our sample to comparable (i.e., occupational) students at the two college types.
2. On the latter two items, we find that while community college students are slightly more likely to help support their parents financially (24% vs. 21%), private college students are more likely to live on their own (30% vs. 26%), so these items yield conflicting messages about which group is "more adult."
3. We measure students' "initial expectations" from their retrospective reports, which they may not remember accurately. However, in this case, such potential distorted recall may not be important. After all, it is students' *perceived* discrepancies which may influence their disappointment, regardless of their actual initial expectations. If students have forgotten their *actual* initial expectations which are discrepant from current expectations, those unperceived discrepancies may not result in disappointment or decreased motivation.

ACKNOWLEDGMENT

The authors would like to thank the Spencer Foundation for their generous support of this research. Opinions expressed are solely those of the authors.

REFERENCES

Apling, R. N. (1993). Proprietary schools and their students. *Journal of Higher Education*, *64*(4), 379–416.

Arnett, J. J. (2000). Emerging adulthood: A theory of development from the late teens through the twenties. *American Psychologist*, *55*, 469–480.

Arnett, J. J. (2004). *Emerging adulthood: The winding road from the late teens through the early twenties*. New York: Oxford University Press.

Bailey, T. (2003). Community colleges in the 21st century: Challenges and opportunities. *CCRC Brief.* Retrieved from Community College Research Center (CCRC) website at http://www.tc.columbia.edu/~iee/ccrc/PAPERS/Briefs/brief15.pdf.

Baltes, P. B. (1997). On the incomplete architecture of human ontogeny: Selection, optimization, and compensation as foundation of developmental theory. *American Psychologist*, *52*, 366–380.

Brewer, D. J. (1999). *How do community college faculty view institutional mission? An analysis of national survey data.* Publication of the Community College Research Center at Teachers College Columbia University. Retrieved from CCRC website at: http://www.tc.columbia.edu/ccrc/PAPERS/brewer1.pdf.

Brint, S., & Karabel, J. (1989). *The diverted dream: Community colleges and the promise of educational opportunity in America, 1900–1985.* New York: Oxford University Press.

Clark, B. (1960). The 'cooling-out' function in higher education. *American Journal of Sociology*, *65*, 569–576.

Clausen, J. S. (1991). Adolescent competence and the shaping of the life course. *American Journal of Sociology*, *96*, 805–842.

Cross, K. P., & Fideler, E. (1989). Community college missions: Priorities in the mid 1980s. *Journal of Higher Education*, *60*, 209–216.

Dannefer, D. (1984). Adult development and social theory: A paradigmatic reappraisal. *American Sociological Review*, *49*, 100–116.

Dannefer, D. (1992). The concept of context in developmental discourse: Four meanings of context and their implications. In: D. Featherman, R. Lerner & M. Perlmutter (Eds), *Life-span development and behavior*, (Vol. 11, pp. 83–110). Hillsdale, NJ: Lawrence Erlbaum Associates.

Deil-Amen, R., & Rosenbaum, J. E. (2003). The social prerequisites of success: Can college structure reduce the need for social know-how? *Annals of the American Academy of Political and Social Science*, *586*, 120–143.

Deil-Amen, R., & Rosenbaum, J. E. (2004). Charter building at low-status colleges: Charters as mechanisms of labor-market access for two-year college students. *Sociology of Education*, *77*, 245–265.

Dougherty, K. (1994). *The contradictory college: The conflicting origins, impacts, and futures of the community college.* Albany, NY: State University of New York Press.

Eccles, J. S., Midgley, C., Wigfield, A., Miller Buchanan, C., Reuman, D., Flanagan, C., & MacIver, D. (1993). Development during adolescence: The impact of stage-environment fit on young adolescents' experiences in schools and in families. *American Psychologist*, *48*, 90–101.

Elder, G. (1998). The life course as developmental theory. *Child Development*, *69*, 1–12.

Grubb, W. N. (1996). *Working in the middle: Strengthening education and training for the mid-skilled labor force.* San Francisco: Jossey-Bass.

Heinz, W. (1999). Job-entry patterns in a life-course perspective. In: W. Heinz (Ed.), *From education to work* (pp. 214–231). Cambridge: Cambridge University Press.

Illinois Board of Higher Education. (2002). *2001 Databook.* Springfield, IL: Illinois State Department of Education.

Kerckhoff, A. (2003). From student to worker. In: J. Mortimer & M. Shanahan (Eds), *Handbook of the life course* (pp. 52–87). New York: Plenum.

Rosenbaum, J. E. (2001). *Beyond college for all.* New York: Russell Sage.

Shanahan, M. J. (2000). Pathways to adulthood in changing societies: Variability and mechanisms in life course perspective. *Annual Review of Sociology*, *26*, 667–692.

FROM OLD TO NEW STRUCTURES: A LONG-TERM COMPARISON OF THE TRANSITION TO ADULTHOOD IN WEST AND EAST GERMANY

Steffen Hillmert

ABSTRACT

This chapter looks at the consequences of political, institutional and market changes for life-course patterns in West and East Germany. It summarizes major institutional changes and developments on the macro level and discusses possible relationships to crucial events which mark the transition to adulthood: leaving education and training, entering the labor market, leaving home and forming unions and families. Empirical analyses draw upon detailed individual-level life-course data from the German Life History Study. The dataset includes birth cohorts between 1919 and 1971 for both West and East Germany and hence covers the historical period for the relevant events between the mid-1930s and the late 1990s.

Life courses in modern societies are not merely a result of individual experiences. They are also shaped by macro-economic conditions and institutional systems located at the local, the nation-state, and even global level.

The Structure of the Life Course: Standardized? Individualized? Differentiated?
Advances in Life Course Research, Volume 9, 151–173

ISSN: 1040-2608/doi:10.1016/S1040-2608(04)09006-9

Germany may be a particularly interesting case in this respect. During the 20th century, both long-term gradual change and radical disruptions can be observed, including two World Wars, division into two states following capitalist and socialist models of society, and eventual re-unification. This development has been accompanied by phases of steady economic growth as well as enduring labor-market problems. What are the consequences of political, institutional and market changes for life-course patterns? Can clear long-term trends be identified?

A fruitful way to analyze the changing structure of life courses is to look at the timing and the order of relevant events for a number of successive cohorts. This chapter concentrates on crucial early events which mark the transition to adulthood: completing education and training, entering the labor market, leaving home and forming unions and families. The aim of the chapter is two-fold: first, to clarify theoretical relationships between macro-social change and individual life-courses, and second, to give an account of long-term developments in the phase of transition to adulthood in East and West Germany. The focus here is not only on incidence and timing of single events but also on interdependencies between them. Empirical analyses draw upon detailed individual-level life-history data for a series of birth cohorts beginning around 1920 and covering the period between (approximately) the mid-1930s and the late 1990s.

TRANSITION TO ADULTHOOD: MACRO-CONDITIONS AND LIFE-COURSE EVENTS

In a role-oriented perspective, the relatively stable status of a full member of society is attained during the phase of the transition to adulthood. This process has two major dimensions. Economic independence is reached by acquiring qualifications and entering gainful employment; emotional and social independence is primarily reached by leaving home and forming own families. During the transition period itself, however, social status is rather indeterminate (Eisenstadt, 1964). Hence, when analyzing societies, it is not only important to know how many people finally make successful transitions but also how extended and complex transitions are. Following a functionalist perspective, social exclusion may include some members of society not making transitions in particular spheres of life who are thus denied access to the role of full members of society. This interpretation particularly applies if the lack of transition is involuntary. More generally, research on the transition to adulthood has the potential to learn more about general trends in society. In the first instance, we can investigate the distinctiveness of age roles. In another instance, the transition to adulthood may be fertile ground to test hypotheses of the 'individualization' of life courses as a number of important life events tend to happen within a relatively short period of time. Finally, it may tell something about links between different spheres of life within a society.

It is therefore no wonder that transition to adulthood has become an important research topic in sociology (for an overview, see Shanahan, 2000). While various ways to study the phenomenon have been proposed, a particularly pragmatic approach involves statistical analysis of the sequence of a whole number of seminal events which mark this transition (Modell, 1989; Billari, 2001). This includes leaving education, entering the labor market, leaving the parental home, forming (marital) unions and becoming a parent. Averages and other aggregate indicators allow comparisons on the collective level.[1]

There is an extensive literature on determinants and historical changes of events which mark the transition to adulthood and the causal processes which drive them, like in the case of leaving the parental home (Goldscheider & Goldscheider, 1993; White, 1994), marriage (Oppenheimer, 1988; Cooney & Hogan, 1991) and fertility (van de Kaa, 2001), to name just a few examples. While some focus on particular events in the process of transition to adulthood separately, this chapter is concerned with links between these events. Such relations have been studied between education and marriage or parenthood (Blossfeld & Huinink, 1991; Marini, 1984), school and work (Kerckhoff, 2003), and between leaving home and educational attainment (White & Lacy, 1997). Interdependent life-course patterns can, however, be rather complex compositions of a number of events, so that relationships between just two events are the norm.

While the basic idea of role transitions can be regarded as universal, it can also be expected that there is considerable variation in their actual forms between societies. This applies to both changes over time and inter-national differences. *Changes over time* reflect long-term developments during the 20th century in the transition to adulthood and have been discussed mainly from a perspective of modernization and under headings like 'institutionalization' (Kohli, 1985) and 'individualization' (Beck, 1992) of life courses. Results from earlier research on the United States suggest increasing standardization in the sense of age grading of events for birth cohorts between 1910 and 1930 (Modell, 1989), but also a growing de-standardization since the late 1960s (Buchmann, 1989; Shanahan, 2000).

Inter-national differences index variation in the order and timing of life-course events and ultimately the general structure of the life course across nation states. It is only recently that studies have investigated to what extent such trends are universal or nation-specific (Corijn & Klijzing, 2001; Blossfeld, Mills, Klijzing, & Kurz, in press). The argument of specific trends follows the idea that systems of welfare and labor-market institutions are distinct and lead to path dependency in outcomes (Esping-Andersen, 1990, 1999; Mayer, 1997; Hall & Soskice, 2001). Results of comparative studies, however, raise the question of explanation. In particular, in what way do nation-based institutional systems, which act as specific 'life-course regimes', determine whether and when the relevant life-course events are experienced and how they are interrelated with each other?

In this chapter, the main interest is on such systematic inter-national and historical differences in the transition to adulthood. Terms like '(de)institutionalization' in the discussion about historical developments highlight the importance of social rules for shaping the patterns of transition to adulthood. Our discussion will follow a model that distinguishes analytically between three kinds of macro-level conditions. The first are *direct rules*. An important sub-group of social rules can be regarded as event-related and the transition to adulthood is probably a part of the life course where a whole number of important rules apply (Parsons, 1942; Modell, Furstenberg, & Hershberg, 1976). Such rules refer both incidence and timing of events and can be informal (like 'everyone should have children until age X'), legal norms (e.g., minimum ages for marriage) or conventions (e.g., age-graded promotions). These rules may also refer to sequences and interrelations between events, as in the case of norms like 'don't have children before marriage'.

A second set of macro-level conditions involves *indirect rules and life-course regimes*. Important as direct rules may be, their number is limited and according to theories of modernization, in particular individualization theory, they lose part of their relevance over time (Beck, 1992). Transition events are, however, also influenced by a whole number of other rules, notably those associated with state institutions and their regulations (cf. Mayer & Müller, 1986), which have consequences far beyond the issues to which they are primarily concerned. For example, formal requirements of qualification may set de-facto age limits, compulsory military service may lead to delayed entries to training and employment for men, and social positions may imply particular life-course patterns and forms of household organization. It is these rather 'latent' consequences which form probably the major part of national life-course regimes shaping individual life courses.

Finally, individual behavior is also a function of available resources, opportunities and constraints. This illuminates the importance of *market conditions*. These conditions may also change over time. For example, it has been argued that volatility of (labor) market conditions has increased over the past few decades, leading to increasing insecurity for choices associated with the transition to adulthood (Blossfeld et al., in press; Kurz, Steinhage, & Golsch, 2001; Tölke & Diewald, 2003).

It should be noted that our distinction between these three forms of macro-level conditions is analytical and not a distinction between various fields of social action like, for example, partnership formation or the labor market. In both examples, there are various macro-level conditions setting constraints for individual behavior. Individuals are confronted with explicit age norms, as well as institutional regulations, which impact on life courses more indirectly, and market fluctuations. Moreover, historical changes may be restricted to one or two of these aspects. For example, basic rules may remain constant while participation rates change. It should be further noted that the three types of determinants are not additive, but stand rather in a hierarchical relation. If explicit rules are salient, then indirect rules are less likely to be effective and accepted norms of behavior have a direct and standardizing effect regardless of specific individual situations and opportunities. It is just when they lose importance that individual cost–benefit calculations based on resources, preferences, and constraints come into play. From the point of view of the individual actor, this could be called a subjective 're-framing' (Esser, 2002) of situations.

In addition to that, all three groups of macro-level factors, but especially institutions and socio-economic conditions, interact with each other. Institutions may influence individual costs and benefits (e.g., by means of different forms of taxation), which form the basis of individual calculations, yet also determine to what degree market principles apply or do not apply in fields like distribution of jobs and houses. This interaction can be interpreted as different degrees of '(de-)commodification' (Esping-Andersen, 1990).

In the beginning of this section, two major dimensions in the transition to adulthood were identified. On the onc hand, economic independence through qualification and labor-market integration, and emotional and social independence through household and family formation on the other. Under the assumption that direct norms of transitions lose relative importance across cohorts, it is likely that institutions and market conditions corresponding to these processes become more important. This highlights the theoretical importance of considering the impact of the educational system, the labor market, the housing market and, finally, the provision of childcare and household-related services. In order to understand historical developments in the transition to adulthood, it is crucial to follow changes in these institutional systems.

INSTITUTIONAL AND MACRO-SOCIAL CHANGES IN WEST AND EAST GERMANY

Institutional developments in Germany have been the result of major political discontinuities during the 20th century. For cohorts born after 1920, there are three important historical periods. The first is World War II and the immediate post-war period (1939–1948) characterized by problems of violence, shortage and a high level of insecurity. A second key period involves *institutionalization* when two German states, the Federal Republic of Germany (FRG) and the German Democratic Republic (GDR), exist as representatives of opposed political systems (1949–1989) and specific institutional infrastructures are built up. A third key period is *transformation*. This followed the German Unification and beyond (1990–1999) when West German institutions were transferred to East Germany within a short period of time, the East German economy underwent a massive restructuring and the welfare state suffered from increasing financial pressure. The following paragraphs give a very short summary of developments in the major areas of state institutions and markets (see also Mayer & Hillmert, 2003).

Educational System

After World War II, the GDR built up a standardized school system, whereas in the West a differentiated system was re-established in each federal state (cf. Cortina, Baumert, Leschinsky, Mayer, & Trommer, 2003). Education and training were characterized by traditional institutions of a tripartite school system and a dual system of vocational training. Educational expansion on the secondary and tertiary level started relatively late in West Germany but overtook the GDR in the early 1970s when further increases in the number of university students was stopped by political decision. Educational expansion was of particular benefit to young women. They passed young men in their share of general education and increased their mean level of occupational qualifications considerably. In the decades after World War II, it became the norm for both genders in both German states to attain a vocational or academic degree after leaving general schooling.

Labor Market

Following World War II, the FRG followed the model of a market economy, albeit highly regulated in many respects. The GDR in contrast had a planned economy where paid employment was both guaranteed and regarded as a duty for both men and women. Particularly for men, differences in chances of employment between systems was likely minimal during the period of post-war economic growth (economic miracle). Labor markets in the 1960s and 1970s were characterized by increasing numbers of jobs. Here, women strongly increased their (full-time) labor market participation, especially in East Germany. Education, training and labor markets were closely matched and coordinated on the basis of formal qualifications (Hillmert, 2002). With recurrent economic difficulties beginning in the early 1970s, high levels of unemployment became a persistent characteristic of the West German labor market, making access to both vocational training and employment more difficult, leading to economic insecurity for individuals and increasing the risks for biographical decisions and long-term commitments. This was even much more the case after the breakdown of communism in the East and the following period of economic crisis.

Housing

After shortages in the immediate post-war period, both states invested in programs for the provision of public housing. In the FRG, most of the available housing was allocated according to market principles. The more regulated system in the GDR made housing an issue more closely connected to the political agenda, giving official privilege to particular groups like (single) parents with children (Bendit, Gaiser, & Marbach, 1999). However, in both systems, there were recurrent periods or regions where demand for (affordable) housing exceeded the supply.

Childcare

It was part of the political program in the GDR to facilitate the participation of mothers in gainful employment. Therefore, a comprehensive system of inexpensive childcare was established. In West Germany, gender-based policies favored men as 'breadwinners' and as a result public childcare was established relatively late and only to a minor extent, especially for children below kindergarten age. Up to the present, there have been differences in the degree of provision between West and East Germany (Frerich & Frey, 1993; Hank, Tillmann, & Wagner, 2001).

In addition to these institutional developments, the example of fertility indicates how individual behavior has also been associated with changes in values. The baby boom of the 1960s can be interpreted as an expression of optimistic expectations based on high rates of economic growth, especially in West Germany. There, traditional family values prevailed at least until the mid-1970s. Then, the women's and the students' movements began to question not only these traditional values but also the binding character of biographical norms in general. Life arrangements, which until then had been rather experimental forms, became a kind of normative new standards, like in the case of non-marital unions as intermediate biographical stages before marriage.

Against this backdrop of major institutional changes and developments in both parts of Germany, possible consequences for the transition behavior of young men and women can be considered. Assuming that there has been a general long-term decline in the salience of norms, specifically a reduction in the binding character of event-related norms including a decline in gender differences, the relationships to state institutions and market conditions have gained relevance across cohorts.

After a period of establishing institutions of the welfare state, institutionalized life-course regimes in both East and West Germany can be expected to have been relatively stable until the late 1980s (in East Germany) or the late 1990s (in West Germany). In addition, however, market conditions have obviously become increasingly important for transitions to adulthood, gradually since the 1970s in the West and rather dramatically (after unification 1990) in the East.

Before deriving hypotheses about the development of life-course patterns, it is important to recognize that structural changes can affect cohorts differently, giving preferences to either intra-cohort mobility or changes across cohorts (Ryder, 1965). Welfare arrangements and labor-market regulations in Germany are likely to have an important consequence in this regard (cf. DiPrete, de Graaf, Luijkx, Tåhlin, & Blossfeld, 1997). Historical changes will not affect the mid-career phase so much but mainly events in the early life course such as labor market entry and family formation and events in the later life course like retirement. This may lead to observable differences across cohorts particularly with regard to the transition to adulthood.

Building upon the above arguments, the guiding theses for long-term historical trends in transition patterns can be grouped into three dichotomies: (1) acceleration versus postponement of transition events; (2) individualization versus standardization of transition patterns; and (3) divergence versus convergence of transition patterns in the two parts of Germany. With respect to acceleration versus postponement, three periods can be distinguished. First, war and post-war insecurities likely lead to delayed transitions and heterogeneous transition patterns (1940s). Second, transitions to adulthood in times of economic prosperity and developing welfare institutions (1950s and 1960s) should happen relatively early. In a third period of the 1970s to the 1990s, important events are shifted along the life course due to longer educational careers, labor market turbulence and other perceived insecurities. This likely began in West Germany well before unification.

In terms of individualization versus standardization, again, a rather U-shaped development can be expected. After the immediate post-war period, transition patterns should be relatively standardized as a consequence of emerging institutional life-course regimes. From the 1970s onwards, however, volatile market conditions should increase inter-individual variation and a decline in interdependencies between the different events. This development is enforced by German unification, particularly in the case of the East Germany which experiences both a rapid institutional transformation and an economic crisis.

Finally, divergence between the life-course regimes in East and West Germany before 1989 should lead to increasingly different consequences for individual transition patterns. Therefore, a non-monotonic development of the transition behavior of young adults in East and West Germany since World War II can be expected. First, we expect divergence while the two German states exist. Second, we expect increased convergence for the transition behavior of young adults in East and West Germany after unification.

DATA, MEASURES, AND ANALYTIC STRATEGY

Empirical analyses draw upon detailed individual-level life-course data (11,280 cases) from the *German Life History Study* (Brückner & Mayer, 1998; Hillmert, 2003; Matthes, 2003). This project consists of a series of retrospective cohort studies which have been conducted at the Max Planck

Table 1. The German Life History Study: Birth Cohorts and Corresponding Historical Periods.

Birth Cohort	Data Collection Year	*N* (Men)	*N* (Women)	Transition to Adulthood (approx.) During Historical Period...	Important Historical Events/Periods
1919–21 (West)	1985–88	559	853	1930/40s	World War II (1939–1945)
1929–31 West	1981–83	347	361	1940/50s	Immediate Post-war period
1929–31 East	1991–92	290	302		
1939–41 West	1981–83	375	355	1950/60s	Two German states: FRG and GDR (1949–1990)
1939–41 East	1991–92	294	292		
1949–51 West	1981–83	364	369	1960/70s	
1951–53 East	1991–92	291	287	1960/70s	
1954–56 West	1989	522	485	1970s	
1959–61 West	1989	512	489	1970/80s	
1959–61 East	1991–92	266	306		
1964 West	1998–99	716	700	1980s	
1971 West	1998–99	716	619	1980/90s	German unification (FRG) and beyond (1990–)
1971 East	1996–98	322	288		

Institute for Human Development since the mid-1980s. The complete dataset (see Table 1) contains monthly life-course information for selected East and West German cohorts born between 1919 and 1971. There is a rather close relationship between cohort and historical period. Taking into account the typical ages of transition, the period between the mid-1930s and the late 1990s is covered. For each data collection, separate samples were drawn. For the East German samples, people were classified according to their place of residence in 1990. Most of the studies included persons of German nationality only, and in order to allow for better inter-cohort comparisons in the following analyses, only persons of German nationality were included for all of the cohorts.

In this chapter, the following crucial biographical events are selected in order to describe the transition to adulthood: (a) leaving education and attaining the first vocational or academic degree; (b) entering the labor market for the first time; (c) leaving the parental home and forming an own household for the first time; (d) forming (marital) unions for the first time;

and (e) becoming a parent for the first time. These singular events are significant biographical markers. For the purposes of analysis, they are defined as non-repeatable. This does not mean that they cannot be repeated in the transition to adulthood and it should be recognized that analyses of standardization and de-standardization on the basis of non-repeatable events are likely to even underestimate the degree of complexity of life courses at this stage of life.

In the following section, there are three major types of empirical analysis. First, historical trends in the *timing* of single events are measured by an inter-cohort comparison of median ages at which these events were experienced. To account for censored observations which exist particularly in cohorts where the 'observation window' between birth and interview was relatively short, all longitudinal analyses have been set up as event–history analyses (cf. Blossfeld & Rohwer, 2002). The analyses of single events use as non-parametric Kaplan–Meier estimates.

Second, while comparison of the timing of the various events gives some account of the standardization of the transition period, a more direct way to investigate this is to look at *sequences* of events. This is done by looking at the relative proportions of selected sequences. Finally, *interdependencies* between events on an individual level are analyzed using semi-parametric Cox models. These transition-rate models do not require a baseline hazard rate. As the aim is to look at interdepencies rather than find a complete list of micro- and macro-level determinants only transition events are used as time-varying predictors for the other events.

Operationalization involved a number of decisions. Concerning entering employment, stable employment requires a 12 months minimum duration to be counted as successful. The assumption here is that short-term jobs are not indicative of and are not experienced as economic independence. With regard to marriage, it is obvious that cohabitation has become more and more a 'functionally equivalent' state (Bumpass, Sweet, & Cherlin, 1991) for younger cohorts so that marriage and cohabitation form a joint category here. As attaining the first degree and entering employment are closely connected, (only) the latter as the main event of transition to economic independence is included in the multivariate models. Of interest here are effects which are in close temporal connection with the respective events. Therefore, possible effects are constrained to a period of 3 years each. Finally, the assumption with regard to becoming a parent is that it is pregnancy rather than actual birth which influences life decisions or is a consequence of them. Hence, the time of conception (calculated as birth date minus 9 months) is included in the models.

LONG-TERM TRENDS IN THE TRANSITION TO ADULTHOOD

Trends for Single Transition Events

The literature on transitions to adulthood in West Germany provides us with results like the prolongation of school-to-work transitions during the last few decades (Hillmert, 1999), a decline in marriage and fertility since the 1960s and a long-term reduction in the age of leaving the parental home (Mayer & Schwarz, 1989). On the other hand, East Germans have left home earlier, but patterns of predictors tend to converge after unification (Juang, Silbereisen, & Wiesner, 1999). Also, the ages for marriage and becoming a parent were considerably lower in the GDR compared with West Germany (Huinink & Wagner, 1995). After the 'Wende' (the breakdown of communism), there was a phase of turbulence which led to a sharp decline in fertility. At least for first births, there has obviously been a catch-up afterwards (Kreyenfeld, 2003).

The following analyses allow both to compare directly between the various transition events and the two parts of Germany on the basis of a standardized dataset. They also make it possible to identify long-term developments as they are represented by differences across a number of birth cohorts. Median ages of the transition events are displayed in Fig. 1 for men and Fig. 2 for women.

For men in the FRG, the median age of the home leaving event shows a long-term decline. In contrast, the median age of marriage, in combination with cohabitation, has remained rather stable across cohorts. In further contrast, leaving education, gaining stable employment, and becoming a parent have been clearly postponed since the 1960s. This can be interpreted as a consequence of educational expansion, as well as the increasing labor-market insecurities that confronted the younger cohorts. Apart from having a first child which has become increasingly unlikely until at least age 30, all the other events are still experienced by the majority of young people.

In contrast to the West, median ages for men in the GDR for all events have been remarkably stable across cohorts. The exception, however, is the youngest cohort, born in 1971, who experienced the relevant events mostly in unified Germany during the 1990s. For them, first marriage or cohabitation tend to happen relatively late, long after leaving education and entering employment. Household formation, in contrast, occurs at a relatively young age for this cohort and resembles the situation in West Germany.

Fig. 1. Median Transition Ages (in Years), by Cohort – Men West Germany (left), East Germany (right). *Note*: Median ages based on Kaplan–Meier estimates. Missing values indicate that less than 50% of the cohort experienced the particular event when the observation window ends.

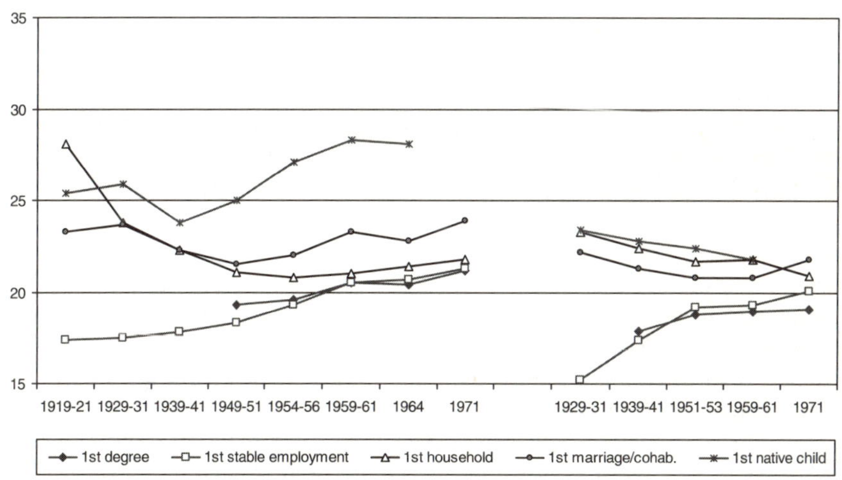

Fig. 2. Median transition ages (in years), by cohort – Women West Germany (left), East Germany (right). *Note*: Median ages based on Kaplan–Meier estimates. Missing values indicate that less than 50% of the cohort experienced the particular event when the observation window ends.

For West German women, the major trends for women were similar to those for men with educational completion and entering the labor market becoming universal and, with leaving the home, the first events in the sequence of transition to adulthood. As leaving home and family-related events such as marriage and parenthood happen at younger ages among women, age dispersion across the complete sequence of events is considerably lower. In East Germany, it is noticeable that the median ages of transition events have become clustered closely together in the three youngest cohorts. Again, there is a trend of leaving home early.

At least for West Germany, the results indicate a convergence in the transition behavior of men and women. This is primarily due to the fact that for both genders the whole set of events, at least completing education, entering the labor market, and forming an own household, has become the standard experience.[2]

Sequences and (De-)Standardization of Transition Patterns

The latter interpretation, based upon the analysis of single events, is confirmed when looking at sequences of events. These are displayed in Fig. 3. This figure shows the proportion of men and women whose transition followed either of two orders: (a) leaving education–entering (stable) employment–getting married or cohabiting–other events or (b) leaving education–entering (stable) employment–forming an own household–other events. The proportion that this selection of sequences accounts for among all sequences is used as an indicator of the standardization – or concentration – of transition patterns. As this share varies across cohorts, changes of this proportion can be regarded as indicators of the standardization and de-standardization of transition patterns as processes. An increase indicates a trend toward standardization, while a decrease refers to de-standardization. Although only a small proportion of (possible) sequences are covered by this definition, it applies to up to 70% of the empirical individual sequences in particular birth cohorts, indicating a high level of standardization there.

Standardization of sequences reached its maximum for men in the birth cohorts around 1940. Afterwards, it decreased steadily but did not fall below the level of the oldest cohorts born around 1920. A major reason for de-standardization is that an increasing number of young people formed their own household before entering the labor market. Probably, they were still dependent on external, in particular parental resources. It should be noted that in the youngest cohort where the observation window is broad enough

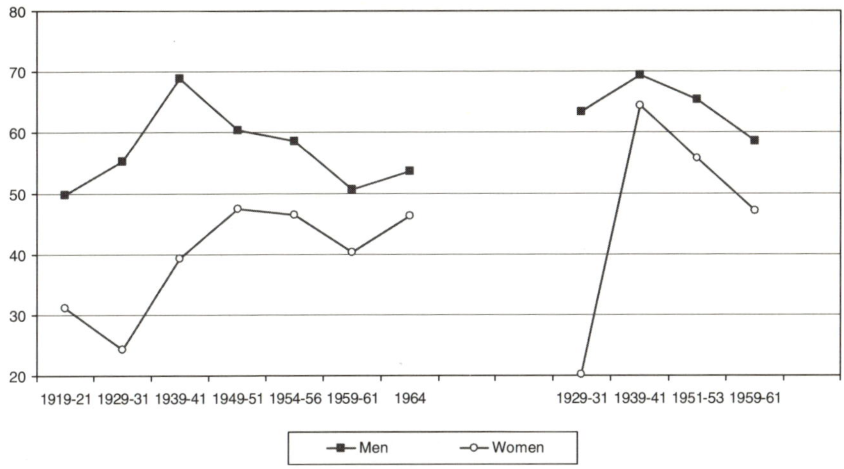

Fig. 3. Proportion of Selected Sequences in % of All Sequences, by Birth Cohort West Germany (left), East Germany (right). *Note*: Selected sequences: A, Degree–Employment–Marriage/Cohabitation–other; B, Degree–Employment–Household–Other.

to apply this analysis (the 1964 West cohort), the decline in standardization appears to have ended.

For women, standardization of transition patterns increased across the older cohorts born until around 1950 due to the rising proportions of women who complete vocational or academic education. It has however remained rather stable since, and there has been an overall convergence between the transition behavior of men and women. Developments in the GDR were different. Already in the birth cohorts around 1940, women's transition sequences reached a high level of standardization which was comparable to men's. This is, primarily, an expression of the high level of female labor-force participation. Still, over the following cohorts, the trends for men and women were parallel with a moderate decline in standardization.

Interdependencies between Transition Events

Finally, we look at general interdependencies between transition events using multivariate event–history analyses (see Table 2 for West Germany and Table 3 for East Germany. There are two kinds of indicators for interdependencies between transition events. The first is the patterns of effects that

Table 2. Interdependencies Between Transition Events (Coefficients and Goodness-Of-Fit Statistics of Semi-Parametric Cox Models), by Birth Cohort, West Germany.

Birth Cohort	1919–21	1929–31	1939–41	1949–51	1954–56	1959–61	1964	1971
First household								
Men								
1st employment				+	+	+		
1st marriage/cohab.	+ +	+ +	+ +	+ +	+ +	+ +	+ +	+ +
1st child	–	+		+		+ +		
Pseudo-R^2	0.14	0.14	0.17	0.13	0.06	0.04	0.09	0.08
Women								
1st employment						+		+
1st marriage/cohab.	+ +	+ +	+ +	+ +	+ +	+ +	+ +	+ +
1st child	–	+	+	+	+			
Pseudo-R^2	0.10	0.14	0.14	0.11	0.07	0.04	0.12	0.13
First marriage/cohabitation								
Men								
1st employment				+	+			
1st household	+ +	+ +	+ +	+ +	+ +	+ +	+ +	+ +
1st child	+ +	+ +	+ +	+ +	+ +	+ +	+ +	+ +
Pseudo-R^2	0.21	0.17	0.21	0.13	0.08	0.07	0.09	0.11
Women								
1st employment					+	+	+	+ +
1st household	+ +	+ +	+ +	+ +	+ +	+ +	+ +	+ +
1st child	+ +	+ +	+ +	+ +	+ +	+ +	+ +	+ +
Pseudo-R^2	0.14	0.17	0.15	0.13	0.10	0.08	0.08	0.11
First child								
Men								
1st employment								
1st household	+	+						
1st marriage/cohab.	+ +	+ +	+ +	+ +	+ +	+ +	+ +	+ +
Pseudo-R^2	0.09	0.07	0.08	0.04	0.03	0.05	0.02	0.05
Women								
1st employment								
1st household	+	+ +						
1st marriage/cohab.	+ +	+ +	+ +	+ +	+ +	+ +	+ +	+ +
Pseudo-R^2	0.08	0.08	0.06	0.03	0.02	0.03	0.02	0.02

Note: +/–, significant effects ($p<0.05$); + +/– –, significant and large effects > +/–0.7 (i.e., odds ratio > 2); Pseudo-R^2 (McFadden), 1-log L (final model)/log L (null model).

Table 3. Interdependencies Between Transition Events (Coefficients and Goodness-Of-Fit Statistics of Semi-Parametric Cox Models), by Birth Cohort, East Germany.

Birth Cohort	1929–31	1939–41	1951–53	1959–61	1971
First household					
Men					
1st employment			+	+	
1st marriage/cohab.	+ +	+ +	+ +	+ +	+ +
1st child	+				
Pseudo-R^2	0.19	0.16	0.15	0.13	0.10
Women					
1st employment	–			+	
1st marriage/cohab.	+ +	+ +	+ +	+ +	+ +
1st child					+ +
Pseudo-R^2	0.22	0.17	0.13	0.11	0.12
First marriage/cohabitation					
Men					
1st employment					
1st household	+ +	+ +	+ +	+ +	+ +
1st child	+ +	+ +	+ +	+ +	+ +
Pseudo-R^2	0.24	0.21	0.15	0.12	0.13
Women					
1st employment					+
1st household	+ +	+ +	+ +	+ +	+ +
1st child	+ +	+ +	+ +	+ +	+ +
Pseudo-R^2	0.19	0.19	0.12	0.11	0.12
First child					
Men					
1st employment					+
1st household					
1st marriage/cohab.	+ +	+ +	+ +	+ +	+ +
Pseudo-R^2	0.04	0.04	0.06	0.06	0.08
Women					
1st employment	–			+	
1st household				–	
1st marriage/cohab.	+ +	+ +	+ +	+ +	+
Pseudo-R^2	0.04	0.03	0.03	0.03	0.01

Note: +/–, significant effects ($p<0.05$); + +/– –, significant and large effects > +/–0.7 (i.e., odds ratio > 2); Pseudo-R^2 (McFadden), 1-log L (final model)/log L (null model).

the time-varying covariates have. The second is the goodness-of-fit measures for the transition-rate models. The first indicator looks qualitatively at the structure of specific relationships between the transition events, while the second is a global measure of interdependency.

In terms of general effect patterns, marriage and household formation more so than parenthood are strongly determined by the other events. This is even seen in the older cohorts where having children was an almost universal experience. Marriage is the major predictor for household formation, while household formation and becoming a parent are the major predictors for marriage. For younger cohorts who grew up in the 1960s or later, becoming a parent is only influenced by marriage, whereas in the older cohorts, household formation was still closely connected with it. This may mean that leaving home has become a separate step in the transition to adulthood which is independent of family formation.

Equally important, there are some notable changes in effects over time. In the case of women, entering employment has become more important as a predictor for household formation and marriage. This can be interpreted as entering employment becoming an increasingly normal step in female life courses. At the same time, having a first child has lost its impact on household formation among women. An interesting exception is the youngest cohort (1971) in East Germany where becoming a (possibly single) parent significantly increases the probability of forming an own household soon. For men, significant effects of first employment on household union formation are visible for the cohorts born around 1950–1960, but are no longer seen among the youngest cohorts.

Finally, the magnitude of interdependencies is seen with respect to the size of the goodness-of-fit statistics. In West Germany, multivariate interdependence is greatest for the birth cohorts around 1940. For the birth cohorts around 1940 and 1960, there was a marked decline. This indicates that the phase of transition became 'looser' in the sense that making one step in the transition to adulthood such as leaving home had less and less impact on the likelihood of making another transition. Since the 1980s, however, this decline has obviously come to an end and appears to have reversed itself, a pattern which resembles the development of the de-standardization of sequences. In contrast, East Germany has seen a rather continuous decline in interdependence across the three oldest cohorts, which again has been followed by relative stability. There is also an important development regarding gender differences in interdependence. For the older birth cohorts until around 1940, men's transition events tended to be more interdependent than women's. Since then, this difference has disappeared, at least in West Germany.

SUMMARY AND CONCLUSION

The analyses of this chapter have investigated long-term changes in the transition to adulthood for young men and women in both parts of Germany. There are a number of important conclusions. In terms of changes in the timing of life-course events, the issue of acceleration versus postponement, there is only one event which has, in the long run and on average, been experienced at younger ages in the younger cohorts than in the older cohorts. This is leaving the parental home. In West Germany particularly since the 1970s, other important transition events have been postponed, shifting along the life course due to longer educational careers, labor market turbulence and perceived insecurities. In the GDR, such market mechanisms were neutralized by the state to a larger extent and transitions to adulthood happened earlier in the life course and remained stable across cohorts. There are a number of indications, however, that German unification has had a major impact on transition patterns in East Germany. This is most clearly seen in the youngest East German cohort in the sample, particularly with respect to the postponement of fertility.

In the long run, there have obviously been an increasing inter-individual variation and a decline in interdependencies between the different events. This echoes arguments about the increasing individualization of the life course. Though this trend is true for both parts of Germany, it has been more significant in West Germany. There is some plausibility for the assumption that this kind of individualization has often relied on 'external' resources provided not by the young people themselves but by their parents. Still, various analyses indicate that the process of de-standardization has come to an end in the youngest cohorts and may actually have reversed.

Different life-course regimes in East and West Germany have also had distinctly different consequences for individual life courses with the consequence of revealing both divergence and convergence in life-course patterns. The timing of transitions to adulthood was much more standardized in the GDR than in the FRG and more stable across cohorts. This led to an increasing divergence of transition patterns between the two German states. However, transition patterns seem to have converged after unification, although the observation window in these analyses may be rather short for definite conclusions.

Another trend is another kind of convergence in the transition behavior of young men and women since World War II. There are still considerable differences in ages at various transitions, but completing education by attaining a degree, entering the labor market, and forming an own household, even if not associated with union-formation and parenthood, have become usual steps in the transition patterns of both young men and women.

Finally, I offer some theoretical conclusions. It is obvious that, with a relative lack of deterministic norms of behavior in modern societies, institutions have tremendous significance for shape and structure of the transition to adulthood. This is especially apparent in the divergent developments of life-course patterns under the conditions of two different German states. On the other hand, institutional rules do not determine individual behavior. In particular, they cannot really explain the trend to postpone life-course transitions in the younger cohorts. Rather, (labor) market conditions have obviously had major consequences for the timing of and the inter-relation between transition events. There are, however, different links to institutions in West and East Germany. In the West, market conditions have gradually become more important over the last few decades while institutions have been relatively stable; in the East, they are clearly a result of massive institutional change after 1989.

NOTES

1. An alternative approach looks at patterns of time-use at different stages of the life course (Gauthier & Furstenberg, 2002).

2. Examination of age dispersion measured by inter-quartile distances and not displayed in this figure shows that the standardization of transition ages has, on average, been higher in East Germany. In the West, the minimum of dispersion was reached in the cohorts born around 1940, and for most events, dispersion increased moderately afterwards.

REFERENCES

Beck, U. (1992). *Risk society: Towards a new modernity*. London: Sage.

Bendit, R., Gaiser, W., & Marbach, J. (1999). *Youth and housing in Germany and the European Union*. Opladen: Leske and Budrich.

Billari, F. (2001). The analysis of early life courses: Complex descriptions of the transition to adulthood. *Journal of Population Research*, *18*, 119–142.

Blossfeld, H. P., & Huinink, J. (1991). Human capital investments or norms of role transition? How women's schooling and career affect the process of family formation. *American Journal of Sociology*, *97*, 143–168.

Blossfeld, H. P., Klijzing, E., Mills, M., & Kurz, K. (Eds) (in press). *Globalization, uncertainty, and youth in society*. London: Routledge.

Blossfeld, H. P., & Rohwer, G. (2002). *Techniques of event history modeling: New approaches to causal analysis*. Mahwah, NJ: Erlbaum.

Brückner, E., & Mayer, K. (1998). Collecting life history data: Experiences from the German life history study. In: J. Giele & G. Elder Jr. (Eds), *Methods of life course research: Qualitative and quantitative approaches* (pp. 152–181). Thousand Oaks: Sage.

Buchmann, M. (1989). *The script of life in modern society: Entry into adulthood in a changing world*. Chicago: University of Chicago Press.

Bumpass, L., Sweet, J. A., & Cherlin, A. (1991). The role of cohabitation in declining rates of marriage. *Journal of Marriage and the Family, 53*, 913–927.

Cooney, T., & Hogan, D. (1991). Marriage in an institutionalized life course: First marriage among American men in the twentieth century. *Journal of Marriage and the Family, 53*, 178–190.

Corijn, M., & Klijzing, E. (Eds) (2001). *Transitions to adulthood in Europe*. Dordrecht: Kluwer.

Cortina, K., Baumert, J., Leschinsky, A., Mayer, K., & Trommer, L. (Eds) (2003). *Das Bildungswesen in der Bundesrepublik Deutschland. Strukturen und Entwicklungen im Überblick*. Reinbek: Rowohlt.

DiPrete, T., de Graaf, P., Luijkx, R., Tåhlin, M., & Blossfeld, H. P. (1997). Collectivist vs. individualist mobility regimes? How welfare state and labor market structure condition the mobility effects of structural change in four countries. *American Journal of Sociology, 103*, 318–358.

Eisenstadt, S. (1964). *From generation to generation: Age groups and social structure*. New York: Free Press.

Esping-Andersen, G. (1990). *The three worlds of welfare capitalism*. Princeton, NJ: Princeton University Press.

Esping-Andersen, G. (1999). *Social foundations of postindustrial economies*. Oxford: Oxford University Press.

Esser, H. (2002). Ehekrisen. Das Re-Framing der Ehe und der Anstieg der Scheidungsraten. *Zeitschrift für Soziologie, 31*, 472–496.

Frerich, J., & Frey, M. (1993). *Handbuch der Geschichte der Sozialpolitik in Deutschland*. München: Oldenbourg.

Gauthier, A., & Furstenberg, F. (2002). The transition to adulthood: A time use perspective. *The Annals of the American Academy of Political and Social Science, 580*, 153–171.

Goldscheider, F., & Goldscheider, C. (1993). Whose nest? A two-generational view of leaving home during the 1980s. *Journal of Marriage and the Family, 55*, 851–862.

Hall, P., & Soskice, D. (Eds) (2001). *Varieties of capitalism: The institutional foundations of competitiveness*. Oxford: Oxford University Press.

Hank, K., Tillmann, K., & Wagner, G. (2001). Außerhäusliche Kinderbetreuung in Ostdeutschland vor und nach der Wiedervereinigung. Ein Vergleich mit Westdeutschland in den Jahren 1990–1999. *Zeitschrift für Bevölkerungswissenschaft, 26*, 55–65.

Hillmert, S. (1999). The 'complex' event of transition to the labour market: Comparing the UK and Germany over time. In: D. Raffe, R. van der Velden & P. Werquin (Eds), *Education, the labour market and transitions in youth: Cross-national perspectives* (pp. 313–331). Edinburgh: University of Edinburgh.

Hillmert, S. (2002). Labour market integration and institutions: An Anglo-German comparison. *Work, Employment and Society, 19*, 675–701.

Hillmert, S. (2003). Ausbildungs- und Berufsverläufe der Geburtskohorten 1964 und 1971 in Westdeutschland: die jüngste Teilstudie der Lebensverlaufsstudie als Scientific Use File. *ZA-Information, 53*, 167–177.

Huinink, J., & Wagner, M. (1995). Partnerschaft, Ehe und Familie in der DDR. In: J. Huinink, K. Mayer, M. Diewald, H. Solga, A. Sørensen & H. Trappe (Eds), *Kollektiv und Eigensinn. Lebensverläufe in der DDR und danach* (pp. 145–188). Berlin: Akademie.

Juang, L., Silbereisen, R., & Wiesner, M. (1999). Predictors of leaving home in young adults raised in Germany: A replication of a 1991 study. *Journal of Marriage and the Family*, *61*, 505–515.

Kerckhoff, A. (2003). From student to worker. In: J. Mortimer & M. Shanahan (Eds), *Handbook of the life course* (pp. 52–87). New York: Kluwer.

Kohli, M. (1985). The world we forgot: A historical review of the life course. In: V. Marshall (Ed.), *Later life: The social psychology of aging* (pp. 271–303). Beverly Hills, CA: Sage.

Kreyenfeld, M. (2003). Crisis or adaptation – reconsidered: A comparison of East and West German fertility patterns in the first six years after the 'Wende'. *European Journal of Population*, *19*, 303–329.

Kurz, K., Steinhage, N., & Golsch, K. (2001). *Case study Germany: Global competition, uncertainty and the transition to adulthood.* GLOBALIFE Working Paper No. 16. University of Bielefeld.

Marini, M. (1984). Women's educational attainment and the timing of entry into parenthood. *American Sociological Review*, *49*, 491–511.

Matthes, B. (2003). Das Teilprojekt 'Ostdeutsche Lebensverläufe im Transformationsprozess (LV-Ost)' der Deutschen Lebensverlaufsstudie. *ZA-Information*, *53*, 156–166.

Mayer, K. U. (1997). Notes on a comparative political economy of life courses. *Comparative Social Research*, *16*, 203–226.

Mayer, K. U., & Hillmert, S. (2003). New ways of life or old rigidities? Recent changes in social structures and life courses in Germany and their political impacts. In: H. Kitschelt & W. Streeck (Eds), *Germany beyond the stable state. West European Politics*, 26, 79–100.

Mayer, K. U., & Müller, W. (1986). The state and the structure of the life course. In: B. Sørensen, F. Weinert & R. Sherrod (Eds), *Human development and the life course: Multidisciplinary perspectives* (pp. 218–245). Hillsdale, NJ: Erlbaum.

Mayer, K. U., & Schwarz, K. (1989). The process of leaving the parental home: Some German data. In: E. Grebenik, C. Höhn & R. Mackensen (Eds), *Later phases of the family cycle* (pp. 145–163). Oxford: Clarendon Press.

Modell, J. (1989). *Into one's own: From youth to adulthood in the United States 1920–1975.* Berkeley: University of California Press.

Modell, J., Furstenberg, F., & Hershberg, T. (1976). Social change and transitions to adulthood in historical perspective. *Journal of Family History*, *1*, 7–32.

Oppenheimer, V. (1988). A theory of marriage timing. *American Journal of Sociology*, *94*, 563–591.

Parsons, T. (1942). Age and sex in the social structure of the United States. *American Sociological Review*, *7*, 604–616.

Ryder, N. (1965). The cohort as a concept in the study of social change. *American Sociological Review*, *30*, 843–861.

Shanahan, M. (2000). Pathways to adulthood in changing societies: Variability and mechanisms in life course perspective. *Annual Review of Sociology*, *26*, 667–692.

Tölke, A., & Diewald, M. (2003). Insecurities in employment and occupational careers and their impact on the transition to fatherhood in Western Germany. *Demographic Research*, *9*, 41–68.

Van de Kaa, D. (2001). Demographic transition, second. In: N. J. Smelser & P. B. Baltes (Eds), *International encyclopedia of the social and behavioral sciences*, (Vol. 5, pp. 3486–3488). Amsterdam: Elsevier.

White, L. (1994). Coresidence and leaving home: Young adults and their parents. *Annual Review of Sociology*, *20*, 81–102.

White, L., & Lacy, N. (1997). The effects of age at home leaving and pathways from home on educational attainment. *Journal of Marriage and the Family*, *59*, 982–995.

AGE NORMS, INSTITUTIONAL STRUCTURES, AND THE TIMING OF MARKERS OF TRANSITION TO ADULTHOOD

Jeylan T. Mortimer, Sabrina Oesterle and Helga Krüger

ABSTRACT

This chapter compares the timing and sequencing of educational completion and first parenthood for samples of U.S. youth and German women. We garner evidence bearing on two hypotheses. The first is that the timing of transition to parenthood is governed by age norms; the second, that this transition is influenced by the incentives posed by distinct school-to-work transition structures. We conclude that strongly institutionalized school-to-work bridges pose greater incentives for the completion of educational programs in Germany and more clearly regulate the timing of parenthood. The more loose connection between school and work in the United States encourages greater variation in the timing of parenthood, and more first births before schooling is complete.

The Structure of the Life Course: Standardized? Individualized? Differentiated?
Advances in Life Course Research, Volume 9, 175–203

ISSN: 1040-2608/doi:10.1016/S1040-2608(04)09007-0

The timing and sequencing of markers of adulthood (e.g., the age of leaving home, full-time labor force participation, marriage, and parenthood) have been closely scrutinized by sociologists and demographers (Mortimer, 1992; Mortimer & Aronson, 2000). In the United States and other post-industrial societies, these events have been delayed as the period of adolescence has become extended in response to multiple, interconnected macrostructural conditions (Arnett, 2000; Buchmann, 1989; Heinz, 1999). Of particular note, years in school have increased and youth are increasingly postponing both marriage and parenthood. While the demographic trends are clearly observable (Shanahan, 2000), the forces underlying these patterns are little understood. Do age norms, and related sanctions, govern the transition to adulthood? Or are transitional events mainly responsive to institutional structures and associated material incentives (Marini, 1984)? We address these questions by examining two important transitional events, the completion of formal education and the initiation of parenthood, in the United States and Germany, societies that have markedly different institutional bridges to adulthood (Mortimer & Krüger, 2000).

According to the "age norm" hypothesis (Neugarten & Hagestad, 1976; Settersten, 1999, 2003), significant life transitions in educational, occupational, and familial realms are regulated by clear, but informal, understandings about the timing at which such events should occur. These understandings are thought to have moral force with a distinct sense of "right" or "ought" attached to them. Norms may be geared to age itself. For example, a student in Germany who receives a university diploma at the age of 25 or 26 and completes the doctoral dissertation by the age of 28 or 29 is considered excellent, while older students with the same accomplishments are not judged as positively.

Age norms may also refer to particular sequential patterns. For example, one should have a promising position in the labor market before marrying, or one should marry before having children. Persons whose actions are in accord with age-based or sequential norms will feel that they are "on time" and others will provide approval or other positive sanctions for normatively timed transitions. Those who do not abide by the timing norms will feel that they are "early" or "late" and may be subject to more or less subtle sanctions that reinforce feelings of deviance or nonconformity.

A plausible alternative hypothesis, specifying "institutional determination," is that the timing of transitions is regulated by the material incentives and other consequences that follow from adherence to structured pathways. Institutional determination is most clear with respect to legal constraints. For example, laws in modern societies govern the age at which persons can enroll in public schools, marry, work, vote, enter a contract, purchase alcohol, and drive an automobile (Mortimer & Aronson, 2000). Furthermore, age grading is often built into standard modes of institutional operation. For example, children in the United States start the first grade of school at age 6 and, if they proceed according to normal schedules, advance one grade each year until they complete grade 12 at age 18. The link between age and grade is not perfect, however, since some children will fail to move from one grade to another with their age mates because of insufficient academic performance, social immaturity, prolonged illness, or other circumstances.

The failure to adhere to institutionally structured educational pathways may have important consequences. For example, if jobs are available only to those who successfully complete educational programs, youth's occupational prospects and the possibility of becoming economically independent of parents may be severely limited by early school-leaving. In view of such constraints, relatively few students would leave school prior to obtaining their degrees. There would also be strong material incentives to delay family formation, especially parenthood, as these would likely jeopardize program completion.

Moreover, if certain educational programs permit entry to more or less rewarding occupations or career lines, the timing of parenthood could be affected accordingly. Especially for women who bear the greater responsibility for child-rearing, parenthood poses a clear threat to occupational success and accompanying economic rewards. In these situations, strong material incentives promote the delay of parenthood. Alternatively, parenthood may be perceived as an escape from stereotypically female dead-end jobs involving poor work conditions, few rewards, and limited prospects for the future.

These two perspectives, emphasizing age-graded norms and institutional structures, are not mutually exclusive. For age norms may arise in response to institutional regulatory schemes and to the rewards and punishments that are linked to compliance. Still, the age norm perspective implies greater stability, as norms are embedded in the wider culture. As such, they would change only slowly in response to changing institutional (or other) pressures. The institutional perspective, in contrast, posits change in the timing of life course transitions in the presence of shifting structural mandates. Change may be rapid, fluid, and without normative force. An institutional perspective also implies greater within-societal variation, given variability in opportunities and constraints resulting from individual differences in social location. For example, in societies where families bear substantial costs for higher education, young people of limited means may have to forego college in favor of a more rapid transition to adulthood: early full-time entry to the labor force, earlier marriage, and parenthood.

Recent prominent conceptions of "destructuring" and "individualization" in the early life course (Buchmann, 1989) emphasize the erosion of traditional institutional pathways, along with a "destandardization" of associated constraints and opportunities (Heinz, 1999). If "destructuration" of the early life course has in fact occurred, one would expect to find increasing variability in both the ages at which markers of the transition to adulthood occur, and in their sequential patterns. Moreover, varying institutional regimes may present different degrees of structuration across societies. Cross-national variability in institutional structures and early life course patterns enables us to address the institutional and age norm hypotheses with respect to the relative timing of educational completion and parenthood.

THE TRANSITION TO ADULTHOOD IN GERMANY AND THE UNITED STATES

In Germany, movement into full-time work is structured by a highly regulated age-graded system of secondary and tertiary education and the institution of the apprenticeship (Hamilton, 1990; Hamilton & Hurrelmann, 1994; Heinz, 1999). This bridging system connects general schooling with occupational qualification and occupational qualification with the labor market. A less clearly differentiated educational structure and a general absence of institutional bridges from school to work characterize the United States, especially for the majority of the youth population that does not receive a baccalaureate degree.

Recent societal transformations in the U.S. heighten the differences in economic and occupational career prospects between those who achieve more and fewer years of formal education. Rapid technological and occupational change has heightened the complexity of some forms of work and generated an increasingly bifurcated labor force. Good "primary" jobs have high pay, ample fringe benefits, and much opportunity (through well-defined internal labor markets). Low-skill service jobs, in contrast, have lower pay, few fringe benefits, and relatively poor career prospects. Declining opportunities available to youth with only a high school diploma have widened the income gap between high school and college graduates (Kerckhoff, 2000). Since jobs in the desirable tier are most often obtained by college graduates, increasing numbers of American youth are seeking higher education.

In the contemporary United States, more than 60% of recent high school graduating classes have entered post-secondary institutions of higher education (Arnett, 2000; Kerckhoff, 2000). However, only about one-fourth of contemporary young people have received 4-year college degrees by the time they have reached 25–29 years of age. College graduates enter the labor market with strong credentials. They also can make use of college career placement services, which link them with prospective employers who offer jobs in the primary labor market involving good pay, fringe benefits, and trajectories of upward advancement. While 8% obtain 2-year degrees, approximately a third of recent cohorts have become college dropouts (Kerckhoff, 2000). College dropouts, as well as those with less education, have little institutional support in making the transition from school to work. For the most part, they must rely on their own resources in finding employment and the jobs available to them are in the secondary sector with relatively low wages, few benefits, and little advancement opportunity. These youth often flounder from one job to another, often quite similar to those they held during high school.

College graduates in Germany have similar support for transitions to advantageous labor market positions. Yet, the vast majority of "sub-baccalaureate" youth in Germany also have access to clear institutional bridges between school and work. Those who do not go to the "gymnasium" (the high school preparatory to university entrance) enter the apprenticeship system after their general schooling. These youth work part-time and their concurrent schooling is geared to the knowledge and skills they must acquire to perform their jobs competently. Required skills and curricula for each occupation are strictly defined on a supra-company level. Each occupational title corresponds to a prescribed "specification of skills" (Berufsbild) that is fixed by the "Bundesinstitut für Berufsbildung," The National Institute for Vocational Education (see also Krüger, 1999). If all goes well, successful completion of the apprenticeship, and associated vocation-specific schooling, enables occupational certification which, in turn, provides access to full-time jobs with quite good wages and benefits and the prospect of stable careers.

This system, however, is far from perfect. In times of economic downturn, not all graduates of the apprentice system will be able to find appropriate placements. Moreover, the system favors males. Certification for female-stereotyped occupations often involves only school-based learning without simultaneous placement in a firm. It is less likely to lead to offers of employment in accord with qualifications (Mortimer & Krüger, 2000). To enhance the likelihood of a successful school to work transition, some German youth attempt to acquire more than one occupation-specific credential, or they upgrade their general education through the "Zweite Bildungsweg" (described below).

According to the institutional hypothesis, clear school-to-work bridges provide material incentives to delay family formation and parenthood. Those subject to such regimes have good reason to delay parenthood until schooling is complete. This hypothesis would predict that few German youth would become parents while still attending school. Both college and noncollege youth have greater material incentive to delay parenting until they finish their education. Without proper occupational certification, a young person in Germany is restricted to unskilled work. In contrast, adolescents in the loose, noninstitutionalized school-to-work transition regime characteristic of the American system would have less to lose by becoming a parent early on. One would expect to find earlier transitions to parenthood in the aggregate and a looser coupling of educational completion and parenthood.

In this chapter, we compare the objective timing and sequencing of educational completion and first parenthood for samples of American youth and German women. While these data sets do not permit observation of age norms directly, we garner substantial evidence relevant to the institutional determination hypothesis. To what extent is there evidence for institutionally governed regularities in transitional behavior? In the U.S., we assess the level of reward – in terms of educational attainment and full-time work – for those who finish their educations at different ages. In Germany, we examine the relation between the type of occupational certification and associated reward differentials, and the timing of family formation. The German case also enables assessment of the stability of parenthood timing in the face of social change in institutional patterns.

DATA SOURCES

Survey data were obtained from the U.S. youth development study (YDS) and from the female life course study of three cohorts of German women with vocational degrees. These data sets allow comparative assessment of the relationships between the completion of education, establishment in an occupational position, and the timing of the first birth.

The Youth Development Study (US)

The YDS is a prospective longitudinal study that began in 1988 with a random sample of ninth grade students in a large Midwestern city (1,010 consented to participate). Its major focus is the impact of early work experience on adolescent development, educational attainment, and the transition to adulthood. Participants annually completed questionnaires. Survey administration took place in high school classrooms in each of the first 4 years and by mailed questionnaires in the fifth through eleventh years. The retention rate through the 11th survey (1999), when most respondents were 25 or 26 years old, was 72%.

Key measures for this analysis are the objective timing of the completion of education and parenthood. Starting in 1992, 1 year after most YDS respondents would have completed high school, panel members were asked whether they had attended school at any time during the past year. The age at which respondents completed their schooling was determined by subtracting the respondents' year of birth from the year for which they last reported attending school (e.g., 1992−1973 = 19). Those who attended an educational institution in the eleventh wave, in 1999, were counted as not having completed their educations.[1] Age at birth of the first child was also computed by subtracting the respondents' birth year from the year when their first child was born.

The objective timing variables had five categories: (1) age 18 and earlier, (2) ages 19–22, (3) ages 23–24, (4) ages 25 and older, and (5) still in school. These correspond roughly to the structure of the American educational system. That is, completion of high school usually occurs at age 17 or 18. Four-year college programs would, for those who enter college at age 18, encompass the age period up to 22 for "traditional" students. However, since many college students in the U.S. take longer to complete their degrees, we separate "later educational finishers" (at age 23–24) and "latest" finishers (at the age of 25 or older). A substantial portion of youth (27.2%) is still in school (for at least some time during the prior year) at the 1999 wave of data collection. These older attendees may be in diverse educational programs (e.g., 4-year or 2-year colleges, vocational schools, technical certification programs, post-graduate degree programs, etc.). Categorizing the timing of educational completion and the age of first parenthood in the same manner enables assessment of the correspondence between them.

In the YDS, educational attainment was measured by the highest degree respondents obtained by ages 25–26 (1999). For 27% of the panel, a high school diploma (25.7%) was the highest degree attained; less than 1% did not attain this degree (0.6%). Twenty-two percent had obtained a certificate from a technical or vocational school or an associate level degree after high school. About a quarter of the panel (24%) attended college for some time without having yet received a degree. Another 27% had a bachelor's, master's, doctoral, or professional degree.

Labor force status is a key marker of the transition to adulthood, as well as a notable "payoff" to educational attainment. It was measured by the months spent, prior to the 1999 data collection, in full-time work, part-time work, unemployment, and full-time homemaking.

The Female Life Course Study (Germany)

The German data were collected as part of a study of social change in life course patterns. During the covered time span (1960–1980), the structure of the apprenticeship system remained much the same, but notable changes occurred in the conception of female family roles and in the extent and patterns of women's labor market participation (Bertram & Hennig, 1996; Seidenspinner & Burger, 1982). The female life course study examined the frictions and contradictions between these forces and how they affected the timing and sequencing of markers of adulthood (Krüger, 2001). The 2,130 participants, all women with apprentice degrees, were equally distributed across three cohorts of school-leavers: those who finished their apprenticeships in 1960, 1970, and 1980 throughout (the former) West Germany. To assess the German apprenticeship system's influence on the timing of marriage and childbirth, and the extent of employment continuity, apprenticeship degrees that are most common for women in all three cohorts were selected.[2] About 70% of all West German women in these cohorts who completed training entered these apprenticeships (Born, 2000).

Information regarding education, work and family careers was collected via mailed questionnaire.[3] Respondents were asked to indicate the dates of family events and durations of work and caregiving on a timeline, starting with the date of completion of the apprenticeship and ending in 1997 (the most recent year of data collection). For the 1960 cohort, the calendar thus covered 37 years; for the 1970 cohort, 27 years; and for the 1980 cohort, 17 years (for a description of the study design, see Bird, Born, & Erzberger, 2000).

The German data provide a retrospective view of how constellations of institutional, normative and economic factors interact with each other across three historical cohorts and how these constellations affect the timing and sequencing of women's initiation of parenthood. The U.S. data offer a more contemporary view of men and women's entry to parenthood, embedded in a distinct institutional and cultural context. Comparing the two data sets enhances understanding of similarities and differences in the coupling of timetables for family formation and the completion of education.

EDUCATIONAL COMPLETION AND ATTAINMENT IN THE U.S. SAMPLE

Pertinent to the institutional hypothesis, Table 1 indicates the attainment benefits associated with delayed educational completion among American youth in the YDS panel. Among those who completed their schooling at or before the age of 22, very few had obtained 4-year college degrees (no one who completed by age 18 and only 16% of those who completed between age 19 and 22). Among the later finishers (aged 23–24), 44% had obtained BA degrees. Of the latest finishers (aged 25–28), 37.5% had obtained college degrees. Nonetheless, the relation between the timing of school completion and educational attainment is far from perfect in the U.S. context. For

Table 1. Highest Degree Obtained by Age of Educational Completion.

Age of Educational Completion	Highest Degree Obtained				
	High school or less (%)	Technical, vocational school or Associate degree (%)	Some college (%)	BA or higher (%)	Row Total (*N*)
By age 18	85.6	7.2	7.2	0.0	100.0 (97)
Traditional college period (age 19–22)	27.6	31.8	24.5	16.1	100.0 (192)
Later finishers (age 23–24)	8.0	23.9	23.9	44.2	100.0 (138)
Latest finishers (age 25–28)	0.0	29.3	33.3	37.5	100.0 (24)
Still in school	12.0	17.9	29.3	40.8	100.0 (184)

Note: Pearson $\chi^2 = 262.48$ ($p<0.001$). Spearman correlation $= 0.49$ ($p<0.001$).

example, among the latest finishers, those who completed their schooling between the ages of 25 and 28, close to 30% were recipients of technical, vocational school or associate degrees and a third had attained some college without receiving the BA. There is similar variability in attainment among the later finishers (age 23–24) and for those who were still attending school in 1999. (Because gender did not condition the relationship between age of educational completion and educational attainment, gender differences are not reported.)

This loose coupling of age of school completion and educational attainment results from a variety of circumstances. Some "older students," that is, older than the "traditional" 18–22 year-old college student, attend school as part of a long-term, uninterrupted progression from high school graduation through college and post-graduate degree programs. Others obtain vocational certification or pursue other kinds of educational programs while working full- or part-time. Still others are educational "returnees," who interrupted their school enrollment one or more times to pursue full-time work, to have children, or for other reasons. The rather weak relationship between age of finishing school and educational attainment in the American context is consistent with an individualization thesis (Buchmann, 1989; Shanahan, 2000). American youth are completing school at different ages, as a result of differences in the structure of their early lives. Because of these individualized patterns, the age of educational completion has varying outcomes with respect to a prominent indicator of socioeconomic status, educational attainment.

We next examine some vocational consequences associated with the timing of educational completion. Table 2 presents average months spent in four work statuses in 1999 by educational completion. These include full-time work (35 h or more), part-time work (less than 35 h), unemployment, and full-time homemaking. At this time, most respondents were 25 or 26 years of age. There is a tendency for those who completed schooling at younger ages to spend more months in full-time work than the oldest finishers (at age 25–26) or those still in school. Those who completed their schooling at older ages and those who were still in school in 1999 spent more months in part-time jobs than their more rapidly completing peers.

Of course, age of school completion and the duration of time since the completion of education are confounded. However, the pattern shown in Table 2 is consistent with the supposition that the more recent educational completers (i.e., latest finishers, aged 25–26) are undergoing some occupational "floundering" and are less likely to have settled into long-term,

Table 2. Months in Four Work Statuses by Age of Educational Completion.

Age of Educational Completion	Mean Months of (0–12)							
	Full-time work		Part-time work		Unemployed		Full-time homemaker	
	Women	Men	Women	Men	Women	Men	Women	Men
By age 18	9.2	9.9	1.4	1.9	0.3	0.5	1.3	0.2
Traditional college period (age 19–22)	9.2	10.0	2.4	1.9	0.6	0.5	1.1	0.3
Later finishers (age 23–24)	8.9	10.9	3.0	2.4	0.4	0.3	1.2	0.0
Latest finishers (age 25–26)	8.8	8.1	3.8	3.5	0.2	0.0	0.5	0.0
Still in school	6.4	7.8	3.9	3.9	0.4	0.1	1.1	0.0
F-test	4.48**	5.67***	3.56**	2.80*	0.66	0.98	0.17	1.03

*$p < 0.05$.
**$p < 0.01$.
***$p < 0.001$.

full-time jobs. Though men are somewhat more likely to be doing full-time work and women to be employed part-time, work patterns, in relation to the age of educational completion, are much the same for both genders. Months of unemployment and full-time homemaking do not differ by age of educational completion for either gender.

Table 3 shows average months in the same four work statuses by highest degree obtained (as of 1999). Men whose highest degree was at an associate level appear at particular risk of part-time work. Young men who have received only a high school degree or less are more prone than other men to unemployment. It is noteworthy how little time, on average, is being spent in full-time homemaking, between 0 and 1.6 months across educational and gender categories. Among women, those who have attained the bachelor's degree spend the least time in full-time homemaking.

Most pertinent to this analysis, the age of educational completion bears a stronger relation than the highest degree obtained to months of full-time work, an important behavioral indicator of work attachment. Surprisingly, months of full-time work are not significantly related to educational attainment for either gender. During the mid-20s, what appears to be more important to the acquisition of stable full-time employment is the amount of time the youth has had, since the completion of schooling, to engage in exploration and job search.

These data provide some indication that American young people do not readily obtain stable, career-type employment after finishing their educations. Because of the weak institutional linkage between school and work in the American context, educational completion does not lead directly to a job in a chosen field. Instead, there is a more or less extended period of exploration, and for some youth, "floundering," in jobs that are not linked to their career goals. As a result, the level of reward associated with the timing of educational completion – in terms of credential level or full-time work – is quite variable and perhaps uncertain for many American young people.

EDUCATIONAL COMPLETION AND ATTAINMENT IN THE GERMAN CONTEXT

In the German educational system (see Fig. 1), school begins at age 6 or 7 and is followed by 4 years of elementary school. At the age of 10 and 11, children are divided into three school types. Those who attend the "Hauptschule" have 5 more years of general schooling. At the age of 15 or

Table 3. Months in Four Work Statuses by Highest Degree Obtained.

Highest Degree Obtained	Mean Months of (0–12)							
	Full-time work		Part-time work		Unemployed		Full-time homemaker	
	Women	Men	Women	Men	Women	Men	Women	Men
High School or less	7.7	9.6	2.3	1.6	0.8	0.7	1.6	0.3
Technical/vocational school	8.6	10.4	2.6	2.0	0.3	0.1	1.2	0.0
Associate degree	9.0	8.3	3.1	4.9	0.0	0.0	1.4	0.0
Some college	8.5	9.1	3.0	2.9	0.4	0.4	1.0	0.2
BA degree or higher	9.2	9.3	3.5	2.9	0.3	0.0	0.3	0.0
F-test	1.13	0.92	1.01	3.11*	1.48	2.50*	2.73*	0.85

*$p < 0.05$.

16, they will seek and start apprenticeships. If their searches are immediately successful, they will finish by the age of 18 or 19. "Realschule" pupils go to school for 6 more years. At age 16 or 17 they will start an apprenticeship and will be age 19 or 20 when they have finished. Those who attend the "gymnasium," leading to higher education at applied or academic universities, attend school for 9 more years following elementary school, to the age of 19 or 20. A program of 4–5 years in duration will follow at applied universities or of 5–6 years at the academic universities. Only 6–8% of German youth enter the labor market as unqualified, without certification (Klammer, Klenner, Ochs, Radke, & Ziegler, 2000, p. 214).

These educational transitions will occur at these ages for youth who successfully navigate standardized pathways. Age variation derives from difficulties in attaining an apprenticeship position in times of labor market downturns. In response to growing labor market risks since the 1970s, two "individualized" educational sequences have gained ground. First, the "Zweiter Bildungsweg""second path," enables graduates from apprenticeships to return to school in order to qualify for entrance to university.

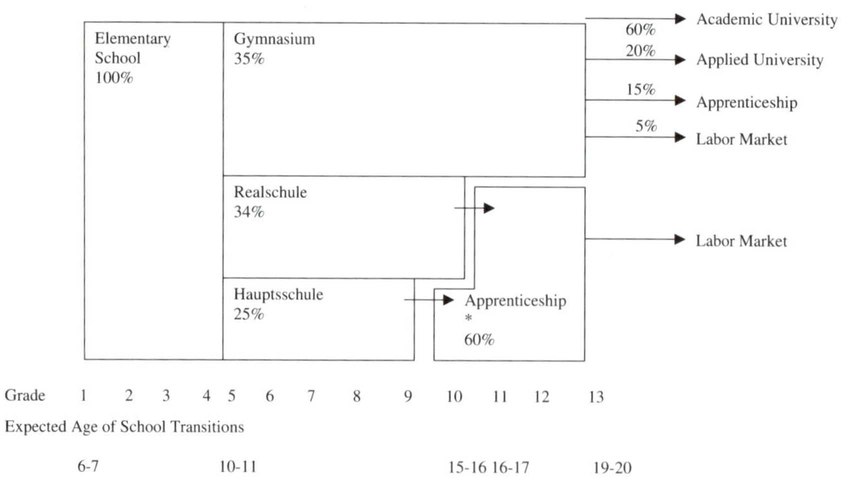

Fig. 1. The German Educational Timeline. *An apprenticeship lasts 3–3.5 years; it can be reduced to 2 years for Gymnasium graduates. *Source*: Federal Ministry for Education and Science (1994). Adapted from: Heinz (1996) Youth Transitions in Cross-Cultural Perspective: School-to-Work in Germany. pp. 2–13 in *Youth Transition: Perspectives on Research and Policy*, edited by B. Gallaway and J. Hudson. Toronto: Thompson Educational Publishing, Inc.

Full-time courses for 1 or 2 years (depending on prior qualification) or evening school qualify participants to attend either an applied or an academic university. If the participants worked for a period of time in the occupational positions for which they trained as apprentices beforehand, they may be considerably older than those who went straight from the "gymnasium" to university. Although they may be considered adults on the basis of their age and labor market participation, they return to the role of student.

Second, in response to labor market downturns in positions requiring academic degrees, an increasing number of youth who complete the "Gymnasium" seek to acquire apprentice certification before proceeding to university. For young people who pursue these alternative routes, the time span between leaving general schooling and entering the labor market will be extended. These individualization processes indicate tensions between institutional pathways and labor market opportunities. The two alternatives function as a buffer against youth unemployment in unstable or shrinking labor markets. They motivate youth to seek more and different qualifications and also disrupt the connection between age grades, educational completion, and early labor market outcomes.

EDUCATIONAL COMPLETION AND PARENTHOOD IN THE U.S. DATA

We now turn to our central focus, the association between the objective timing of educational completion and first parenthood in the two societies. If the timing of parenting were largely institutionally regulated, one would expect most cases of parenthood to follow educational completion in both countries. Table 4 presents a cross-tabulation of educational completion and age of first birth for youth in the YDS. Generally, parenthood occurs at later ages as the age of educational completion increases. Young people who complete their education by the age of 18 are especially likely to have children by their mid-twenties. At this age, almost two-thirds have had a child. Youth who complete their education after age 18 are less likely to have children by this age. For example, the latest finishers (completing at ages 25–28) and those still in school are least likely of all to have children (79% and 72%, respectively, had not yet become parents).

Table 5 shows the relation between the highest degree completed and the age of transition to parenthood. Consistent with the pattern shown in

Table 4. Age of Educational Completion by Age at Birth of First Child.

Age of Educational Completion Row Total (N)	Age at Birth of First Child					
	During high school (age 15–18) (%)	During college period (age 19–22) (%)	After college period (age 23–24) (%)	Age 25–28 (%)	No	children yet (%)
By age 18	16.2	33.3	8.1	6.1	36.4	100.0% (99)
Traditional college period (age 19–22)	4.0	18.6	8.0	10.1	59.3	100.0% (199)
Later finishers (age 23–24)	7.6	13.1	5.5	12.4	61.4	100.0% (145)
Latest finishers (age 25–28)	0.0	4.2	8.3	8.3	79.2	100.0% (24)
Still in school	6.2	13.9	2.1	5.7	72.2	100.0% (194)

Note: Pearson $\chi^2 = 64.10$, df $= 16$ ($p<0.001$). Spearman correlation $= 0.22$ ($p<0.001$).

Table 5. Highest Degree Completed by Age at Birth of First Child.

Highest Degree Completed	Age at Birth of First Child					
	During high school (age 15–18) (%)	During college period (age 19–22) (%)	After college period (age 23–24) (%)	Age 25–28 (%)	No children yet (%)	Row Total (N)
High school or less	13.3	33.0	8.5	6.4	38.8	100.0% (188)
Vocational/technical school or associate degree	10.1	19.6	4.7	11.5	54.1	100.0% (148)
Some college	4.8	15.5	7.1	9.5	63.1	100.0% (168)
BA degree or higher	0.5	2.2	3.2	6.5	87.6	100.0% (185)

Note: Pearson $\chi^2 = 122.34$, df $= 12$ ($p<0.001$). Spearman correlation $= 0.40$ ($p<0.001$).

Table 4, those whose highest degree was high school or less were the most likely to have become parents. Fully 61% of this group were parents. The difference between the least and the most highly educated young people is striking. Fully 88% of those who had obtained a bachelor's degree or higher had not yet become parents. Table 6 presents the results of a finer breakdown, taking into account the exact year of educational completion and age of first parenthood. It shows that most members of the U.S. panel do not have children before completing their educations. This is true of 73% of the women and 88% of the men. Very few young people (<3%) complete their educations and become parents in exactly the same year. Notably, males are less likely to have children prior to the completion of their educations (24% of women versus 11% of men have children before they finish schooling).

The patterns observed in the U.S. data are consistent with a dynamic of institutional structuring. Young people who are still in school are less likely to be economically self-sufficient and less capable of providing economic support for children. However, the data are also in accord with a normative rule that prescribes educational completion prior to parenthood. While some in the "not yet completed education" group may never become parents, we can still conclude that they have breached neither an institutionally regulated nor a normatively regulated sequence of finishing school prior to parenting. For young men, institutional regulation may be more pronounced as a result of their better opportunities in the labor market. For them, there are greater material incentives to complete their education before becoming parents. Since males are traditionally expected to be the primary breadwinners, there may also be stronger normative constraints for males to complete their educations prior to having children.

Table 6. Birth of First Child in Relation to Educational Completion, by Gender.

Birth of First Child	Total (%)	Women (%)	Men (%)
Prior to completion of education	18.5	23.8	10.9
In the same year	2.1	2.8	1.1
After completion of education	58.2	53.6	64.7
No children and still in school	21.2	19.7	23.3
Total	100.0	100.0	100.0
N	661	386	275

EDUCATION–WORK BRIDGES AND FAMILY FORMATION IN GERMANY

The German female life course study documents striking effects of occupation-specific apprentice qualification on the female trajectories of work and family life. The strong effect of the educational/occupational qualification system on the subsequent female life course is evident in the timing of marriage and childbirth. It is especially noteworthy that only 11 of 1,665 women (0.7%) had children prior to the completion of their apprenticeships. Unlike in the American context, where almost one of four women did so, having a child while still in school is a rare event in Germany.

In contrast to the American pattern of exploration and sometimes "floundering" after finishing education, all women in the German study obtained their occupational entry positions within 6 months of finishing their apprenticeships. This constitutes strong evidence that the highly institutionalized school-to-work bridge was capable of "delivering" to those who completed their educational certifications. However, there is also a strong relationship between type of occupational certification and the capacity to sustain stable employment in the field for which one is qualified. This is apparent by dividing the accumulated years of employment for women into two categories: the years employed in the occupation corresponding to the woman's apprentice qualification (which implies commensurate wages) and in other jobs (usually providing lower wages).

In the aggregate, the women spent about 60–75% of their potentially available time actually employed during the time span between the end of their apprenticeship training and the time this study was conducted.[4] Irrespective of cohort, occupations are clearly differentiated in their capacity to maintain qualified employment (and pay). A comparison of means (results not shown) reveal significant differences across occupations in accumulated years of work experience in the field of initial qualification and in other occupations.[5] In the aggregate we can differentiate between two groups of occupations: those with a high retentive power and those with a low retentive power. Women in the first group spent a greater portion of their total time in the labor force employed in the fields for which they were trained (see Table 7).

Before addressing the effects of occupational retention capacity on birth rates, it is important to situate German family formation in its historical context. The postponement of parenthood is a general cohort trend throughout the post-war period (Nave-Herz, 1988). Delayed family

Table 7. Percent of Time in the Labor Force Employed in the Occupation Trained for and in Other Occupations.

	Percent of Time in Initial Occupation	Percent of Time in Other Occupations	Total
High-retention occupations			
Nurse	62	13	75
General office worker	53	17	70
Bank office worker	49	16	65
Industrial office worker	48	20	68
Low-retention occupations			
Hairdressers	34	30	64
General retail	34	31	65
Food retail	34	25	59
Hotel specialist	30	34	64

formation has also been linked to historical periods of labor market downturn and the fact that employment discontinuity is quite disadvantageous in the German occupationally structured labor market (Brinkmann, 1980; Engelbrech, 1991). In our data, the age of first childbirth increased by roughly 2 years between the 1960 and the 1980 cohorts, among women who had children (see Table 8). Furthermore, in all three cohorts the occupational groups whose certifications lead to longer tenure in qualifying occupations are also those in which marriage and childbirth occur later. The women who worked in high retention occupations in the 1960 cohort gave birth to their first child at age 25.9, on average. In contrast, those in low retention occupations in the 1960 cohort gave birth to their first child at age 23.8. Thus, differences between occupational groups in the timing of family formation remain the same across historical time.

Consistent with the "institutional determination hypothesis," the timing of marriage and childbirth occurs in response to particular occupational qualifications and career trajectories. Differences in the timing of the first birth result from two specific developments. The first is the change in opportunities in access to apprenticeships since the mid 1970s. The reduced supply and increasing competition for apprenticeship placement have made it advantageous to have greater educational qualifications to obtain entry to the more attractive apprentice positions. For example, in our data, only very few members of the oldest cohort (3%) or the middle cohort (1.9%), but nearly one-fifth (19.2%) of the 1980 cohort attained the "Abitur," the

Table 8. Average age at first birth (years).

	Apprenticeship Completion Cohort					
	1960		1970		1980	
	Mean	Standard deviation	Mean	Standard deviation	Mean	Standard deviation
Age at first birth (all women)	25.2	3.9	25.8	4.7	27.0	4.2
High-retention occupations	25.9	3.8	26.8	4.6	27.6	4.1
Low-retention occupations	23.8	3.6	25.3	5.3	25.7	3.9
Difference between high- and low-retention	2.1***		1.5**		1.9***	

$^{**}p<0.01$.
$^{***}p<0.001$. ($H_o : \mu_{high} = \mu_{low}$, $H_a : \mu_{high} > \mu_{low}$).

secondary degree following completion of the gymnasium, which – despite their older age – gave them a competitive advantage over school-leavers with lower qualifications.[6]

It is no coincidence that the office occupations with higher entrance requirements are also among those with a high retentive power, whereas the occupations in a retail environment have low retentive power. The Abitur has become an important criterion for gaining an apprenticeship in an attractive office occupation (in our sample the general office workers, industrial office workers, bank office workers, and wholesale and export office workers), but it has remained largely irrelevant for less attractive occupations (the general merchandise and food retail saleswomen, and hairdressers).[7] The link to family formation is clear: because it takes longer to obtain the Arbitur and then to complete an apprenticeship than to take the traditional apprenticeship route right after secondary school, women who have obtained this qualification will be older upon completing their educations and both labor force entry and family formation will be delayed (See Table 9).

The second development derives from labor market incentives. Most women attempt to achieve a secure and stable position in the labor market before they have a child (Horstmann, 1996). In this way, they better their

Table 9. Percent of Apprentices Who Attained Abitur before Entering an Apprenticeship.

	Cohort		
	1960	1970	1980
Office occupations	5.3	3.1	32.7
Hairdressing, retail occupations	1.9	0.0	5.0
Cohort average (all occupations)	3.0	1.9	19.2

chances for further promotion and increase the likelihood that they will be able to return to their jobs after taking time off for motherhood. Considering all occupational groups together, the median duration between entering the labor market and first childbirth rose from 6.2 years (1960 cohort) to 7.1 years (1980 cohort). The proportion of women without children (by the time of the 1997 survey administration) also rose in the same period from 18.9% to 23.4%. Still, much variation was observed between occupations. Those who had qualified for occupations with opportunities for career advancement within their firms delayed parenthood, or remained childless. In contrast, those women who had poorer occupational prospects, or poor working conditions, were more likely to opt for earlier parenting. The effects of the escalation of educational credentials and occupational career structures on family formation can be illustrated by three occupations with contrasting push and pull factors: the bank office workers (high retentive power, high educational investment, and high career opportunities), the hairdressers (low retentive power, low educational investment, and low career opportunities), and the nurses (high retentive power, middle educational investment, and changing career opportunities over historical time). The differences in family formation for these three vocational groups are summarized in Fig. 2. The bars show the median duration in years between completing the apprenticeship and the birth of the first child in each cohort. The superimposed line shows the proportion of women who trained in each occupation but did not have any children by the date of the 1997 survey.

Consider first the bank office worker. In the 1980 cohort, a very large proportion of the bank office workers started their apprenticeship with the Abitur (63.9%) and delayed family formation or did not enter motherhood. This occupation provides clear incentives to postpone children to take advantage of occupational advancement opportunities and good long-term prospects. Furthermore, heavy investment in work is expected for those who

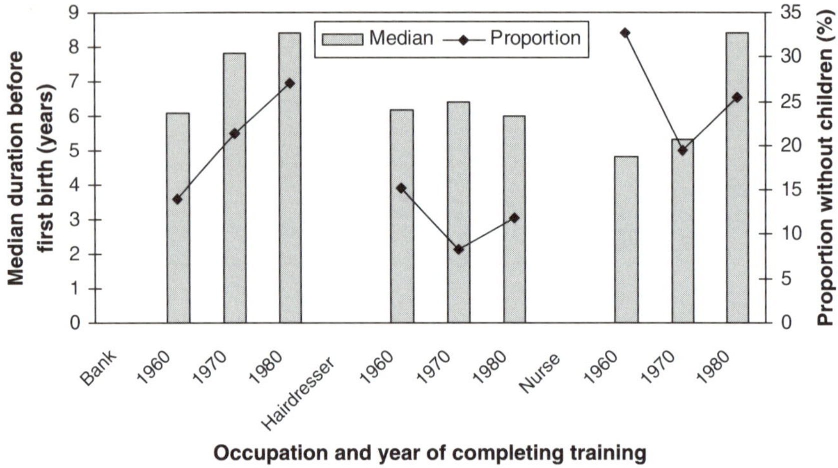

Fig. 2. A Comparison of the Median Duration between Ending Apprenticeship and First Birth, and Proportion of Women Without Children for Three Occupations.

wish to be considered for advancement. This includes expectations of overtime, geographic mobility, and travel for jobs and conventions. Here, motherhood is not a desirable alternative role but rather an occupational risk that threatens advancement. Not only has motherhood been increasingly postponed within this group (the median duration between completing training and first birth is 8.4 years in the 1980 cohort), but the proportion of women in banking without children has risen from 14% in the 1960 cohort to 27% in the 1980 cohort.

The second example is the hairdresser, representative of low-retentive occupations with little advancement opportunity. In all three cohorts there was only one hairdresser with the Abitur (in the 1960 cohort); hairdressers take a mid-level position in their speed of family formation, but are more likely to become mothers than most other women in the sample. For hairdressers, the escape to early motherhood after 6–8 years employment is a good alternative to poor long-term prospects in the labor market. The birth of a child is seen as a preferable alternative to paid work. This results in the low proportion (8–15%) of women trained in this occupation who do not have children.

The third illustration is nurse, the occupation with the highest retentive power and a lower than average proportion of women with the Abitur (13.1% in the 1980 cohort). In the 1960 and 1970 cohorts, nurses started their families most quickly after receiving their vocational certification, and then returned to work as nurses.[8] This is an occupation that had rather little advancement opportunity, but its variable and negotiable work schedule made it possible to integrate the work and mother roles. Motherhood was not perceived as a conflicting role as both work and parenting could be combined. In the 1960 and 1970 cohorts, nurses tended to have children relatively quickly (between 5 and 6 years after qualifying). With the intensification of work, new options for advancement in this field, and changes in the ratio of doctors and nurses, family formation patterns changed rapidly during the 70s. We observe in the 1980 cohort that nurses exhibited a much more lengthy duration of time between completing education and the first birth, typical of bank office workers.

These patterns, across occupations and over time, show the extent to which institutional structures, specifically the qualification system and the availability of occupationally structured career paths, influence the timing of family formation in the German context. The general historical trend, along with the maintenance of differences by type of apprentice qualification, do not seem attributable to cultural (or subcultural, occupationally specific) change in age norms. Instead, they are plausibly viewed as responses to change in educational/vocational training and occupational opportunities and constraints.

Changes in fertility timing after the unification of the former East and West Germany provide further support for the institutional thesis. In the former East Germany, everyone had the legal right to work; therefore, obtaining a job after completing education was guaranteed by the State. Women had children while they were still attending training and educational programs. State welfare programs, providing housing for families with children and full-time child care, made it possible for young mothers to continue their educations, while moving out of the parental home into their own residences. In fact, giving birth was an almost exclusive means of getting one's own apartment. Thus, there was a clear material incentive in favor of early fertility.

In 1980, prior to reunification, the mean age of first birth was 21 years in the former East Germany, while it was 26 in the West (Bundesministerium für Familie, Senioren, Frauen, Jugend, 1997; Dobritz & Gärtner, 1995). Ten years following reunification with West Germany, and the elimination of the State's employment guarantee along with support for housing in the former East, the age of birth of the first child in the former East had reached the mark of 28.4. Initially, it appeared that East German women were engaging in a "birth strike." It soon became clear, however, that the decline resulted from a pervasive postponement of marriage and childbirth (Kreckel &

Schenk, 2001). After reunification, East German women quickly adopted West German family patterns (Klammer et al., 2000). This recent historical trend supports an institutional interpretation, rather than one featuring age norms.

SUMMARY AND DISCUSSION

It is apparent that in both the United States and Germany the timing of educational completion (and vocational training) is rewarded by objective educational and occupational attainments. In general, youth in both societies who prolong their educations receive higher-level, or more diverse, credentials, yielding higher socioeconomic attainment later in life. Thus, in accord with the institutional structuring hypothesis, there are substantial incentives to avoid interrupting educational programs by childbirth in both societies.

Since the economic advantages accruing to college-trained young people in the United States are great, youth who obtain bachelor's degrees are especially apt to postpone family formation, especially parenthood, until their educations are complete. In general, youth who stay in school longer have children at older ages. Weak school-to-work bridges, however, combined with the considerable individualization of early educational trajectories in the U.S., generate much variability in the YDS in educational attainment among young people who complete their educations at the same age. Moreover, in the contemporary American context, many youth hold jobs that are unrelated to their career goals following the completion of their educations. Many American youth require a period of exploration before they locate work that they see as having long-term career prospects. In fact, among YDS participants, time since educational completion was more strongly related than the level of educational attainment to a behavioral measure of career establishment, months in full-time work. Such patterns could undermine the incentive to complete educational programs without disruption by parenthood.

Given the strong institutional structuring of age-graded educational pathways, there appears to be a stronger tendency to delay parenting until education and labor market establishment is complete in Germany than in the American context. Postponement of parenthood is observed in Germany, like the United States, as more young people extend their educations. Though age grading in educational trajectories is more firmly regulated in Germany, increasingly individualized patterns are also emerging. The

"Zweite Bildungsweg" and the pathway from the "Gymnasium" to the apprenticeship, both of which enable young people to heighten their vocationally relevant credentials in the face of increasing economic uncertainty, are key examples.

With stronger institutional bridges from school to work, occupational rewards and labor market placement are more closely tied to the completion of education in Germany than in the U.S. The German women in the three cohorts studied were uniformly successful in obtaining work following their apprenticeships. While almost none became parents prior to completing their apprenticeship programs, variation in the timing of parenthood was clearly linked to occupational prospects linked to particular apprentice certifications. Those in the "better" apprenticeships, more closely linked to long-term employment in the field of qualification and opportunities for advancement, waited longer to have children than those who completed less promising apprentice certifications.

This study indicates that the structures which bridge education and work during the transition to adulthood are highly consequential for life course decision-making regarding family formation. This is evident in the U.S. sample, since only a minority of youth became parents prior to completing their educations (about one-fourth of women and one of ten men), and, in the German sample, where almost no one did so. Educational programs, the timing of career consolidation, and the availability of alternate pathways (such as the "Zweite Bildungsweg" and the apprenticeship following "Gymnasium" completion) are most clearly reflected in the timing of family formation in Germany.

Given the increasingly prolonged educational pathways to gainful and promising labor market positions, one might expect to find historical trends in the timing of childbearing in relation to educational completion. But the normative and institutional hypotheses would yield different predictions about the patterns of change. For example, in Germany, if age norms primarily regulated the timing of family formation, one would expect to find increasing numbers of births occurring prior to the completion of apprenticeship programs, and prior to establishment in the career. That is, if age norms are rather stable, age of first birth would increasingly "lag behind" institutional pressures toward longer schooling and more extensive educational qualification. The fact that the age of family formation has instead moved upwards in an historical era marked by educational extension as well as periodic, tightening labor markets, disconfirms the notion that age-based norms govern the transition to parenthood. If, in contrast, norms were linked to sequencing and not timing (i.e., one should not have children prior to educational completion irrespective of age), the observed pattern would be predicted.

Still, the observed trends in family formation are clearly consistent with an institutional argument. Women (and men) increasingly pursue alternative education–career sequences in adaptation to economic uncertainties. These new institutional patterns come to govern the timing of entry to marriage and parenthood, just as the more traditional school to work patterns did in a prior era. Finally, the observed pattern of change in fertility patterns among the former East German women is difficult to explain by adherence to age norms, as these are likely to remain rather stable over time.

While we cannot firmly adjudicate between these perspectives in assessing the German historical trends, the cross-national comparison reveals the significance of social structure for family formation. While parenthood follows educational completion in both countries, the link is much stronger in Germany, with its strong school-to-work bridge, than in the loose coupling regime characteristic of the United States. This research thus indicates the interdependence of institutional structures, individual aging, and the objective timing of key life events – educational completion, acquisition of full-time work, and parenthood – as young people make the transition to adulthood.

NOTES

1. Some respondents did not complete the survey every year and some had missing data for the school attendance question; those cases with at least five data points (1992–1999) were included. Respondents without valid data for the years after our last record of school attendance were not included because it was not possible to determine whether they were in school in subsequent years. (Although "completing school" in any 1 year does not rule out subsequent re-entry, for the purposes of this analysis those who finish schooling and do not return to the classroom during the period of observation are considered to be "completers.") Using these decision rules, the age of educational completion (including those classified as "not yet" having completed) was computed for 717 respondents (71% of all cases); 65 cases had no valid data in any year (from 1992 onwards), and for 228 cases, the record was incomplete.

2. These include medical/dental assistant ("Arzt-/Zahnarzthelferin"), bank office worker ("Bankkauffrau"), office worker ("Bürokauffrau"), general merchandise retail saleswoman ("Einzelhandelskauffrau"), saleswoman in food retail ("Nahrungsmittel-Fachverkäuferin"), hairdresser ("Frisörin"), nurse ("Krankenschwester"), wholesale and export trade office worker ("Gross - und Aussenhandelskauffrau"), hotel industry specialist ("Hotelfachfrau"), and industrial office worker ("Industriekauffrau").

3. The addresses of potential respondents were randomly selected from occupational board organizations ("Kammern") throughout West Germany (before unification) and updated through city registration offices. The board organizations (industry and trade, crafts, medical, etc.) provide final board certification and keep records of the graduates' gender, name, and address.

4. Reductions occur from periods of nonemployment between the completion of apprenticeship and the attainment of occupational entry positions and interruptions in employment due to family, caregiving, unemployment, illnesses, household moves, etc.

5. The nurses, general office workers, bank office workers, and industrial office workers, worked significantly longer in their initial occupation than the hairdressers, general merchandise retail sales personnel, food retail saleswomen, and hotel industry specialists ($p < 0.01$) or the doctor's assistants and wholesale and export office workers ($p < 0.05$).

6. Women with Abitur would have entered the labor market at ages 21–23, whereas those who started their apprenticeships after completing "Realschule" would have entered the labor market at the ages of 18–20.

7. Restrictions on opening hours in the retail sector translate into rigid working times for employees that are perceived as detrimental to family life (evenings and Saturdays).

8. The high proportion of nurses without children in the 1960 cohort (32.7%) can largely be accounted for by the norm of celibacy that nurses were expected to adhere to at that time.

ACKNOWLEDGMENTS

The Youth Development Study is supported by grants (titled "Work Experience and Mental Health: A Panel Study of Youth") from the National Institute of Child Health and Human Development (HD44138) and the National Institute of Mental Health (MH42843). The inspiration for this chapter came from work initiated while the first author was a Fellow at the Center for Advanced Study in the Behavioral Sciences. She is grateful for support provided by the Center, as well as by the Hewlett Foundation and the W.T. Grant Foundation (Grant 95167795). Sabrina Oesterle was supported by a Fellowship from the University of Minnesota Graduate School while working on this project. The Bremen Female Life Course Study was sponsored by the Deutsche Forschungsgemeinschaft and is part of the Bremen Sonderforschungsbereich 186 on Status Passages and Risks in the Life Course.

REFERENCES

Arnett, J. (2000). Emerging adulthood: A theory of development from the late teens through the twenties. *American Psychologist*, *55*, 469–480.

Bertram, H., & Hennig, M. (1996). Das katholische Arbeitermädchen vom Lande: Milieus und Lebensführung in regionaler Perspektive. In: A. Bolder, H. Heid, W. Heinz, G. Kutscha, H. Krüger, A. Meier & K. Rodax (Eds), *Die Wiederentdeckung der Ungleichheit: Aktuelle Tendenzen in Bildung für Arbeit. Jahrbuch '96 Bildung und Arbeit* (pp. 229–251). Opladen: Leske und Budrich.

Bird, K., Born, C., & Erzberger, C. (2000). *Ein Bild des eigenen Lebens zeichnen.* Der Kalender als Visualisierungsinstrument zur Erfassung individueller Lebensverläufe. Arbeitspapier Nr. 59, Sonderforschungsbereich 186, Universität Bremen.

Born, C. (2000). Erstausbildung und weiblicher Lebenslauf: Was nicht nur junge Frauen bezüglich der Berufswahl wissen sollten. In: W. R. Heinz (Ed.), *Übergänge – Individualisierung, Flexibilisierung und Institutionalisierung des Lebensverlaufs* (pp. 50–65). Weinheim: Juventa Verlag.

Brinkmann, C. (1980). Erwerbsbeteiligung und Arbeitsmarktverhältnisse: Neue empirische Ergebnisse zur "Entmutigung" und zusätzlichen "Ermutigung" von weiblichen Erwerbspersonen. *Beiträge zur Arbeitsmarkt- und Berufsforschung*, Bd. 44. Nürnberg: IAB.

Buchmann, M. (1989). *The script of life in modern society: Entry into adulthood in a changing world.* Chicago: University of Chicago Press.

Bundesministerium für Familie, Senioren, Frauen, & Jugend (Eds) (1997). *Die Familie im Spiegel der amtlichen Statistik.* Bonn: Kohlhammer.

Dobritz, J., & Gärtner, K. (1995). Bericht 1995 über die demographische Lage in Deutschland. *Zeitschrift für Bevölkerungswissenschaft, 20*, 339–448.

Engelbrech, G. (1991). Frauenspezifische Restriktionen des Arbeitsmarktes – Situationsbericht und Erklärungsansätze zu Phasen des Berufsverlaufs anhand von IAB-Ergebnissen. In: K. U. Mayer, J. Allmendinger & J. Huinink (Eds), *Vom Regen in die Traufe: Frauen zwischen Beruf und Familie* (pp. 91–118). Frankfurt and New York: Campus.

Hamilton, S., & Hurrelmann, K. (1994). The school-to-career transition in Germany and the United States. *Teachers College Record, 96*, 329–344.

Hamilton, S. F. (1990). *Apprenticeship for adulthood: Preparing youth for the future.* New York: Free Press.

Heinz, W. R. (1999). Job-entry patterns in a life-course perspective. In: W. R. Heinz (Ed.), *From education to work: Cross-national perspectives* (pp. 214–234). New York: Cambridge University Press.

Horstmann, S. (1996). *Kindererziehung und Alterssicherung: Verteilungspolitische Aspekte ausgewählter Reformvorschläge zu einer familienorientierten Ausgestaltung der gesetzlichen Rentenversicherung.* Grafschaft: Vektor-Verlag.

Kerckhoff, A. (2000). The transition form school to work. In: J. T. Mortimer & R. Larson (Eds), *The changing adolescent experience: Societal trends and the transition to adulthood* (pp. 52–87). New York and Cambridge: Cambridge University Press.

Klammer, U., Klenner, C., Ochs, C., Radke, P., & Ziegler, A. (Eds) (2000). *WSI – Frauen daten report.* Berlin: Edition Sigma.

Kreckel, R., & Schenk, S. (2001). Full time or part time? The contradictory integration of the East German female labour force in unified Germany. In: V. W. Marshall, W. R. Heinz, H. Krüger & A. Verma (Eds), *Restructuring work and the life course* (pp. 159–176). Toronto: University of Toronto Press.

Krüger, H. (1999). Gender and skills: Distributive ramifications of the German skill system. In: P. Culpepper & D. Finegold (Eds), *The German skills machine: Sustaining comparative advantage in a global economy* (pp. 189–227). New York and Oxford: Berghahn Books.

Krüger, H. (2001). Social change in two generations: Employment patterns and their costs for family life. In: V. W. Marshall, W. R. Heinz, H. Krüger & A. Verma (Eds), *Restructuring work and the life course* (pp. 401–423). Toronto: University of Toronto Press.

Marini, M. (1984). Age and sequencing norms in the transition to adulthood. *Social Forces*, *63*, 229–244.

Mortimer, J. T. (1992). Adulthood. In: E. Borgatta & M. Borgatta (Eds), *Encyclopedia of sociology* (pp. 1–12). New York: Macmillan.

Mortimer, J. T., & Aronson, P. (2000). Adulthood. In: E. F. Borgatta & M. Montgomery (Eds), *Encyclopedia of sociology*, (2nd ed.) (pp. 25–41). New York: Macmillan.

Mortimer, J. T., & Krüger, H. (2000). Pathways from school to work in Germany and the United States. In: M. Hallinan (Ed.), *Handbook of the sociology of education* (pp. 475–497). New York: Plenum Publishers.

Nave-Herz, R. (1988). *Kinderlose Ehen: Eine empirische Studie über die Lebenssituation kinderloser Ehepaare und die Gründe für ihre Kinderlosigkeit*. Weinheim: Deutscher Studienverlag.

Neugarten, B. L., & Hagestad, G. (1976). Age and the life course. In: R. H. Binstock & E. Shanas (Eds), *Handbook of aging and the social sciences* (pp. 35–55). New York: Van Nostrand Reinhold.

Seidenspinner, G., & Burger, A. (1982). *Mädchen '82*. München: Verlag Deutsches Jugendinstitut.

Settersten, R. A. (1999). *Lives in time and place: The problems and promises of developmental science*. Amityville: Baywood Publishing Company.

Shanahan, M. A. (2000). Pathways to adulthood in changing societies: Variability and mechanisms in life course perspective. *Annual Review of Sociology*, *26*, 667–692.

TRIALS AND TRIBULATIONS IN COUPLING CAREERS

Shin-Kap Han

ABSTRACT

Focusing on how men and women's work and family careers are coupled, I put forth an analytical framework that integrates the seemingly disparate empirical evidence on the relationship between married women's employment and divorce into a more layered picture. The "coupled careers interface" framework incorporates three features. First, it is anchored at the couple level. Second, it specifically incorporates contingencies in outcome. Third, it focuses on the dynamics of the careers and their connections from the life course perspective. Combining sequence analysis and log-linear analysis techniques, I illustrate this framework using life history data from married couples in upstate New York. The results show that, depending on the ways in which careers are coupled, some marriages benefit from the dual-earner configuration, while others suffer. This counters the presumption of uniform trade-off. Also found is the asymmetric interdependence between men and women's work careers, especially in the context of the parallel family careers involving children.

Trends in female labor force participation rate and divorce rate effectively capture the direction and magnitude of the transformations in work and

The Structure of the Life Course: Standardized? Individualized? Differentiated?
Advances in Life Course Research, Volume 9, 205–236

ISSN: 1040-2608/doi:10.1016/S1040-2608(04)09008-2

family that have occurred over the last few decades. Noting the strong correlation between the two, many have concluded that the two are connected, which bears theoretically upon the fundamental interdependence between the two life domains (e.g., Davis, 1984; Cherlin, 1992; Popenoe, 1993).

In light of this, the discord of late around the present and future states of work and family is puzzling. The diagnoses vary widely and are often sharply at odds with each other. They range from buoyant optimism in "Two-Income Families Are Happier, Healthier, and Better Off" (Barnett & Rivers, 1996) to the gloom-and-doom scenario in "The Great Disruption" (Fukuyama, 1999) and from debunking the anti-marriage myths in "The Case for Marriage" (Waite & Gallagher, 2000) to suggesting a radical reconfiguration in "When Work Becomes Home and Home Becomes Work" (Hochschild, 1997). Mass media accounts are equally polarized, ranging from depictions of "relatively happy and stable" dual-doctor marriages, "Doctor Marries Doctor: Good Medicine" (Jauhar, 1999) to the woe brought on by combining two academic jobs in marriages, "Strange Bedfellows" (Boufis, 1999).

This chapter focuses on how men and women's work and family careers are coupled. It develops and illustrates a framework, in which these seemingly disparate observations could be arrayed to form a more layered picture. The question being asked is not *whether* women's employment hinders or helps marriage, but *when* and *where* it helps and *when* and *where* it hinders. Three features of the framework are worth noting. First, the analysis is anchored at the couple level rather than at the aggregate or individual level. Second, the connection is examined from the life course perspective, emphasizing the underlying dynamics of the careers and their linkages. Third, I pay attention to the various ways in which those careers could be linked and their varied consequences. The framework articulates these features by utilizing the models and methods of log-linear and sequence analysis techniques.

COUPLING WORK AND FAMILY: A FRAMEWORK

The need to focus on the specific intersections between work and family as connected organizers of experience, social relations, and life chances is greater than ever today (Kanter, 1977; Moen & Wethington, 1992; Abbott, 1993; Parcel & Cornfield, 2000). Recent reviews of the field hence strongly emphasize the issue as a priority. Menaghan (1991), for instance, draws attention to perspectives that link individual behavior and outcomes in one set of roles to experiences in other roles and to their historical and social contexts. Rosenfeld (1992) too takes up the issue, pointing out the neglect of how one's work career, especially women's, is affected by other life domains and the life course of other family members. There have been considerable effort to theoretically and empirically identify this connection (e.g., Elder, 1974, 1995; Oppenheimer, 1974, 1977; Felmlee, 1982; England & Farkas, 1986; Mayer & Tuma, 1990; Bielby & Bielby, 1992; Brines, 1994; Blossfeld, Drobnič, & Rohwer, 1996; Bernasco, de Graaf, & Ultee, 1998; Brines & Joyner, 1999).

The framework developed here extends prior work by directly confronting the complexity of the work–family connection and heterogeneity in its outcomes. It draws on three ideas. First, I anchor the analysis at the couple level, where the core dynamics of work–family interface operate and its distinct and emergent features can be more readily observed (England & Farkas, 1986; Waite & Lillard, 1991; Blossfeld et al., 1996; Bernasco et al., 1998; Brines & Joyner, 1999; Quick & Moen, 1999; Han & Moen, 1999a, 2001; Sørensen, 1994).[1] Second, the temporal dimension underlying the connection between work and family is explicitly considered from the life course perspective (Kanter, 1977; Moen & Wethington, 1992; Rosenfeld, 1992; Abbott, 1993; Elder, 1995). Oppenheimer's "life-cycle squeeze" (1974; Wilensky, 1963) and Waite's "family life cycle" (1980), for example, clearly illustrate that the ways in which the connection is structured render it highly contingent on the timing and synchronization of the transitions and trajectories in both work and family. The concept of career, as operationalized by the sequence analysis technique, charts the overall patterning and the connection between the two as they unfold over time (Wilensky, 1961; Spilerman, 1977; Abbott & Hrycak, 1990; Abbott, 1995; Han & Moen, 1999b; South, 2001). Third, the framework takes into account the possibility that the connection between work and family may not be uniformly structured across all couples (Tilly & Scott, 1978; Bielby & Bielby, 1992; Abbott, 1993). Also, both positive and negative effects might be at work side by side, which requires a model that allows heterogeneous effects (Oppenheimer, 1977; Gerson, 1985; England & Farkas, 1986; Menaghan, 1991; Moen, 1992; Moen & Wethington, 1992; Rosenfeld, 1992). These elements are articulated below to form an integrated framework.

Coupled Careers Interface

I draw largely from the prior work that emphasizes differentiated roles and the ways in which those roles are related to each other in the substantive context of work and family (Goode, 1964; Oppenheimer, 1977, 1988; England & Farkas, 1986; Moen, 1992). Placed at the core of the connection between work and family is a *two-person-two-role-domain interface* within which a man and a woman are engaged in role negotiations with each other over time (see Fig. 1).

First, each of the four interfaces – ①:②, ②:③, ③:④, and ④:① – is postulated as an active interaction between the two sides, not just as a "spillover" from one to the other (England & Farkas, 1986; Bernasco et al., 1998). These interfaces, however, are not required to be competitive zero-sum games. Second, it takes into account the differences between men and women and between work and family domains. The interface between work role and family role for men need not be identical with that for women and the interface between two work roles may not necessarily be symmetric with that between two family roles (e.g., Sørensen & McLanahan, 1987; Bielby & Bielby, 1992; Brines, 1994). Third, it explicitly recognizes that the interfaces are interrelated, conditioning each other directly and indirectly. Some links may be more important than the others, but in the end it is the chain as a whole that matters in shaping the connection between work and family. Hereafter, this whole complex is referred to as the "coupled careers interface."

There are numerous ways in which these roles and interfaces can be configured. Configuration A in the bottom panel of Fig. 1, for instance,

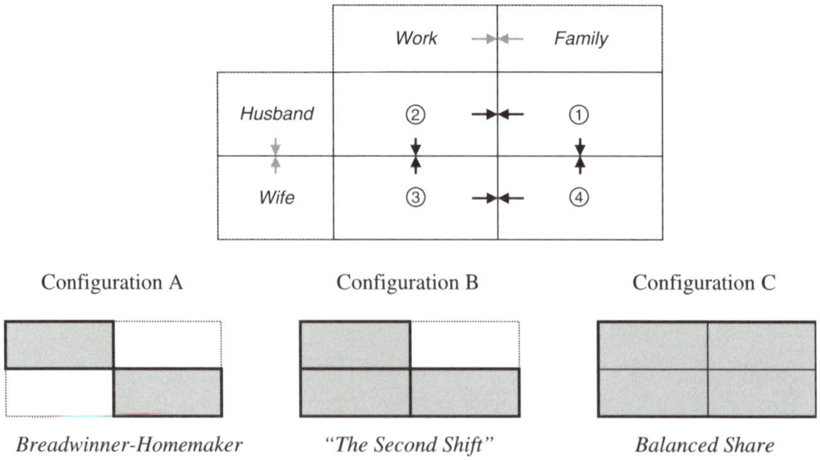

Fig. 1. The Interface Structure and Its Various Configurations.

shows the traditional husband-breadwinner (②) and wife-homemaker (④) interface structure where the two domains of work and family are sharply separated along gender lines (Kanter, 1977; Moen, 1992). The key element simplifying role allocation is segregation. This sexual division of labor is, for some, a way to deal with the tension between the two "greedy institutions" (Coser, 1974) and is presumed to be an "efficient" solution, "maximizing" the benefits of division of labor and comparative advantage (Becker, 1981).

However, the rapid growth in the female labor force participation, especially among married women, has significantly undermined the normalcy of this configuration (Bernard, 1982 [1972]; Blumstein & Schwartz, 1983; Sørensen & McLanahan, 1987; Skolnick, 1991). The "dual-earner couples" overtook traditional breadwinner–homemaker families in number as early as the mid-1970s (Hayghe, 1990) and this trend has continued through the 1980s and 1990s. By 1998, there were over 30 million dual-earner households in the United States, outnumbering the so-called "traditional" households nearly three-to-one (Clarkberg & Moen, 2001; U. S. Bureau of the Census, 1998).

Not all dual-earner couples are alike though. Configuration B illustrates the circumstance of many married employed women who, bearing a disproportionate share of domestic work in addition to their paid work, confront the strains imposed by a new work–family interface. This is the predicament that Hochschild (1989) calls "the second shift," which is at the basis of "the myth of the miserable working woman" (Barnett & Rivers, 1996; Walsh, 1995).[2] Finally, configuration C is what one of the respondents in Hood's (1983) study of two-job families describes as: "It's more of a 50/50 deal. We're both providers and we're both homemakers and we're both parents." At least in terms of equity, this might be an ideal situation, variously called, "new families" (Waite & Gallagher, 2000), "the collaborative couple" (Barnett & Rivers, 1996), and "the egalitarian system" (Davis, 1984).[3]

A Double-Edged Sword: Compatible or Incompatible Coupling?

As discussed earlier, the parallel increases in the employment of married women and in divorce strongly suggest a link between the two. The question is how to describe and explicate this link. There are a few plausible specifications for the relationship. One line of reasoning is to see it as a direct cause-and-effect link.[4] Few believe, however, that women's employment is a direct cause of marital dissolution. Presently, the dominant view is that women's growing economic independence is the major factor in the rise in marital instability (Espenshade, 1985; Goldscheider & Waite, 1986).[5] It is indeed a highly plausible case, "stronger and more suggestive than that linking any other concurrent trend with the rise in divorce" (Cherlin, 1992, p. 53).

Still the evidence remains mixed. On the one hand, wives' employment can strain the marriage by itself or bring about its dissolution when other conditions are present. On the other, the earnings she brings home, for example, can ease her family's financial burden and reduce the tensions of economic hardship or help equalize the balance of power at home and thereby reduce the likelihood of marital dissolution. Her work might also bring her increased personal satisfaction, which would, in turn, improve her relationship with her husband (Cherlin, 1992; Barnett & Rivers, 1996; South, 2001).

In order to sort out and reconcile these mixed expectations, I suggest a two-step model with the coupled careers interface as a mediator between the two. The model illustrated in Fig. 2 postulates that the effect of wife's employment on marital stability is in part contingent on the interface. The observed association between wife's employment and marital stability (α) can then be decomposed into two parts ($\beta\gamma$ and α'), where α', β, and γ denote the paths in the figure. The value of $\beta\gamma$ will depend on what kind and how much of a role the coupled careers interface plays. The value of α' can be equal to that of α only if $\beta\gamma = 0$; that is, if the interface is irrelevant.

Although it is possible to specify the variation as gradational, I assign for the sake of parsimony two discrete values for the coupled careers interface: incompatible ($r = j$) and compatible ($r = k$). The mediated effect will be positive for the compatible interface ($\beta_k\gamma_k > 0$), whereas it will be negative for the incompatible interface ($\beta_j\gamma_j < 0$). At the aggregate level, the observed

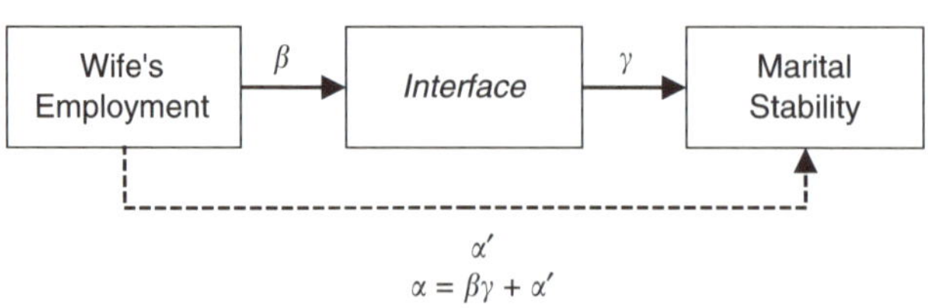

Fig. 2. Specifying the Relationship between Wife's Employment and Marital Stability.

α can be thought of as a weighted sum of the two as follows:

$$\alpha = \omega_j(\beta_j\gamma_j + \alpha'_j) + \omega_k(\beta_k\gamma_k + \alpha'_k) \quad (1)$$

where the subscripts index the two possibilities in the interface (j and k), ω_i denotes the weight for each, and $\omega_j + \omega_k = 1$ $(0 \leqslant \omega_r \leqslant 1)$. When $\omega_k > \omega_j$, thus, one is likely to see a more positive relationship at the aggregate level, and when $\omega_k < \omega_j$, a more negative one (Elster, 1998).

The present formulation, in short, takes into account the possibility that some couples may benefit from the wife's employment and strengthen the marriage, while others may suffer and weaken the marriage. That is, married women's employment may increase as well as decrease the likelihood of divorce, depending on the interface mediating the two. The model thus can account for the seemingly conflicting pieces of empirical evidence within a single explanatory scheme. The last – and, perhaps, the most crucial – piece of the model is then to ascertain which couples are compatible and which incompatible.

DATA

To illustrate the framework, I examined 546 married couples from the data collected in the first and second waves of the Cornell Retirement and Well-Being Study (Wave I-1994 and Wave II-1996/1997). The initial respondents are from six large manufacturing and service companies in four cities of upstate New York who were aged between 50 and 72 at the time they were first interviewed in 1994 (see full description in Han & Moen, 1999b). It is, admittedly, limited in external validity and by its retrospective design, but does offer an extensive collection of detailed life history data with regard to work and family for both the respondents and their spouses.

The most serious concern, though, is that the sample consists only of currently married couples. It is primarily a limitation imposed by the practicality of data collection. The selection bias, however, can be dealt with within the frame of the Bayes Theorem as shown in the Appendix (Manski, 1995, pp. 73–87). Based on the reasoning provided there, the following analysis supposes that, if an appropriate baseline can be ascertained for the initial, time of marriage, distribution of couples, the difference between the observed and baseline distributions can be attributed to the effects of the interface (cf. Bernasco et al., 1998). The likelihood of marital dissolution would be lower for the couples with compatible interface, whereas it would be higher for the couples with incompatible interface.

CHARTING WORK AND FAMILY CAREERS

The interface does not operate as a one-time event. Rather it is a continuous process spanning the couple's life course. To trace its path as it unfolds over time, a sequence analysis technique, also known as "optimal matching" or "optimal alignment," is employed. I consider the whole sequence and its overall patterning of career transitions and trajectories, taking into account the *incidence*, *timing*, and *duration of diverse events*, and *their sequence* in work and family domains. References to sequence analysis procedures can be found in Abbott and Tsay (2000).

I examine the life histories from ages 30 to 50, where the tension between work and family is likely to be high and the connection between the two brittle. With regard to work, I use data on the employment histories of respondents and their spouses ($n = 543$ couples $\times\ 2 = 1,086$), which provide information on transitions and trajectories over the life course in occupation, work status, and interorganizational mobility.[6] For the family domain, two separate measures are constructed. One is marital history, which is based on the number of marriages up to age t and the marital status for each year ($n = 741$). The other is based on the number of children in the household under 6 years of age for each year ($n = 738$). The clustering analyses based on the distance matrices obtained from the optimal matching algorithm identify sets of sequentially equivalent groups called "career tracks" or just "tracks" for short (see Han & Moen, 1999b, for details). Considered in tandem, they provide operational handles in taking into account the temporal dimension underlying the connection between work and family. For employment history, seven distinct career tracks are identified. With regard to marriage and children, five tracks are identified, respectively.[7] The major characteristics of each are described below.

Work Careers: Employment History

Panel (a) of Table 1 shows the basic distribution by gender across work career tracks, and Fig. 3 provides two age-profile charts for each track. The top panel charts the average cumulative interorganizational mobility. The bottom one presents work status profile, where the bottom layer indicates proportion full-time, the middle layer part-time, and the top layer out of the labor force or unemployed.[8]

The first track, comprised predominantly of men (83.5%), shows a continuous full-time trajectory with a medium rate of organizational mobility – about 2 companies by age 50. By and large, it consists of typical professional occupations (i.e., highly skilled white-collar occupations). This is called, "orderly-I." The second track is "high mobility," which is also

Table 1. Work and Family Career Tracks.

(a) Work

Cluster No./Track	*N*	% Female	% Male	Column % F	Column % M
1. Orderly-I track	79	16.5	83.5	2.6	13.6
2. High-mobility track	54	22.2	77.8	2.4	8.7
3. Orderly-II track	92	28.3	71.7	5.3	13.6
4. Orderly-III track	423	31.7	68.3	27.1	59.7
5. Delayed-entry, high-status track	36	83.3	16.7	6.1	1.2
6. Delayed-entry, low-status track	68	86.8	13.2	11.9	1.9
7. Minimum-work track	227	97.4	2.6	44.6	1.2
Total	979	50.6	49.4	100.0	99.9

(b) Marriage

Cluster No./Track	*N*	% Female	% Male	Number of Marriages by Age 50
1. No-interruption track	602	47.2	52.7	1.03
2. Early-interruption track	43	60.5	39.5	2.05
3. Mid-interruption track	24	62.5	37.5	2.00
4. Late-interruption track	30	43.3	56.7	2.00
5. No-marriage track	34	73.5	26.5	0.00
Total	733	49.7	50.3	1.12

(c) Parenting

Cluster No./Track	*N*	% Female	% Male	Total Number of Children
1. Early-parenting track	199	62.8	37.2	2.62
2. Modal-parenting track	328	47.0	53.0	3.37
3. Late-parenting track	55	30.9	69.1	3.49
4. Delayed-parenting track	24	37.5	62.5	2.21
5. No-children track	76	65.8	34.2	0.00
Total	682	52.1	47.9	3.09

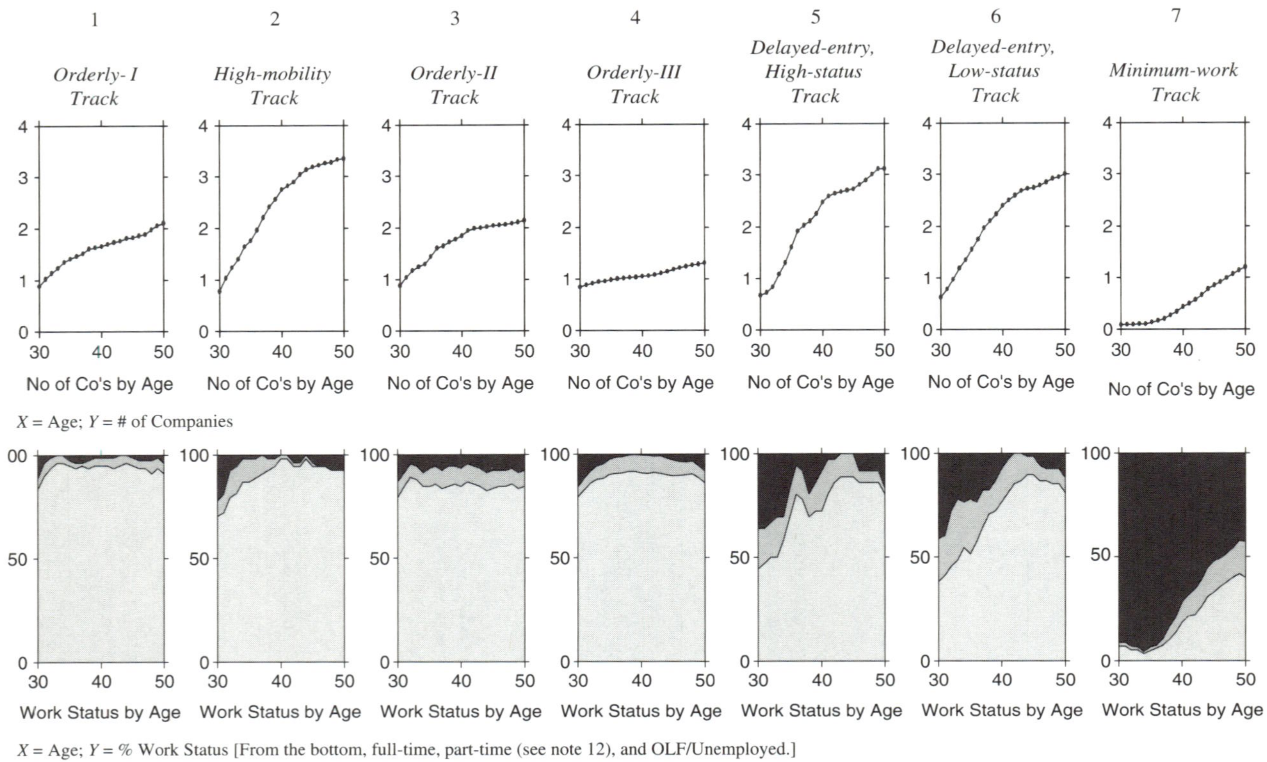

Fig. 3. Work Career Tracks: Profiles of Organizational Mobility and Employment Status across Age.

predominantly male (77.8%) and with a continuous full-time trajectory. It is made up of white-collar (i.e., managerial, executive, and entrepreneurial) occupations and shows a high rate of organizational mobility, more than 3 companies by age 50 on average. The third track, "orderly-II," shares the male dominance (71.7%) and the continuous full-time work status, but organizational mobility is lower – about 2 companies by age 50. It consists, however, mostly of skilled blue-collar occupations. The next track, "orderly-III," includes single-company men and women with virtually no organizational mobility. It is less male dominated (68.3%) and involves continuous, full-time work. The men on this track typically work in blue-collar occupations, and the women in pink-collar occupations.

The next two tracks are predominantly female, 83.3% and 86.8%, and share many characteristics. Most important is delayed entry or reentry to the work force. Both consist of women starting or returning to work in their thirties. The proportion of part-time employment is also sizable and organizational mobility is relatively high, about 3 companies on average by age 50. The fifth track consists mostly of teachers and nurses, while the sixth consists mostly of secretaries. Hence, I label these, "delayed-entry, high-status" and "delayed-entry, low-status." The last track, "minimum-work," is overwhelmingly female (97.4%). They are mostly out of the labor force in their 30s and enter the labor force partially and intermittently around age 40 with a significant portion of their employment in part-time work. On average, they spend 15 years not working, with 62 out of 227 never working during the period from age 30 to 50.

Overall, there seem to be distinct and separate career tracks for men and women ($L^2_{(6)} = 450.35$, $p<0.001$), showing a gendered pattern in work careers (Moen, 1985, 1992). Equally important, women seem to have traveled quite diverse paths, whereas men's work careers tend to be much more standardized, falling primarily into the first four tracks.

Family Careers: Marriage and Children

Panel (b) of Table 1 presents the basic distribution with regard to marriage tracks. Fig. 4 plots the age profiles of marital status (the dashed line; 0, if not married, and 1, if married) and cumulative number of marriages (the gray line). The majority, 602 out of 733 (82.1%), of the sample fall into the first track. It is evenly split between men and women. This consists of those who remain married to his or her first spouse, labeled "no-interruption."

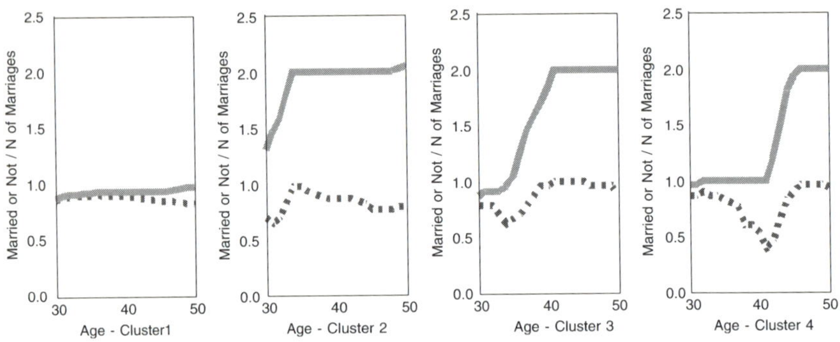

Fig. 4. Age Profiles of Marriage Career.

Note: 1. No-interruption track; 2. Early-interruption track; 3. Mid-interruption track; 4. Late-interruption track; and 5. No-marriage track. Profile for 5, which is not shown above, can be drawn by two flat lines at 0.

The second one, "early-interruption," is more female (60.5%) than male. By age 50, on average, they are in their second marriage. Typically, the first marriage breaks up before 30 and they resettle by 35. Next, "mid-interruption," is also more female (62.5%) and consists of those who also experience two marriages on average in which a first marriage ends during their mid-30s and a second begins by age 40. The fourth, "late-interruption," has more men (56.7%) than women. The average number of marriages by age 50 is also two, but the first marriage breaks up around the 40s and they resettle by 45. Among these three tracks, the difference is primarily in the timing of interruption, from early to late. The last track is the most gender-skewed, where women account for 73.5%. It is made up of those who have not married by age 50 or those who have not been married between ages 30 and 50, thus, "no-marriage." In terms of the overall pattern, the sample divides into three: [1], [2, 3, and 4], and [5]. Gender differences are present but somewhat weak ($L^2_{(4)} = 13.55$, $p = 0.009$), with the major difference in the last track, where women show much higher likelihood of remaining single than men. While most in the sample follow the typical script, those who diverge vary greatly in their paths.

Panel (c) of Table 1 shows the basic distribution of parenting tracks. The first one, "early-parenting," is characterized by the relatively small number of children and the timing of the last child. By age 35, all the children are over 6 years old. The second and third are similar in many respects. "Modal-parenting" is the largest group and evenly split between men and women. They have the second largest number of children with the peak between 30 and 35. The third track, with the largest number of children, is mostly male. Its peak is between ages 35 and 40 and thus is characterized as "late-parenting." The next track is the smallest, consisting mostly of young men. It has the smallest number of children on average. There is an early spike followed by a larger one later, which lasts from late 30s to late 40s, thus, "delayed-parenting." The fifth and last track is "no-children," which is mostly female. Fig. 5 plots the age profiles of parenting status. The labeling follows the order of peak age as shown in the figure, with the last one being an exception. Gender differences ($L^2_{(4)} = 30.67$, $p<0.001$) are mainly due to the age gap between men and women at marriage. The preponderance of women in the last track seems to be related to their being more likely to be not married as described earlier.

The Pattern of Association

The sequence analysis summarized above has identified sets of typical careers in both work and family domains. The next step in fleshing out the interface is to put together the work and family careers of both men and women, examining the linkages between them. I apply log-linear models for

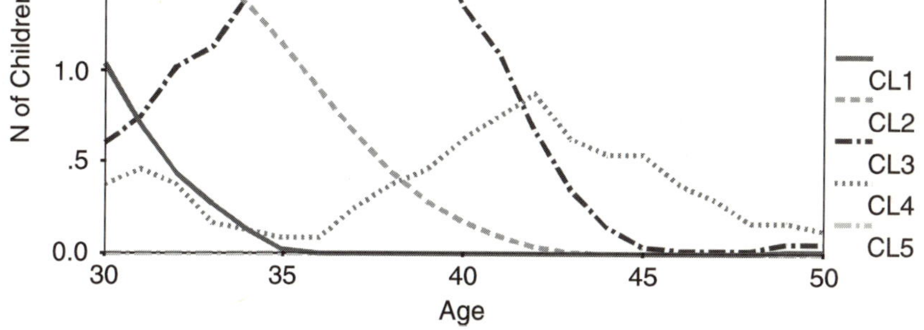

Fig. 5. Age Profiles of Parenting Career.
Note: 1. Early-parenting track; 2. Modal-parenting track; 3. Late-parenting track; 4. Delayed-parenting track; and 5. No-children track.

cross-classified data to characterize the association. Panel (a) of Table 2 shows the distribution of couples in observed counts with respect to two work careers, respondent's and his/her spouse's. Note that the units of analysis in the table are couples, *not* individual respondents (Kalmijn, 1998, p. 419).

The table can be constructed either for the stock of marriages at a given point in time – *prevalence measures* – or for people who marry in a given period of time – *incidence measures*. Incidence measures are generally preferred in the study of intermarriage. If the stock of marriages is used, one can analyze characteristics at the time of survey or at the time of marriage. Typically, the latter measures are considered more suitable because some characteristics change after marriage. On the one hand, partners may become more alike during marriage. They may, for example, switch faith which tends to bias current measures of homogamy upwardly (Kalmijn, 1998). On the other, prevailing marriages may have been affected by differential rate of marital dissolution due to selective attrition, which can be related to the degree of homogamy (Kalmijn, 1998; Mare, 1991). These are precisely what the present study is concerned with. In other words, once the effect of assortative mating (i.e., the initial condition) is taken into account, the remaining effect would be due to the interactions, what I term the "interface," between spouses during marriage (Kalmijn, 1994; Bernasco et al., 1998). The couples with compatible interface are more likely to stay married, while the opposite will hold for those with incompatible interface. The differential rate of marital disruption, then, would be reflected in the difference between the initial distribution, the baseline distribution expected by the effects of assortative mating, and the observed distribution.

Several log-linear models are formulated. In brief, these models assume that the expected frequencies for the cell in row i and column j (m_{ij}s) are a log-additive function of the table total (u), the ith row ($u_{1(i)}$), the jth column ($u_{2(j)}$), and the interaction between rows and columns ($u_{12(ij)}$). The latter measures patterns of association independent of the marginal row and column distributions (Agresti, 1990; Haberman, 1979; Sobel, Hout, & Duncan, 1985). The most general model can thus be written as:

$$\ln m_{ij} = u + u_{1(i)} + u_{2(j)} + u_{12(ij)} \tag{2}$$

Except in the model of *independence* (i.e., $u_{12(ij)} = 0$ for all i and j), the $u_{12(ij)}$s are interpreted as deviations from row and column averages. Significantly positive values of $u_{12(ij)}$ indicate that persons in career track i marry and, once married, stay married to those in career track j more than what the chance would expect. Negative values of $u_{12(ij)}$ suggest the opposite.

A general class of models that has intuitive appeal in the present context is the set of models assuming *symmetry*. In particular, the model of *quasi-symmetry* posits symmetry in the two-variable interaction, but not the additional requirement of marginal homogeneity specified in the model of symmetry. This seems well-suited, given the design of the survey from which Table 2 is obtained. It is not, however, sufficiently restrictive to permit one to characterize the pattern of differential association with much specificity.

Hence, I consider two more restricted versions that are developed in reference to the notion of homogamy (Johnson, 1980; Hout, 1982; Mare, 1991; Kalmijn, 1994). First, homogamous association, or assortative mating, can be operationalized as persons of a career track preferring others like themselves as spouses. But, if such persons are unavailable, they do not differentiate among the other tracks in terms of desirability. This means that getting and staying married between different career tracks should occur at chance levels, given the levels of "ingroup" preference exhibited by the different tracks. It further asserts that the latter tendencies are the same for all tracks, thus "*constant incoupling*."[9]

The second alternative model, which is called the "*differential incoupling*" model, removes the set of restrictions that specifies the strength of ingroup preference measured by $u_{12(ii)}$ to be the same for all career tracks i (cf. Goodman, 1968). In other words, it permits the tendencies toward incoupling to differ across career tracks and allows the likelihood of getting and staying married between different career tracks to vary.

Table 3 reports values of the likelihood-ratio test statistics for several models fitted to the data in Table 2. First of all, the G^2 for the *independence* model indicates that it does not fit the data well. In other words, the association between the respondents' and their spouses' work career tracks is not random. While the model of *symmetry* is also not consistent with the data, the model of *quasi-symmetry* fits the data well ($p = 0.614$). About 90% of the association left unexplained by the model of independence is explained by the model. In fact, due primarily to the very general nature of the model, it fits the data too well to permit much specific inference on the pattern of association.

In contrast, the *constant incoupling* model does not yield an adequate fit to the data. The *differential incoupling* model is however the most general model for differential association that can be formulated without introducing distinctions among outgroups. While providing a much better fit to the data, the general fit to the data, strictly speaking, is not adequate. However, the difference in G^2 between the constant and differential incoupling models ($\Delta G^2_{(6)} = 41.47$) conditionally tests the constancy

Table 2. Coupling Work Careers: Observed Frequencies, Expected Frequencies, and Residuals.

Respondent's Work Career	*Spouse's* Work Career						
	Orderly-I	High-mobility	Orderly-II	Orderly-III	Delayed-entry, High-status	Delayed-entry, Low-status	Minimum-work
(a) *Observed frequencies*							
Orderly-I	8	2	2	8	4	1	25
High-mobility	2	1	4	7	3	3	10
Orderly-II	1	3	4	14	0	1	24
Orderly-III	5	7	16	67	7	17	96
Delayed-entry, high-status	0	2	1	14	0	0	1
Delayed-entry, low-status	0	3	5	27	1	0	5
Minimum-work	3	3	4	27	1	1	1
(b) *Expected frequencies from quasi-symmetry model (the baseline model)*							
Orderly-I	*8.00*	2.59	2.12	9.06	2.61	0.58	25.04
High-mobility	1.41	*1.00*	3.98	7.81	2.54	2.57	10.69
Orderly-II	0.88	3.02	*4.00*	14.68	0.44	2.18	21.80
Orderly-III	3.94	6.19	15.32	*67.00*	9.46	16.41	96.67
Delayed-entry, high-status	1.39	2.46	0.56	11.54	*0.00*	0.42	1.64
Delayed-entry, low-status	0.42	3.43	3.82	27.59	0.58	*0.00*	5.16
Minimum-work	2.96	2.31	6.20	26.33	0.37	0.84	*1.00*
(c) *Expected frequencies from differential incoupling (quasi-independence) model*							
Orderly-I	*8.00*	1.87	3.11	17.11	1.46	2.22	16.23
High-mobility	0.75	*1.00*	2.19	12.04	1.03	1.56	11.43
Orderly-II	1.14	2.02	*4.00*	18.41	1.57	2.39	17.47
Orderly-III	6.05	10.69	17.72	*67.00*	8.34	12.65	92.55
Delayed-entry, high-status	0.46	0.81	1.34	7.40	*0.00*	0.96	7.02
Delayed-entry, low-status	1.06	1.88	3.12	17.17	1.47	*0.00*	16.29
Minimum-work	1.54	2.73	4.52	24.87	2.13	3.22	*1.00*
(d) *Standardized residuals*, (a) – (c)							
Orderly-I	*0.00*	0.09	−0.63	−2.20	2.10	−0.82	2.18
High-mobility	1.45	*0.00*	1.22	−1.45	1.94	1.15	−0.42
Orderly-II	−0.13	0.69	*0.00*	−1.03	−1.25	−0.90	1.56
Orderly-III	−0.43	−1.13	−0.41	*0.00*	−0.46	1.22	0.36
Delayed-entry, high-status	−0.68	1.32	−0.30	2.43	*0.00*	−0.98	−2.27

Table 2. *(Continued)*

Respondent's Work Career	*Spouse's* Work Career						
	Orderly-I	High-mobility	Orderly-II	Orderly-III	Delayed-entry, High-status	Delayed-entry, Low-status	Minimum-work
Delayed-entry, low-status	−1.03	0.82	1.06	2.37	−0.39	*0.00*	−2.80
Minimum-work	1.18	0.17	−0.24	0.43	−0.77	−1.24	*0.00*
(e) *Standardized residuals*, (b) − (c)							
Orderly-I	*0.00*	0.52	−0.56	−1.94	0.95	−1.10	2.19
High-mobility	0.77	*0.00*	1.21	−1.22	1.49	0.81	−0.22
Orderly-II	−0.24	0.71	*0.00*	−0.87	−0.90	−0.14	1.04
Orderly-III	−0.86	−1.38	−0.57	*0.00*	0.39	1.06	0.43
Delayed-entry, high-status	1.37	1.83	−0.68	1.52	*0.00*	−0.55	−2.03
Delayed-entry, low-status	−0.62	1.13	0.40	2.51	−0.73	*0.00*	−2.76
Minimum-work	1.14	−0.25	0.79	0.29	−1.21	−1.33	*0.00*

Table 3. Models of Association for Work Careers and Test Statistics.

Model[a]	G^2	*df*	*p*-value[b]	PRI[c]	*BIC*
Independence	128.104	36	<0.001		—
Symmetry	102.335	21	<0.001	—[d]	−25.535
Quasi-symmetry	12.853	15	0.614	0.900	−78.482
Constant incoupling	116.696	35	<0.001	0.089	−96.421
✔Differential incoupling[e]	75.225	29	<0.001	0.413	−101.357

Note: Valid $N = 441$ (couples). ✔ marks the preferred, selected model.

[a]Models are referred to by the set of constraints placed on the *u*-terms in them.

[b]The *p*-values associated with G^2 statistics may be treated only as heuristic though, for they are calculated as if each table is based on a straightforward multinomial sampling scheme (Marsden, 1981).

[c]The measure indicates proportional reduction in G^2, thus, relative improvement in fit, vis-à-vis the model of independence or null association.

[d]Comparison of the value of G^2 in the symmetry model with that in the independence model is inappropriate, because the latter is not a special case of the former, nor is the former a special case of the latter.

[e]It is also called as the model of *quasi-independence*.

hypothesis and shows that it is not tenable ($p<0.001$). It also has the best *BIC* among the models, indicating the best balance of parsimony and substance (Raftery, 1986), and explains more than 40% of the unexplained association of the independence model ($\Delta G^2_{(7)} = 52.88$, $p<0.001$). This is the preferred model that describes the observed distribution of the couples in the data.

In order to ascertain effects due to the compatibility and incompatibility of the interface during marriage, a baseline needs to be established against which the model selected above would be compared. The baseline as such would reflect the conditions at the time of marriage. Since the information is not directly available from the data, an ersatz baseline is constructed. For starters, prior research suggests that the association is likely to be symmetrical. Further insights for the specification can be obtained by examining the associations in other substantive dimensions that are more reflective of the initial condition. Education fits the bill well, for it is a more reliable status indicator for women, changes little after marriage (Kalmijn, 1998), and correlates highly with occupational career. From the likelihood ratio test for several models fitted to the distribution with respect to respondent's and his/her spouse's education (not shown), the model of *quasi-symmetry*, a model of symmetry without the marginal homogeneity restrictions, is selected for the baseline, corroborating the prior studies.

The differences in the estimated frequencies between these two models, the model of differential incoupling for the current distribution and the model of quasi-symmetry for the baseline distribution, are then to be attributed to the interface effects (Bernasco et al., 1998).[10] Note that the differential incoupling model implies quasi-symmetry. They differ from each other, however, in that the differential incoupling model places no restrictions on the diagonal elements of the interaction term u_{12}, permitting the tendencies toward incoupling to differ across tracks. Also, as discussed earlier, unlike under the quasi-symmetry model, the off-diagonal terms $u_{12(ij)}$ $(i \neq j)$ need not be equal under this model, allowing the likelihood of pairing different career tracks to vary.

Making Sense of the Pattern: Compatible Careers, Incompatible Careers

To facilitate the discussion and make the gender identification explicit, let us establish the following notational conventions. First, the seven work career tracks will be referred to by the abbreviated labels as follows: orderly-I track – *ORD1*, high-mobility track – *HMOB*, orderly-II track – *ORD2*, orderly-III track – *ORD3*, delayed-entry, high-status track – *DEHS*, delayed-entry, low-status track – *DELS*, and minimum-work track – *MINW*. Second, an operator $\underline{C}$ is defined as follows: $\underline{C}(a, b) \equiv$ A couple with the male on career track *a* and the female on career track *b*.

Table 4 reports parameter estimates of the two-variable interaction effects for the differential incoupling model along with parameter estimates of the row (respondent's work career) and column (spouse's work career) effects. Panel (c) in Table 2 presents the expected frequencies based on these parameters. The other panels in Table 2 are for (a) the observed frequencies, (b) the expected frequencies from the baseline model of quasi-symmetry, (d) the standardized residuals of the preferred model vis-à-vis the observed frequencies, and (e) the standardized residuals of the preferred model vis-à-vis the expected frequencies from the baseline model.[11]

Only one of the incoupling effect parameters is substantial and significant, orderly-I track (2.021). When odds ratios are calculated, they show that this pairing is about 6–7 times (6.58) of what is to be expected. This seems to be a novel yet sustainable form of coupling, which enjoys a positive interface effect.[12]

Table 4. Parameter Estimates of the Differential Incoupling Model.

Work Career	Interaction/ Incoupling Effect	Odds Ratio	Respondent/ Row Effect	Spouse/Column Effect
Orderly-I track	2.021***	6.58	−0.374†	−2.728***
High-mobility track	−0.278	0.67	−0.725**	−2.159***
Orderly-II track	0.179	1.05	−0.301	−1.653***
Orderly-III track	−0.376†	0.60	1.367***	0.053
Delayed-entry, high-status track	−21.163	—[a]	−1.212***	−2.407***
Delayed-entry, low-status track	−21.476	—[a]	−0.370	−1.990***
Minimum-work track	−3.161**	0.04	0.000	0.000
Intercept	3.161***			

Note:
†$p<0.10$.
**$p<0.01$.
***$p<0.001$.
[a]No incoupling couples were observed.

The other pairings can be partitioned into six sectors as shown in Fig. 6, given the residual values in the panels (d) and (e). Table 5 codes the pairings by the criteria. Since both models fit the diagonal cells exactly, these residuals are informative only in the off-diagonal cells.[13] Of interest are the pairings that show the movement away from the baseline model (i.e., quasi-symmetry) toward the preferred model (i.e., differential incoupling). On the one hand, in sectors II and IV, there are more couples than what the baseline model expects. In sector IV, in particular, the actual counts even exceed what the preferred model expects ($m_{qs} < m_{di} < x_{ij}$). On the other, in sectors III and V, there are fewer couples than expected by the baseline model, suggesting a more than expected decrease. In the former, sectors II

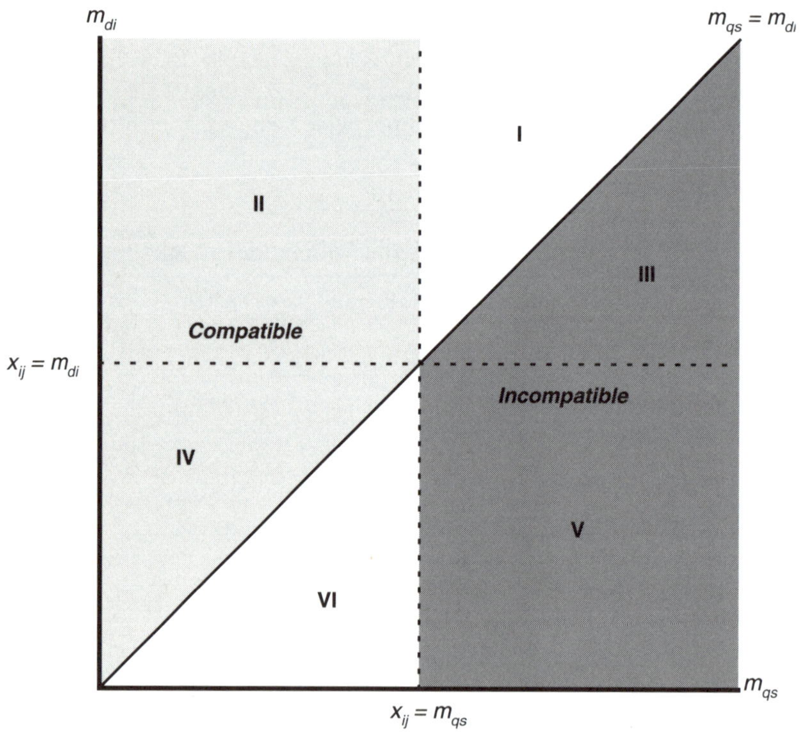

Fig. 6. Partitioning by the Residuals.
Note: The pairings can be partitioned into six sectors as shown above, given the residual values in the panels (d) and (e) of Table 2. For instance, sector I is the area, where $x_{ij} - m_{di} < 0$, $x_{ij} - m_{qs} < 0$, and $m_{qs} - m_{di} < 0$, thus $x_{ij} < m_{qs} < m_{di}$.

Table 5. Pairings Coded by the Residual Criteria.

Respondent's Work Career	*Spouse's* Work Career						
	Orderly-I	High-mobility	Orderly-II	Orderly-III	Delayed-entry, High-status	Delayed-entry, Low-status	Minimum-work
Orderly-I	0	**V**	I	I	**VI**	II	V
High-mobility	**VI**	0	**VI**	I	VI	VI	I
Orderly-II	**II**	V	0	**I**	I	I	VI
Orderly-III	**II**	II	II	0	**III**	VI	V
Delayed-entry, high-status	**III**	V	II	VI	0	**I**	I
Delayed-entry, low-status	**I**	V	VI	V	II	0	**I**
Minimum-work	**VI**	**IV**	III	VI	**II**	II	0

Note: The residuals for the diagonal cells are 0 by definition for both quasi-symmetry and differential incoupling (quasi-independence) models.

and IV, there may be a tendency for those pairings to stay intact at a level greater than expected. In other words, these are the compatible pairings with positive interface effect, thus enjoying lower likelihood of marital dissolution. The couples falling into the latter, sectors III and V, on the contrary, suffer from incompatible interface and negative effect, thus experiencing higher likelihood of marital dissolution.

Taking into account the observed frequency, I selected several pairings for further note and mapped them onto the two tables in panel (a) of Fig. 7, which are separately constructed by respondent's gender. The distribution of the pairings across the tables suggests a sort of block structure. There seems to be a split between male-dominated (*ORD1*; *HMOB*; and *ORD2*) and female-dominated career tracks (*DEHS*; *DELS*; and *MINW*) with *ORD3* track serving as a demarcation line. This could be read in the context that work career tracks are highly gender-specific, as discussed earlier.

However, the distribution of compatible and incompatible pairings does not fit the conventional notion of sexual division of labor. Couples with the husband on male-dominated career tracks and the wife on female-dominated tracks are not necessarily compatible. In fact, panel (b) of Fig. 7 shows that there is no clear pattern that divides these couples into compatible and incompatible pairings. There are couples, for instance, for which traditional male-breadwinner/female-homemaker model seems to work. But the couples with similar configurations are also found on the opposite side. The couples with reversed gender-role configurations can be found both above as well as

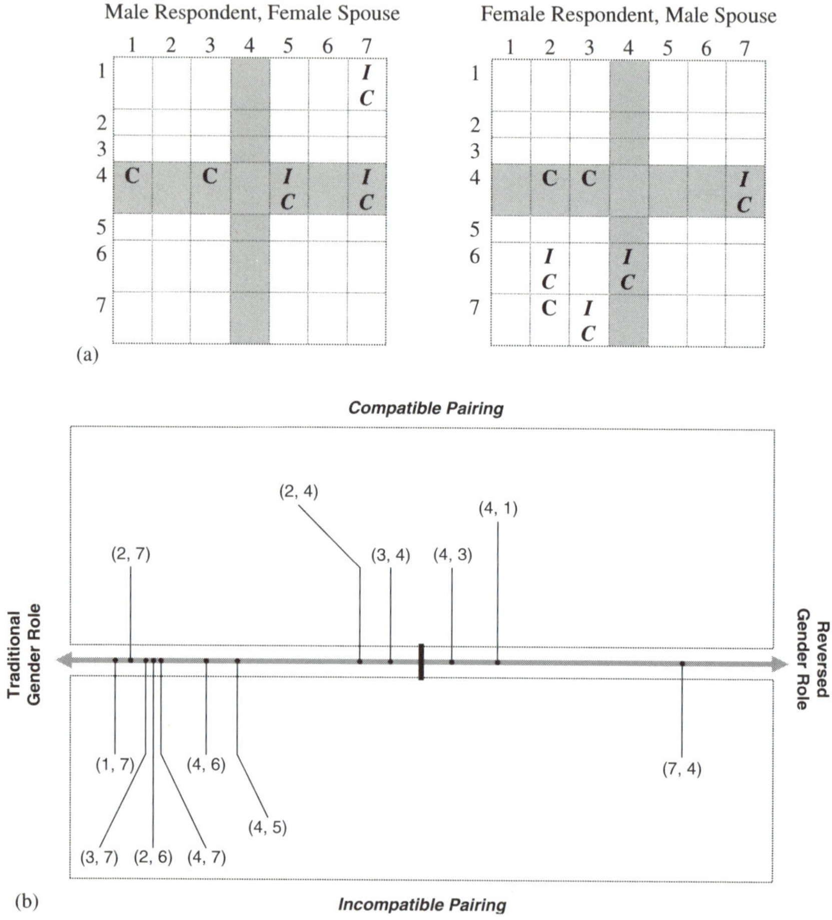

Fig. 7. Compatible and Incompatible Careers: Selected Pairings Arrayed by (a) Respondent's Gender and (b) Gender-Role Configuration.

Note: Pairs are noted by (men's career track, women's career track), where career tracks are numbered as 1 – Orderly-I, 2 – High-mobility, 3 – Orderly-II, 4 – Orderly-III, 5 – Delayed-entry, High-status, 6 – Delayed-entry, Low-status, and 7 – Minimum-work track. Horizontal locations are determined by relative gender composition of men's and women's career track. For example, the locations for (1, 7) – 80.9 – and (7, 4) – −65.7 – are obtained from the proportions of male in career tracks 1 (83.5%) and 7 (2.6%) and those in career tracks 7 (2.6%) and 4 (68.3%), respectively.

below the line too. The deviation from the traditional gender-role division in and of itself does not seem to be much of a factor in determining on which side of the divide a couple would fall.

The results thus far do not suggest any clear single dimension along which these couplings can be arrayed. Yet, when family careers are incorporated, the compatible couples show distinct paths from the incompatible couples. To highlight the difference, the couples in sectors II and IV are put together with $\underline{C}(ORD1, ORD1)$ as compatible couplings ($n = 45$), while those in sectors III and V are put together with $\underline{C}(MINW, MINW)$ as incompatible couplings ($n = 170$). They are then cross-classified against the marriage and parenting clusters obtained earlier (see Table 1 and Figs. 4 and 5). Panels (a) and (b) in Table 6 report the observed frequencies along with the adjusted standardized residuals.

The association in panel (a) between the pair characteristic, whether a coupling is compatible or incompatible, and the marital career is not significantly different from the null association ($p = 0.143$). Panel (b), however, shows a strong association between the pair characteristic and the profile of parenting ($p = 0.005$). Compatible couples are more likely to not have any children at all. When they do, they are more likely to have them either early (early-parenting track), or, in a few cases, very late (delayed-parenting track). Between ages 30 and 40, as a result, they are relatively unburdened by pre-school children at home (see Table 1 and Fig. 5). In addition to *whether* a couple has any child or not, *when* they have children seems to matter a great deal with respect to the interface compatibility (Cherlin, 1977; Waite & Lillard, 1991; Quick & Moen, 1999).

DISCUSSION AND CONCLUSIONS

The ways in which we organize two of the most fundamental institutions, work and family, and the connection between them have gone through radical transformations over the last few decades. These transformations have left a great deal of disruption and uncertainty in their wake and present a challenge. The need for a new perspective is more pressing than ever, given the ongoing changes in the composition of the labor force, institutional arrangements at work as well as at home, and gender roles (Riley & Riley, 1994). How then should, and could, the careers of men and women who find themselves negotiating the two spheres of life with each other over the life course be conceptualized?

Table 6. Compatible and Incompatible Pairings.

(a) By Marriage Cluster	No-Interruption Track	Early-Interruption Track	Mid-Interruption Track	Late-Interruption Track	No-Marriage Track	Sum
Compatible pairings	41	0	2	2	0*	45
	(−0.3)	(−1.4)	(1.4)	(0.8)		
Incompatible pairings	156	7	2	4	0*	169
	(0.3)	(1.4)	(−1.4)	(−0.8)		
Sum	197	7	4	6	0*	214

Note: *These are structural zeros by definition. In parentheses are adjusted standardized residuals; $L^2_{(3)} = 5.44$, $p = 0.143$.

(b) By Parenting Cluster	Early-Parenting Track	Modal-Parenting Track	Late-Parenting Track	Delayed-Parenting Track	No-Children Track	Sum
Compatible pairings	18	17	1	2	6	44
	(2.2)	(−2.4)	(−1.7)	(0.4)	(2.7)	
Incompatible pairings	38	92	16	5	5	156
	(−2.2)	(2.4)	(1.7)	(−0.4)	(−2.7)	
Sum	56	109	17	7	11	200

Note: Adjusted standardized residuals are in parentheses; $L^2_{(4)} = 14.92$, $p = 0.005$.

In this paper, I developed an analytical framework of "coupled careers interface" to address the concern. It incorporates three major features. First, it is anchored at the couple level. Second, it emphasizes the process aspect of the connection between work and family, embedding the interface in temporal context. The framework, finally, factors in the complex, contingent operation of the interface structure and its varied, unequal consequences across couples. The key is a two-person-two-role-domain interface, a system of interaction consisting of careers in the two life domains of both spouses, placed at the intersection between work and family. The interface acts as a differentiating mechanism on which how the connection will operate, whether positively (compatible coupling) or negatively (incompatible coupling), depends.

Although largely illustrative, the model usefully accounts for the conflicting evidence in the relationship between married women's employment and divorce for dual-earner couples within the present data. By and large, the findings corroborate existing explanations. Yet, there are some significant new insights. Of most importance, despite the fact that there are more boundaries to be negotiated, and thus, more potential tensions and conflicts, some couples do benefit from the dual-earner configuration (Hood, 1983; Hochschild, 1989; Popenoe, 1993). The effect of married women's employment on divorce is found to be contingent on the interface, instead of being uniformly negative.

The asymmetric interdependence between men and women's work careers in the interface, especially in the context of the parallel family careers involving children, is one of the findings that closely echo previous research (also see Han & Moen, 1999a, 2001). The interface seems to operate in such a way as to produce a clear differentiation into primary and secondary careers, in which the husbands are given the priority most of the time (Sørensen & McLanahan, 1987; Bielby & Bielby, 1992; Pavalko & Elder, 1993). There are, however, a few exceptions to the pattern of asymmetry. The results indicate, for example, that the distribution of compatible and incompatible couplings does not necessarily reflect the conventional notion of sexual division of labor (cf. Becker, 1981; Willis, 1987). The question of *when* seems to be a crucial discriminating factor, of how to time, schedule, and synchronize the transitions and trajectories in both work and family careers (Oppenheimer, 1974; Waite, 1980; Waite & Lillard, 1991; Rosenfeld, 1992; Moen & Wethington, 1992; South, 2001).

On the whole, the notion of coupled careers and its direct formulation in the interface framework deepens the current understanding of the connection between work and family and brings the debate to a new level (Menaghan, 1991; Rosenfeld, 1992; Abbott, 1993). More definitive studies, of course, will need a longer time span to examine historically situated changes in the interface and a more representative sampling scheme to accommodate its whole spectrum (Goldin, 1990). Yet, the present framework seems to be flexible enough and its findings robust enough to be readily extended to such investigations.

NOTES

1. In the end, these levels would have to be integrated, preferably in a multilevel framework that is firmly grounded empirically (Abbott, 1993). The framework provides a crucial nexus in that direction.

2. Those who focus on this configuration typically locate the main source of the tension between ③ and ④. Yet also present are the tensions between ② and ③ – whose

job is more important, more lucrative, and more absorptive – and between ① and ④ – who is doing what and how much at home.

3. Of course, these three configurations are not discrete categories. Not all breadwinner–homemaker couples, for example, would be able to sustain such complete segregation. And most of the dual-earner couples would fall somewhere between 100–0 (configuration B) and 50–50 (configuration C) in dividing family responsibilities. There can be, and are, numerous variations. Nor are these successively ordered stages; rather, they coexist side by side (Tilly & Scott, 1978; Davis, 1984; cf. Hochschild, 1989). Moreover, some couples may shift from one configuration to another over time. One crucial structural difference is to be kept in mind though: For the dual-earner couples (configurations B and C), there are more boundaries to be negotiated, and thus, more potential tensions and conflicts.

4. Another line of reasoning is that both trends could be the result of a third factor, such as changes in opportunity structure or cultural norms. This argument is not without merits: Those exogenous variables certainly have played important roles (Tilly & Scott, 1978; Cherlin, 1992; Goldin, 1990; Barich & Bielby, 1996; Hochschild, 1989; cf. Huber & Spitze, 1981). Yet, the argument suffers from the problem of not being able to provide concrete behavioral mechanisms for the link (England & Farkas, 1986).

5. The institution of marriage persists, though, despite its decline as a cultural imperative and as an economic necessity, even more preferred by women (Cherlin, 1992, pp. 128–130).

6. Occupation is coded with a nominal coding scheme used in the Cornell Retirement and Well-Being Study ($k = 64$). Work status is coded into five categories – employed full time, part time, alternating between the two, sporadic or seasonal, and not employed. And, lastly, (inter)organization mobility is coded by counting the number of companies or employers one has worked for up to age t. These three dimensions of work career are later merged for the clustering analysis, producing one set of tracks.

7. The solution I adopt here for work careers is based on an initial solution with 17 clusters. The seven largest clusters, which account for 90.1% of the sample, are used and the rest are dropped from the analyses reported below due to the small Ns. The same procedure is used for marriage and parenting careers. For the former, I use 5 out of 9 (98.9%), and for the latter 5 out of 23 (92.4%). The substantive effect of this procedure is to rule out the small number of severely deviant cases.

8. Two other employment statuses, "alternating between full-time and part-time" and "seasonal or sporadic employment," are placed together with the part-time employment status in order not to clutter the graphs.

9. Typically, the model is referred to as "constant *inbreeding*" model in the literature, as the model that follows is referred to as "differential *inbreeding*" model (see Marsden, 1981). I adapted the labels to suit the context.

10. A supplementary analysis on the dissimilarities between the two distributions (not shown) finds considerable differences that are neither random nor uniform, which is presumably, and at least partly, due to the interface effects.

11. For panels (d) and (e), the standardized residuals are obtained by taking $(x_{ij} - \hat{m}_{ij})/\sqrt{\hat{m}_{ij}}$.

12. $\underline{C}(MINW, MINW)$ appears significant and is at the opposite end of the continuum. Without any gainful employment for either spouse, it would be practically very difficult, if not impossible, to sustain the marriage in this case. Yet, the observed frequency is too small to put much weight on this case ($n = 1$).

13. These inferences should be taken cautiously for a couple of reasons. First, several of them are based on very small observed frequencies x_{ij}. Second, the implied notion of "social distance," on which the following inferences are based, assumes the existence of systematic tendencies for marriages between certain work career tracks to be more likely to stay intact than others. Under the preferred model, although the expected frequencies in the off-diagonal cells will differ from each other, they differ only as a function of the differential levels of incoupling parameters. Thus, if the differential incoupling model closely accounts for the data, then analyses using the concept of "social distance," as is done below, might run the risk of confounding the two.

ACKNOWLEDGMENTS

The research was funded in part by the Alfred P. Sloan Foundation (#96-6-9) and by the National Institute on Aging (#IT50-AG11711), through the Cornell Careers Institute and the Gerontology Research Institute at Cornell University, respectively. I would like to thank Phyllis Moen for her continuing support and Heather Quick, Shinok Lee, and Chun-Hao Fan-Chiang for their able assistance. I also thank Tim Liao, Phyllis Moen, Gillian Stevens, Pamela Tolbert, Linda Williams, and especially, Ross Macmillan for their comments and suggestions on the earlier drafts. Direct correspondence to Shin-Kap Han, Department of Sociology, 326 Lincoln Hall (MC-454), 702 S. Wright Street, University of Illinois at Urbana-Champaign, Urbana, IL 61801. E-mail: skh@uiuc.edu.

REFERENCES

Abbott, A. (1993). The sociology of work and occupations. *Annual Review of Sociology*, *19*, 187–209.

Abbott, A. (1995). Sequence analysis: New methods for old ideas. *Annual Review of Sociology*, *21*, 93–113.

Abbott, A., & Hrycak, A. (1990). Measuring resemblance in sequence data: An optimal matching analysis of musicians' careers. *American Journal of Sociology*, *96*, 144–185.

Abbott, A., & Tsay, A. (2000). Sequence analysis and optimal matching methods in sociology: Review and prospect. *Sociological Methods and Research*, *29*, 3–33.

Agresti, A. (1990). *Categorical data analysis*. New York: Wiley.

Barich, R., & Bielby, D. (1996). Rethinking marriage: Change and stability in expectations, 1967–1994. *Journal of Family Issues*, *17*, 139–169.

Barnett, R., & Rivers, C. (1996). *She works/he works: How two-income families are happier, healthier, and better off*. San Francisco: HarperSanFrancisco.

Becker, G. (1981). *A treatise on the family*. Cambridge: Harvard University Press.

Bernard, J. (1982 [1972]). *The future of marriage*. New Haven: Yale University Press.

Bernasco, W., de Graaf, P. M., & Ultee, W. C. (1998). Coupled careers: Effects of spouse's resources on occupational attainment in the Netherlands. *European Sociological Review*, *14*, 15–31.

Bielby, W., & Bielby, D. (1992). I will follow him: Family ties, gender-role beliefs, and reluctance to relocate for a better job. *American Journal of Sociology*, *97*, 1241–1267.

Blossfeld, H.-P., Drobnič, S., & Rohwer, G. (1996). Employment patterns: A crossroad between class and gender. Arbeitspapier Nr. 33 (Sonderforshungsbereich 186). Bremen, Germany: Der Universitat Bremen.

Blumstein, P., & Schwartz, P. (1983). *American couples: Money, work, sex*. New York: Morrow.

Boufis, C. (1999). Strange bedfellows: Does academic life lead to divorce? *Salon* (March 24).

Brines, J. (1994). Economic dependency, gender, and the division of labor at home. *American Journal of Sociology*, *100*, 652–688.

Brines, J., & Joyner, K. (1999). The ties that bind: Principles of cohesion in cohabitation and marriage. *American Sociological Review*, *64*, 333–355.

Cherlin, A. (1977). The effect of children on marital dissolution. *Demography*, *14*, 265–272.

Cherlin, A. (1992). *Marriage, divorce, remarriage*. Cambridge: Harvard University Press.

Clarkberg, M., & Moen, P. (2001). Understanding the time-squeeze: Married couples' preferred and actual work-hour strategies. *American Behavioral Scientist*, *44*, 1115–1136.

Coser, L. (1974). *Greedy institutions: Patterns of undivided commitment*. New York: Free Press.

Davis, K. (1984). Wives and work: The sex role revolution and its consequences. *Population and Development Review*, *10*, 397–417.

Elder, G. (1974). *Children of the great depression*. Chicago: University of Chicago Press.

Elder, G. (1995). The life course paradigm: Social change and individual development. In: P. Moen, G. H. Elder Jr. & K. Lüscher (Eds), *Examining lives in context: Perspectives on the ecology of human development* (pp. 101–139). Washington, DC: American Psychological Association.

Elster, J. (1998). A plea for mechanisms. In: P. Hedström & R. Swedberg (Eds), *Social mechanisms: An analytical approach to social theory* (pp. 45–73). New York: Cambridge University Press.

England, P., & Farkas, G. (1986). *Households, employment, and gender: A social, economic, and demographic view*. New York: Aldine.

Espenshade, T. J. (1985). Marriage trends in America: Estimates, implications, and underlying causes. *Population and Development Review*, *11*, 193–245.

Felmlee, D. H. (1982). Women's job mobility processes within and between employers. *American Sociological Review*, *47*, 142–151.

Fukuyama, F. (1999). The great disruption. *The Atlantic Monthly*, *283*(5), 55–80.

Gerson, K. (1985). *Hard choices: How women decide about work, career, and motherhood*. Berkeley: University of California Press.

Goldin, C. (1990). *Understanding the gender gap: An economic history of American women*. New York: Oxford University Press.

Goldscheider, F. K., & Waite, L. J. (1986). Sex differences in the entry into marriage. *American Journal of Sociology*, *92*, 91–109.

Goode, W. J. (1964). *The family*. Englewood Cliffs: Prentice-Hall.

Goodman, L. A. (1968). The analysis of cross-classified data: Independence, quasi-independence, and interaction in contingency tables with or without missing cells. *Journal of American Statistical Association*, *63*, 1091–1131.

Haberman, S. (1979). *Analysis of qualitative data*. New York: Academic Press.

Han, S., & Moen, P. (1999a). Work and family over time: A life course approach. *The ANNALS of the American Academy of Political and Social Science*, *562*, 98–110.

Han, S., & Moen, P. (1999b). Clocking out: Temporal patterning of retirement. *American Journal of Sociology*, *105*, 191–236.

Han, S., & Moen, P. (2001). Coupled careers: Pathways through work and marriage in the United States. In: H. Blossfeld & S. Drobnič (Eds), *Careers of couples in contemporary society: From male breadwinner to dual-earner families* (pp. 201–231). London, UK: Oxford University Press.

Hayghe, H. (1990). Family members in the work force. *Monthly Labor Review*, *113*, 14–19.

Hochschild, A. (1989). *The second shift: Working parents and the revolution at home*. New York: Avon Books.

Hochschild, A. (1997). *The time bind: When work becomes home and home becomes work*. New York: Henry Holt and Co.

Hood, J. (1983). *Becoming a two-job family*. New York: Praeger.

Hout, M. (1982). The association between husbands' and wives' occupations in two-earner families. *American Journal of Sociology*, *88*, 397–409.

Huber, J., & Spitze, G. (1981). Wives' employment, household behaviors, and sex-role attitudes. *Social Forces*, *60*, 150–169.

Jauhar, S. (1999). Doctor marries doctor: Good medicine. *The New York Times* (March 23).

Johnson, R. A. (1980). *Religious assortative marriage in the United States*. New York: Academic Press.

Kalmijn, M. (1994). Assortative mating by cultural and economic occupational status. *American Journal of Sociology*, *100*, 422–452.

Kalmijn, M. (1998). Intermarriage and homogamy: Causes, patterns, trends. *Annual Review of Sociology*, *24*, 395–421.

Kanter, R. M. (1977). *Work and family in the United States: A critical review and agenda for research and policy*. New York: Russell Sage Foundation.

Manski, C. F. (1995). *Identification problems in the social sciences*. Cambridge: Harvard University Press.

Mare, R. D. (1991). Five decades of educational assortative mating. *American Sociological Review*, *56*, 15–32.

Marsden, P. V. (1981). Models and methods for characterizing the structural parameters of groups. *Social Networks*, *3*, 1–27.

Mayer, K. U., & Tuma, N. B. (1990). Life course research and event history analysis: An overview. In: K. U. Mayer & N. B. Tuma (Eds), *Event history analysis in life course research* (pp. 3–20). Madison: University of Wisconsin Press.

Menaghan, E. G. (1991). Work experiences and family interaction processes: The long reach of the job? *Annual Review of Sociology*, *17*, 419–444.

Moen, P. (1985). Continuities and discontinuities in women's labor force participation. In: G. H. Elder Jr. (Ed.), *Life course dynamics: 1960s to 1980s* (pp. 113–155). Ithaca: Cornell University Press.

Moen, P. (1992). *Women's two roles: A contemporary dilemma.* Westport: Greenwood.

Moen, P., & Wethington, E. (1992). The concept of family adaptive strategies. *Annual Review of Sociology*, *18*, 233–251.

Oppenheimer, V. (1974). The life-cycle squeeze: The interaction of men's occupational and family life cycles. *Demography*, *11*, 227–245.

Oppenheimer, V. (1977). The sociology of women's economic role in the family. *American Sociological Review*, *42*, 387–406.

Oppenheimer, V. (1988). A theory of marriage timing. *American Journal of Sociology*, *94*, 563–591.

Parcel, T., & Cornfield, D. (2000). *Work and family: Research informing policy*. Thousand Oaks: Sage.

Pavalko, E., & Elder, G. (1993). Women behind the men: Variations in wives' support of husbands' careers. *Gender and Society*, *7*, 548–567.

Popenoe, D. (1993). American family decline, 1960–1990: A review and appraisal. *Journal of Marriage and the Family*, *55*, 527–555.

Quick, H., & Moen, P. (1999). Careers in competition? An analysis of U.S. couples' employment trajectories. A Paper Presented at the Annual Meeting of the American Sociological Association, Chicago, IL.

Raftery, A. (1986). Choosing models for cross-classifications. *American Sociological Review*, *51*, 145–146.

Riley, M., & Riley, J. (1994). Structural lag: Past and future. In: M. Riley, R. Kahn & A. Foner (Eds), *Age and structural lag: Society's failure to provide meaningful opportunities in work, family, and leisure* (pp. 15–36). New York: Wiley and Sons.

Rosenfeld, R. (1992). Job mobility and career processes. *Annual Review of Sociology*, *18*, 39–61.

Skolnick, A. (1991). *Embattled paradise: The American family in an age of uncertainty*. New York: Basic Books.

Sobel, M., Hout, M., & Duncan, O. (1985). Exchange, structure, and symmetry in occupational mobility. *American Journal of Sociology*, *91*, 359–372.

South, S. (2001). Time-dependent effects of wives' employment on marital dissolution. *American Sociological Review*, *66*, 226–245.

Sørensen, A. (1994). Women, family and class. *Annual Review of Sociology*, *20*, 27–47.

Sørensen, A., & McLanahan, S. (1987). Married women's economic dependency, 1940–1980. *American Journal of Sociology*, *93*, 659–687.

Spilerman, S. (1977). Careers, labor market structure, and socioeconomic attainment. *American Journal of Sociology*, *83*, 551–593.

Tilly, L., & Scott, J. (1978). *Women, work, and family*. New York: Holt, Rinehart and Winston.

U.S. Bureau of the Census. (1998). *Current population survey*. Washington, DC: U.S. Department of Commerce.

Waite, L. (1980). Working wives and the family life cycle. *American Journal of Sociology*, *86*, 272–294.

Waite, L., & Gallagher, M. (2000). *The case for marriage: Why married people are happier, healthier, and better off financially*. New York: Doubleday.

Waite, L., & Lillard, L. (1991). Children and marital disruption. *American Journal of Sociology*, *96*, 930–953.

Walsh, E. (1995). *Divided lives: The public and private struggles of three accomplished women.* New York: Simon and Schuster.

Wilensky, H. L. (1961). Orderly careers and social participation: The impact of work history on social integration in the middle class. *American Sociological Review*, *26*, 521–539.

Wilensky, H. L. (1963). The moonlighter: A product of relative deprivation. *Industrial Relations*, *3*, 105–124.

Willis, R. J. (1987). What have we learned from the economics of the family? *American Economic Review*, *77*, 68–81.

APPENDIX A. OBTAINING THE ODDS RATIO FROM TRUNCATED DATA

Consider a population of couples, each described by some covariates x and by a binary criterion variable y, with $y = 1$ if a couple breaks up and $y = 0$ if it stays married. Let $x = (w, r)$, where w denotes some covariates and r denotes other covariates specifically referred to as risk factors for marital dissolution. In the model, wife's employment status, along with other characteristics of the couple, will be considered as w. The interface compatibility is the risk factor, which is indexed by r, where $r = j$ for incompatible interface and $r = k$ for compatible interface. The relationship between interface compatibility/incompatibility (r) and divorce (y) can be summarized by a 2-by-2 table as follows:

	Interface: *Incompatible* ($r = j$)	Interface: *Compatible* ($r = k$)
Divorce: *Yes* ($y = 1$)	$P(y = 1\|w, r = j)$	$P(y = 1\|w, r = k)$
Divorce: *No* ($y = 0$)	$P(y = 0\|w, r = j)$	$P(y = 0\|w, r = k)$

If $\beta\gamma \neq 0$ – that is, if the interface indeed operates as a selective, and differential, attrition mechanism, the odds ratio defined below should be larger than 1:

$$OR \equiv \frac{P(y = 1|w, r = j)P(y = 0|w, r = k)}{P(y = 0|w, r = j)P(y = 1|w, r = k)} \tag{A.1}$$

Due to the selection bias, though, what is observable in the data at hand bears only on the second row of the table, the "surviving" couples, and thus does not allow obtaining the odds ratio directly. However, it can be calculated indirectly and can be shown as below that the value of *OR* would be larger than 1, if $P(y = 0|w, r = j) < P(y = 0|w, r = k)$.

Eqs. (A.1) and (A.2) hold by definition.

$$P(y = 0|w, r = k) + P(y = 1|w, r = k) = 1 \tag{A.2}$$

$$P(y = 0|w, r = j) + P(y = 1|w, r = j) = 1 \tag{A.3}$$

The odds ratio (*OR*), therefore, can be rewritten by substituting $[1 - P(y = 0|w, r = k)]$ for $P(y = 1|w, r = k)$ and $[1 - P(y = 0|w, r = j)]$ for $P(y = 1|w, r = j)$ as in (A.4):

$$\frac{[1 - P(y = 0|w, r = j)]P(y = 0|w, r = k)}{P(y = 0|w, r = j)[1 - P(y = 0|w, r = k)]} \tag{A.4}$$

The *OR*, hence, can be obtained from the information given in the second row of the table above. It can further be simplified as in (A.5):

$$\frac{P(y = 0|w, r = k) - P(y = 0|w, r = j)P(y = 0|w, r = k)}{P(y = 0|w, r = j) - P(y = 0|w, r = j)P(y = 0|w, r = k)} \tag{A.5}$$

Since the second term in the denominator and that in the numerator are identical, the odds ratio will depend on the relative magnitude of $P(y = 0|w, r = k)$ and $P(y = 0|w, r = j)$. Given (A.2) and (A.3), it also follows that, if $P(y = 0|w, r = j) < P(y = 0|w, r = k)$, then $P(y = 1|w, r = j) > P(y = 1|w, r = k)$.

EMBEDDED CAREER CLOCKS: THE CASE OF RETIREMENT PLANNING

Phyllis Moen, Stephen Sweet and Raymond Swisher

ABSTRACT

We investigate employees' expectations and planning about a key later life course transition, retirement. Drawing on an organizationally derived sample of workers in dual-earner households in upstate New York, we find that personal mastery, along with health, income, and job conditions, are key predictors of planning. Also important are prior biographical pacing, gender, and relational contexts (at home and at work). Members of today's mostly baby boom cohort tend to plan more financially than for life after retirement, and most anticipate retiring earlier than the conventional age of 65.

When institutionalized clocks govern role transitions, actors approaching them vary little in either their expectations or their plans. But today's American workers experience two conflicting trends regarding a key later life course status passage, retirement. On the one hand, the retirement transition remains entrenched in established social and organizational policies and practices that, by the middle of the 20th century, were normatively defined. These normative expectations, in conjunction with the institutionalization of income supports in the form of Social Security benefits and

The Structure of the Life Course: Standardized? Individualized? Differentiated?
Advances in Life Course Research, Volume 9, 237–265

ISSN: 1040-2608/doi:10.1016/S1040-2608(04)09009-4

private pensions set retirement apart from unemployment as a work exit that can be planned for, anticipated, and positively defined (Costa, 1998; Graebner, 1980).

On the other hand, changes in the employer/employee contract associated with the restructuring of corporations and a global economy increasingly mean that seniority no longer guarantees job security. Mergers and downsizing have destroyed traditional career patterns, making early and mid-career prospects increasingly uncertain (Hardy, Hazelrigg, & Quadagno, 1996; Kotter, 1995). Moreover, federal policies such as the prohibition of mandatory retirement and age discrimination and delays in Social Security eligibility make continued employment attractive for older adults. Complicating retirement plans even further is the common perception that Social Security may no longer provide a secure safety net for future retirees. This is accompanied by the reality of pensions, savings, inheritance, and escalating real estate values that make a large segment of the population feel they can afford to retire far earlier than did their fathers or grandfathers. Still others feel they can never "afford" to stop working. The mismatch between these shifting circumstances and outdated but entrenched norms (cf. Riley, Kahn, & Foner, 1994) means that retirement for contemporary workers has become an increasingly incomplete role transition, one without a taken-for-granted script. In this regard, the variability of retirement may contribute to increased concerns over the "individualization" (Shanahan, 2000) and "de-standardization" (Bruckner & Mayer, *this volume*) of the life course. American workers in this context increasingly must make strategic selections around this key status passage, developing their own plans and assessing their own risks and prospects (see Moen & Altobelli, in press).

The growing number of women in the workforce also influences changing patterns of retirement planning and timing. Women as a status group have historically experienced retirement as their husbands' transition, not their own. Women's workforce experiences have traditionally been intermittent and frequently part-time, meaning that previous generations of women, especially those who are married, have not spent sufficient time in full-time jobs with single employers to accrue pension benefits. Moving in and out of the workforce meant that women often viewed their exits as simply another "leaving," rather than an official retirement. But as women's workforce participation has become the norm (U.S. Bureau of Census, 2001), more women are now confronting retirement than ever before. As a result, most couples now experience *two* retirements, his and hers, and must now take their spouses' careers and retirement plans into account in planning their own retirements (Blossfeld & Drobnic, 2001; Han & Moen, 2001; Henretta,

O'Rand, & Chan, 1993a, b; Moen & Han, 2001; Moen, Kim, & Hofmeister, 2001). How individuals in today's workforce – both women and men – plan for retirement and how couples coordinate retirement expectations in these times of flux remain poorly understood.

Uncertainty about the retirement transition also reflects a larger overall trend toward greater heterogeneity in the timing, duration, and sequencing of all life course transitions – such as completing schooling, moving into full-time work, marriage, and parenting (Han & Moen, 1999; Henretta, 1994; Hogan & Astone, 1986; Moen, 1985; Moen, Dempster-McClain, & Williams, 1989; Mortimer, 2003; O'Rand, 1996; Rosenfeld, 1992; Settersten, 1999; Settersten & Mayer, 1997). Dislocation in the structure and culture of work, occupational careers, and retirement are producing a climate of ambiguity and uncertainty.[1] Understanding workers' planfulness and expectations can shed light on retirement as a key element in the life course, a personal and family transition occurring upon a moving platform of social, demographic, and organizational change. Such macro-level transformations also shape the more proximate environments in which actors develop plans and expectations at home and at work. Workers' retirement expectations and plans have enormous organizational, economic, and policy ramifications.

In this study, we draw on a life course-role context approach (e.g., Moen, Dempster-McClain, & Williams, 1992; Musick, Herzog, & House, 1999; Spitze, Logan, Joseph, & Lee, 1994), along with other relevant theoretical strands, to investigate the planning implications gender, age/cohort, and location in particular workplace and family environments. We develop and test a multilevel model of retirement planning, drawing on data from a survey of two-earner couples in which at least one spouse is employed in one of 10 large organizations in upstate New York. We assess retirement planning along three dimensions: incidence, timing, and degree. We begin by considering the nature of planning, locating it in a broader theoretical frame.

PLANFULNESS AS PRAGMATIC DECISION-MAKING AND CONTROL

Scholars recognize the active shaping of the life course by planful actors (Clausen, 1991; Elder, 1995, 1998; Giele & Elder, 1998; Marshall, Heinz, Krüger, & Verma, 2001; Shanahan, Hofer, & Miech, 2002), but there has been comparatively little work on career planning in general, much less retirement planning. Lachman and Burack (1993) and Prenda and Lachman

(2001) posit that planning may well be an outgrowth of a sense of mastery, feeling in charge of one's life. In turn, planning may facilitate mastery, structuring events that are both predictable and manipulable (see also Rodin, 1990). This suggests that, "the extent to which people see themselves as being in control of the forces that importantly affect their lives" (Pearlin, Menaghan, Lieberman, & Mullan, 1981, p. 340), having a sense of mastery or control, will be positively related to the incidence, timing, and degree of workers' retirement planning.

According to rational choice theory (e.g. Becker, 1981; Coleman, 1990; Homans, 1950; Lindenberg, 1985), individuals weigh various life course transitions in light of perceived costs and benefits. Sociological approaches emphasize the pragmatic embedding of such decisions within the contexts of opportunities, meanings, and constraints (e.g. Breiger, 1995; Giddens, 1984). Much research on retirement similarly assumes that individuals are active, purposive agents in planning their retirements but that they do so in a cultural and organizational environment constraining their options.

Prospect theory was developed to explain decision-making under risky conditions (Kahneman & Tversky, 1983). We contend that decisions about retirement timing represent risky choices, made without advance knowledge of the consequences of future conditions, particularly with regard to health, security of pension investments, and downsizing. Those who value the income, status, and purposeful activity their jobs provide may envision retirement as a role loss. By contrast, workers in jobs with high demands and little control, along with those in downsizing environments, may envision retirement as a gain of autonomy and reduced stress. Thus the age at which workers choose to retire is often in response to changes in incentives, disincentives, and perceived gains or losses (Burtless & Quinn, 2001; Hayward & Hardy, 1985).

Prior studies support this conclusion. For example, older men who are well educated and in professional jobs are more likely to continue working (Hayward, Hardy, & Grady, 1990) and to be better off financially (Siegel, 1993). But a good income permits workers to retire early and to plan to do so and some workers may also face financial penalties if they continue to work beyond the normative retirement age of 65 (Burtless & Quinn, 2001; Quadagno & Quinn, 1997). Those who see their jobs as demanding or unrewarding are also likely to leave the workforce early (Herzog, Kahn, Morgan, Jackson, & Antonucci, 1989; see also Streib & Schneider, 1971). Good health can also be the rationale for planning to retire early in order to do other things, including taking on "bridge" jobs following retirement from one's primary career (Quinn, 1998). Yet poor health of family members may also focus workers attention on retirement planning, whether to provide care, or as a signal of the possibility of workers' own eventual poor health. By and large, we believe that people who see themselves as very healthy may well put off planning, plan less, and expect to retire later.

This leads to specific, testable premises. First, we propose that both a personal sense of mastery and job conditions providing a high degree control will be associated with earlier and more planning for retirement, as well as an earlier expected age of retirement. Second, perceived financial adequacy and actual household income will similarly be associated with earlier planning, higher levels of planning, and earlier anticipated exits. Restrictions, such as having a demanding job or poor health (of self or spouse), should also predict earlier and more planning and earlier expected exits, although for different reasons.

RETIREMENT PLANNING IN MULTILAYERED CONTEXTS

We hold that retirement planning cannot be understood apart from the multilayered contexts in which lives unfold. These include individual biographies, families, households, workplaces, and the structures and cultures of broader institutional arrangements shaping the life course. Such contexts shape the nature of decision-making throughout life. Becker (1981) developed a model of household decision-making, embedding individual choice in couples' rational assessments of their circumstances. Reference group theory (e.g., Merton, 1968; Williams, 1975) gives even greater attention to the relational aspects of decision-making processes (see also Cook & Levi, 1990). It suggests that individuals model their own behavior in relation to the values and behaviors of the groups to which they belong or aspire (Shibutani, 1961). This approach fits well with symbolic interaction theory's emphasis on shared and socially constructed meanings and definitions of situations that develop through interaction over time (Stryker & Serpe, 1994; see also Reitzes, Mutran, & Fernandez, 1998). It also conforms to the life course emphasis on meaning and linked lives (e.g., Moen, 2003).

Historical and Biographical Contexts

Until recently, the American experience of retirement was a single, irreversible exit that was timed in relation to biographical exigencies (such as illness or disability), mandatory retirement policies, and strong social norms. By the middle of the 20th century, retirement was a scripted and clearly demarcated status transition, especially for men who spent most of their adulthood in full-time, uninterrupted employment, before moving to full-time "leisure." By contrast, retirement today no longer necessarily means a final exit from the workforce, as growing numbers of people take on post-retirement jobs. In fact, the age at which people actually make their final exit from the workforce may be increasing (Quinn, 2002; but see also Gendell, 2001). O'Rand and Henretta (1999) and Han and Moen (1999) more generally document the increasing "fuzziness" of the retirement life stage. For example, growing proportions of older workers are occupying "bridge jobs." Others are employed in post-retirement work that is either scaled-back versions of their primary career jobs or else something entirely different. There is growing variability in career paths and retirement timing, as well as an increasing duration of time individuals can expect to spend "retired" – whether from their primary career jobs or from the workforce altogether (Fullerton & Toossi, 2001; Gendell, 2001). Such heterogeneity in retirement possibilities may well lead workers to actively engage in planning their own retirements in order to maintain some semblance of control in navigating their life courses. Han and Moen (1999) found that the age workers began planning their retirement was progressively earlier for successive cohorts. As their sample consisted of pre-baby boom retirees, it is not clear whether their findings also apply to contemporary workers, members of the baby boom cohort and those following in their wake. As Uhlenberg and Miner (1996) show, historical changes in the economic, policy, and cultural climates affect both labor force participation and retirement exits. We suspect that workers today begin to plan for retirement, at least financially, even earlier than the retirees in Han and Moen's study. A recent analysis of baby boomers' saving patterns reveals that almost half (46%) save regularly, with another three in ten (30%) saving occasionally (Gist, Wu, & Ford, 1999).[2] Yet cohort is typically confounded with age, meaning younger workers may well be less apt to plan for (a distant) retirement. In light of growing retirement ambiguities and uncertainties, we expect younger workers who do plan will begin doing so earlier and expect to retire earlier than their older coworkers.

An emerging issue in life course research is the effect of variation in the pacing of life events. "Biographical pacing" refers to the age at which individuals undergo key status passages (Han & Moen, 1999). Such pacing shapes not only life pathways but the subjective side of life as well, coloring workers' expectations and goals. The intention to retire "on" or "off" time, earlier or later than the conventional norm (Brim & Ryff, 1980; Neugarten & Hagestad, 1976; Settersten & Hagestad, 1996), is likely influenced by other "on" or "off" time transitions (e.g., finishing school, marriage, parenthood). Retirement planning and expectations may reflect not only contemporaneous circumstances, but also the imprints of the timing of prior life course transitions. As there has been little research to date on the effects of biographical pacing as a predictor of retirement planning or expectations, we directly examine this issue, expecting that delays in earlier life course transitions (such as completing education, getting married, and having children) will similarly delay both workers' initiation of retirement planning and their expected age of retirement.

Gender and Gendered Household Contexts

Most of the classic studies of retirement have focused on men. And yet reference group, life course, and feminist theoretical approaches suggest the whole process of retirement planning may well be a different experience for women.[3] Opportunities and constraints regarding social roles, career paths, and passages at all life stages are heavily gendered, as are the differentiated responses to family exigencies among men and women. The traditional (male) career model presumes an unbroken, full-time attachment to the labor force throughout adulthood, culminating in retirement (Moen, 2003; Moen & Roehling, 2005). When men leave their jobs they are exiting from a role that has typically dominated their adult years (Weiss, 1997). Women, on the other hand, commonly experience greater discontinuity, moving in and out of the labor force, in and out of part-time jobs in tandem with shifting family responsibilities (e.g., Han & Moen, 1999; Rosenfeld, 1980; Sorensen, 1983). Consequently, they are less likely to have the same duration of employment or the same accumulation of work experience as men. Given occupational segregation and their less stable employment histories, women are also less likely to be covered by pensions and even those with pensions typically have potential incomes far lower than men's (O'Rand & Henretta 1999).

Since women as a group are less experienced in retirement, we expect that women will plan less and begin planning later. In the past, women who have retired have traditionally done so earlier than men. In fact, this has been institutionalized in terms of women's earlier age eligibility for Social Security benefits, grounded in the assumption that wives are younger than their husbands and that both will want to retire at about the same point in time.

We anticipate that contemporary women workers also expect to retire earlier than men, in part to synchronize with their husband's retirement. Women's retirement plans should be especially tied to life course exigencies around family responsibilities (such as the need to care for an ailing or infirm family member, or having children later in life).

Spouses in dual-earner households may constitute another frame of reference, with each considering their own retirement timing in the context of their partner's. As a consequence, although large gender differences in the amount of retirement planning activities may occur in comparisons among workers, we anticipate fewer differences within couples. Moreover, given their gendered life experiences, we propose that women are more apt to accommodate to their husbands' planning and timing than vice versa. Thus, we expect women to expect to retire earlier and to be more likely to "time" their retirements in relation to their spouses' plans.

Workplace Contexts

Another potentially important frame of reference is the workplace. Coworkers may both create and perpetuate particular workplace cultures with implicit rules and routines regarding retirement planning and timing. Accordingly, we propose that organizational demographics, customs, and norms about retirement timing influence the expectations of individual workers, particularly in terms of their anticipated retirement timing. Weiss (1990) describes workplaces as *relationships of community* that furnish friends and workmates and provide a sense of place and social meaning. In such contexts, whether and when their coworkers expect to retire can shape workers' own plans and expectations.

There is great diversity in the types of relationships cultivated in the workplace and these likely play a role in shaping how employees plan for their retirement exits. Our theory draws upon insights generated through the study of workplaces as social contexts (see, for example, Wharton, Rotolo, & Bird, 2000). In terms of retirement planning, workplace characteristics such as age demographics, workplace policy, and job security likely influence retirement planning efforts in important ways. For instance, younger employees in demographically older or more diverse age organizations may well have an accelerated pattern of retirement planning in comparison to employees working with people who are all about the same age. Age structures of workplaces may be especially important in shaping the culture of retirement planning. Since retirement issues are apt to be more salient to older workers, organizations with older workers are likely to create workplace cultures in which retirement issues are "in the air." Employment in such workplaces should predict individuals' retirement planning efforts, with even younger workers beginning to think about their future retirement. However, the very fact of being around coworkers in their late 50 s and 60 s could reinforce the notion of retiring "on time" or even later. Those in organizations with older workforces may begin planning earlier, engage in higher levels of retirement planning, yet expect to retire later.

DATA AND MEASURES

We draw on data from a sub-sample of the Ecology of Careers Study, involving interviews of both members of dual-earner households and designed such that at least one spouse works for one of 10 participating organizations. This study is well suited to the examination of the contingent and coinciding relationships between spouses' experiences and plans, as well as the organizational contexts shaping them. Approximately hour-long telephone interviews were conducted with both spouses (in separate interviews) in dual-earner, mostly middle-class households in upstate New York. Respondents were asked to report their family, work, and biographical experiences and expectations, including extensive life histories ($N = 1{,}283$ couples).[4] The ten strategically selected organizations from which we obtained random samples of married workers, and subsequently interviewed workers' spouses, represent both manufacturing and service (utilities, health care, and higher education) industries. The sampling is random within organizations but the selection of organizations is not. Although this influences generalizability, these data still offer a rare opportunity to examine individual-, couple-, and organizational-level influences on retirement plans and expectations. We limit our analysis to the retirement planning efforts of the "referent" respondent, that is, the spouse who currently works at one of the ten selected organizations. This produced a final sample of 1,063 workers in dual-earner, non-retired households. Each of the participating organizations offers employees access to pension programs.

Individual-Level Variables

We analyze the retirement planning process along three dimensions, the incidence, timing, and degree of retirement planning. *Incidence* relates to whether respondents report having done any retirement planning. *Timing* involves the age at which planning began, as well as the anticipated age of exit from one's career job.[55] To capture couple effects, we examine responses to the question, "What influence do you expect your (spouse/partner's) retirement or retirement plans to have on your retirement – do you think it will speed up your retirement, delay it, or have no effect?" asked of respondents over age 40 ($N = 720$).

The design of the computerized survey instrument was such that respondents reporting that they had not engaged in any retirement planning ($N = 89$) or who refused to answer the question ($N = 22$) were not asked when they started planning or intended to retire and are thus excluded from those analyses. Due to concerns about sample selection bias and potential censoring created by the loss of these respondents, we include incidence as a component of our analysis, estimating multilevel logistic models of whether or not respondents engaged in any retirement planning. With the exception of being somewhat younger and tending to be female, the 10.4% of respondents who did no planning do not differ in significant ways. Although the potential for some sample selection effect remains, it is unclear in which direction it would bias our results.

Degree of retirement planning is subdivided into two measures: financial and lifestyle planning. For financial planning, respondents rate the degree to which they have engaged in financial planning for retirement on a hundred-point scale, with 0 indicating no planning and 100 indicating a lot of planning. Similarly, we constructed an index of "lifestyle" planning, combining the degree to which respondents have developed hobbies and interests for retirement, have thought of a second or third career after retirement, or have thought about volunteer work after retirement ($\alpha = 0.60$).

We used the MidLife Development Inventory (MIDI) perceived constraints scale as an indicator of *personal control* or *mastery*. This index was constructed from four questions assessing the degree to which subjects feel (a) helpless in dealing with problems in life, (b) a lack of control over what happens, (c) out of control in responding to life's events, and (d) unable to solve problems ($\alpha = 0.78$).

Family income is the log of respondents' and spouses' combined salaries. *Job prestige* is calculated according to the method described by Nakao and Treas (1990) using BLS three-digit job classifications. Prestige in this sample ranges from a low of 25.73 to a high of 73.51. *Job tenure* is indicated by the number of years working in the present job. *Job security* is gauged by respondents' assessments on a 100-point scale where 0 indicates that one expects to lose one's job in the next couple of years and 100 indicates certainty of keeping one's job. The *job schedule autonomy* scale consists of eight 5-point items indicating control over work hours, timing of work, working at home, amount of work at home, vacation time, receipt of personal calls, use of personal email, and ability to take a few hours off ($\alpha = 0.75$). *Workload* is an index of three questions that assess the degree to which respondents' jobs require working hard, fast, or involve excessive amounts of work ($\alpha = 0.64$). We also include a measure of self-reported *physical health* ranging from 0 to 10 with 10 indicating the very best health.

We assess biographical pacing along several dimensions. One is the *educational career pathway*, as indicated by the number of years spent in higher education and whether respondents returned to school after age 25. *Marital pacing* is measured by the age at first marriage. Movement into *parenthood* is measured as whether respondents became parents by age 27 (the sample mean) or later, with non-parents serving as the reference category.

Organizational Variables

To develop indicators of workplace context we aggregate the responses of individuals within each organization, yielding mean levels of workforce age, schedule autonomy at work, and job security. Heterogeneity in the age structure is represented by the standard deviation of the *mean age of workers* within each employing organization. Mean workforce age across organizations varies from 38.4 to 45.4 years, with a standard deviation of 2.3 years. *Mean levels of job security* vary quite widely across organizations, with a mean of 72.7 (on a 0–100 scale), and a range of 58.0–91.9. *Mean schedule autonomy* varies from a low of 2.5 to a high of 3.8 (sample mean is 3.5). Intra-class correlations, which capture the percent of total variation that is between rather than within organizations, indicate that 8%, 23% and 26% of the variance is between organizations for mean age, mean job security, and mean schedule autonomy, respectively.[6]

ANALYTIC STRATEGY

Hierarchical models (HLM) are used to account for the clustered nature of our sample, with individuals nested within organizations and to test our hypotheses about the effects of organizational-level variables on the retirement planning process (Bryk & Raudenbush, 1992). For continuous outcomes, we use hierarchical linear models. Using financial planning as an example, at level 1:

$$Financial_{ij} = \beta_{0j} + \beta_{1j}Age_{ij} + \cdots + \beta_{kj}X_{kij} + r_{ij} \quad (1)$$

within-organizational variation in financial planning ($Financial_{ij}$) is modeled as a function of a level-1 intercept (β_{0j}), individual-level independent variables (X_k) such as age, gender, family income, etc., and an error term (r_{ij}) capturing the unique disturbance for individual i working in organization j. Variation in financial planning between organizations, is captured at level 2:

$$\beta_{0j} = \gamma_{00} + \gamma_{01}MeanAge_j + \gamma_{02}W_{2j} + \cdots + \gamma_{0S}W_{Sj} + u_{0j} \quad (2)$$

with the mean amount of financial planning in organization j (β_{0j}) a function of a level-2 intercept (γ_{00}), level-2 independent variables (W_S) such as the mean age of workers within that organization, and an error term unique to the organization (μ_j).

Linear models are inappropriate for dichotomous outcomes, such as whether or not an individual has engaged in any retirement planning. For this outcome, we apply the hierarchical generalized linear model (Raudenbush, Bryk, Cheong, & Congdon, 2000). Following Guo (2000) we model the probability that an individual has done any retirement planning as $p_{ij} = Pr(y_{ij} = 1)$ using a logit link function and assume p_{ij} follows a Bernoulli distribution. The model at level 1 then is:

$$\log[p_{ij}/(1 - p_{ij})] = \beta_{0j} + \beta_1 Age_{ij} + \cdots + \beta_{kj}x_{kij} \quad (3)$$

with variation in any retirement planning modeled again as a function of an intercept and individual level covariates. Coefficients can then be interpreted as in a standard logit model (Liao, 1994). At level 2, the model is the same as in the hierarchical linear model (2) above.

RESULTS

Table 1 presents descriptive statistics for the individual and couple variables in our analysis. On an average, this (mostly baby-boom age or younger) sample begins to plan for retirement at just under 32 years of age and expects to retire a full 5 years ahead of the traditionally institutionalized retirement age of 65. Subtracting these two figures reveals that at least some planning for retirement now typically takes place for more than 30 years of the adult life course. We also see more financial than lifestyle planning (mean of 68 versus 48 on a 100-point scale).

Table 1. Descriptive Statistics of Individual and Couple Variables by Gender.

Variables	Men		Women		Total	
	Mean/%	SD	Mean/%	SD	Mean/%	SD
1. No retirement planning	7.67***		13.87		10.44	
2. Age began planning	33.10***	(8.52)	30.43	(7.69)	31.95	(8.28)
3. Financial planning (0–100 scale)	69.68**	(23.72)	66.18	(25.19)	68.11	(24.44)
4. Lifestyle planning (0–100 scale)	49.51**	(22.61)	46.44	(23.65)	48.14	(23.12)
5. Age expects to retire	60.36***	(4.49)	59.18	(5.8)	59.83	(5.14)
6. Age	44.59***	(7.84)	40.77	(7.11)	42.88	(7.76)
7. Health (0–10 scale)	8.26	(1.26)	8.29	(1.3)	8.27	(1.27)
8. Perceived constraints	1.93	(0.46)	1.92	(0.46)	1.93	(0.46)
9. Post high school years of education	4.91***	(0.71)	4.20	(2.5)	4.59	(2.64)
10. Returned to school	42.00		43.00		43.00	
11. Age at first job	22.01+	(2.8)	21.71	(2.86)	21.88	(2.83)
12. Age at first marriage	24.93**	(4.12)	24.27	(4.3)	24.64	(4.21)
13. First birth < Age 28	40.40		39.33		40.00	
14. First birth Age 28 +	45.17*		39.54		43.00	
15. No children	14.43		21.13		7.00	
16. Husband/wife age gap	1.62***	(3.78)	−2.10	(4.46)	−0.04	(4.5)
17. Salary (log)	4.82***	(0.2)	4.65	(0.2)	4.75	(0.22)
18. Job prestige	55.51***	(9.93)	52.47	(9.88)	54.15	(10.01)
19. Job tenure (0–100 scale)	6.66***	(6.49)	5.39	(5.72)	6.09	(6.19)
20. Job security (0–100 scale)	71.59+	(23.31)	74.09	(24.34)	72.71	(23.8)
21. Work schedule autonomy	3.64**	(0.65)	3.52	(0.77)	3.59	(0.71)
22. Work load	2.88***	(0.49)	2.99	(0.49)	2.93	(0.49)
23. Spouse's salary (log)	4.36***	(0.39)	4.69	(0.25)	4.51	(0.37)
24. Spouse's health (0–10 scale)	8.37	(1.31)	8.29	(1.26)	8.34	(1.29)
25. Influence of spouse on retirement						
Delay	7.95***		16.40		11.30	
Speed up	13.86***		28.20		19.40	
No effect	78.18***		55.40		69.30	
N	587		476		1,063	

Note: Significance tests assess gender differences.
***$p<0.001$.
**$p<0.01$.
*$p<0.5$.
+$p<0.10$.

Given our interest in the gendered nature of the retirement process, we note both similarities and differences by gender among workers in dual-earner households. As expected, men are more likely than women (92–86%) to have engaged in any retirement planning and to report higher levels of both financial and lifestyle planning. Among workers who have planned, men tend to begin their planning efforts nearly 3 years later than women. While our sample is restricted to those employed and in a dual-earner household arrangement, we attempt to untangle such planning differences from gender variations in salaries, education and other confounding factors in multivariate models below.

Incidence and Timing of Planning and Expected Age of Retirement

We begin with analyses of the incidence and timing of retirement planning. Table 2 presents coefficients from hierarchical models of incidence, that is, whether or not respondents have engaged in *any* retirement planning (logistic model in first set of columns), when they first started to plan (linear model in second set), and when they intend to retire (linear model in third set).[7]

As expected, retirement planning is related to a person's sense of control over their lives. People who perceive more constraints (and hence, less control) tend to both delay planning ($\beta = 1.15$, $p<0.01$) and anticipate retiring later ($\beta = 0.57$, $p<0.10$). Both greater financial resources ($\beta = -0.68$, $p<0.10$) and perceived income adequacy ($\beta = -0.03$, $p<0.01$) predict earlier expected ages of retirement, while spouses' (good) health predicts later retirement expectations. Having a demanding workload is associated with expectations for earlier retirement ($\beta = -0.71$, $p<0.01$). Workers in jobs with very demanding workloads may well associate retirement with a release from time pressures and the beginning of a more leisurely lifestyle.

Age has multiple meanings: it is a marker for cohort, a proxy for proximity to retirement, a gauge of life course location and prior biographical pacing. Older workers are significantly more likely to have done any retirement planning ($\beta = 0.06$, $p<0.01$). Among those who have begun to plan, age shapes when they began to do so, as well as their expected ages of retirement. Consistent with our "cohort" hypothesis, we find that older workers began their planning later than younger ones ($\beta = 0.69$, $p<0.001$). We believe this likely reflects the growing destandardization and uncertainty around retirement, with younger workers having the greater concern.[8] Older workers also expect to retire later than younger workers ($\beta = 0.10$, $p<0.001$). This may reflect societal trends, seen in the expectations of younger workers, toward anticipating retiring from one job only to take up another. It may also, however, be the result of sample selection (i.e., survival) bias, as some older workers who expected to retire "early" may have already retired from these companies and thus are not in our sample.

Workers' current retirement planning and expectations also reflect prior biographical decisions and transitions. Those with longer educational

Table 2. Hierarchial Models of Any Retirement Planning, Age Began Retirement Planning, and Age Intending to Retire.

Variables	Any Planning for Retirement (Logistic Model)		Age Began Retirement Planning (Linear Model)		Age Intended to Retire (Linear Model)	
	Coefficient	(SE)	Coefficient	(SE)	Coefficient	(SE)
Individual Level (fixed effects)						
Intercept	2.38***	(0.27)	32.14***	(0.21)	60.09***	(0.21)
Age	0.06**	(0.02)	0.69***	(0.03)	0.10***	(0.02)
Age at first job	−0.01	(0.03)	0.15+	(0.08)	0.15*	(0.06)
Age at first marriage	−0.03	(0.03)	−0.06	(0.06)	−0.04	(0.04)
Age at first child (< 27)	0.05	(0.34)	1.81**	(0.64)	0.02	(0.48)
Age at first child (27+)	−0.31	(0.30)	−0.32	(0.60)	0.47	(0.46)
Years of higher education	−0.05	(0.05)	0.19*	(0.10)	0.11	(0.07)
Gap in schooling	0.05	(0.21)	0.12	(0.42)	0.62*	(0.31)
Gender (women = 1)	−0.46+	(0.26)	0.04	(0.49)	−0.77*	(0.38)
Age gap with spouse	−0.02	(0.04)	0.01	(0.08)	0.06	(0.06)
Age gap * gender	0.03	(0.05)	0.08	(0.10)	0.09	(0.08)
Own health rating	0.08	(0.08)	−0.44**	(0.16)	0.10	(0.12)
Spouse's health rating	−0.05	(0.08)	0.01	(0.16)	−0.27*	(0.12)
Perceived constraints	−0.27	(0.23)	1.15*	(0.45)	0.57+	(0.34)
Family income (logged)	0.02	(0.31)	−0.37	(0.54)	−0.68+	(0.41)
Income adequacy	0.01	(0.00)	−0.03*	(0.01)	−0.03*	(0.01)
Job tenure	−0.02	(0.02)	−0.03	(0.03)	−0.03	(0.03)
Job security	0.00	(0.01)	0.02	(0.01)	0.01	(0.01)
Schedule autonomy	0.17	(0.16)	−0.18	(0.32)	−0.22	(0.23)
Work load	−0.04	(0.22)	−0.05	(0.42)	−0.71*	(0.32)
Organizational level (random intercept)						
Mean age	−0.13	(0.12)	−0.30*	(0.12)	0.27*	(0.11)
Age spread	−0.10	(0.29)	0.27	(0.27)	1.06**	(0.25)
Schedule autonomy	0.73	(0.71)	−0.30	(0.87)	−2.08*	(0.72)

Table 2. (*Continued*)

Variables	Any Planning for Retirement (Logistic Model)		Age Began Retirement Planning (Linear Model)		Age Intended to Retire (Linear Model)	
	Coefficient	(SE)	Coefficient	(SE)	Coefficient	(SE)
Percent variation explained						
Individual level	—		43.40		9.80	
Organizational level	—		96.90		90.80	
Total variation	—		44.40		19.40	

Notes: Number of respondents is 1,063 for any planning, and 952 for age began retirement planning and age intended to retire number of organizations is 10 for all models.
***$p<0.001$.
**$p<0.01$.
*$p<0.05$.

careers ($\beta = 0.19$, $p<0.01$) or who began their first jobs later ($\beta = 0.15$, $p<0.10$) begin planning for retirement at later ages. Parenthood also suggests the importance of biographical pacing, but not in the ways we anticipated. Compared to respondents who have not had children or had children after age 27, those who had their first child early delayed any retirement planning by almost 2 years ($\beta = 1.81$, $p<0.01$). This suggests that those who remain childless or who postpone having children may be more "planful" in comparison to workers who have their children earlier than average. Other aspects of biographical pacing relate to the anticipated age of retirement in more expected ways. Starting one's first full-time job later ($\beta = 0.15$, $p<0.05$) and returning to school ($\beta = 0.62$, $p<0.05$) both predict delays in expected age of retirement.

In line with our theoretical perspective about the gendered nature of retirement and the life course more generally, gender and gendered relationships influence the retirement planning process. Women in our sample are less likely to have done any retirement planning than the men ($\beta = -0.46$, $p<0.10$) and expect to retire three-quarters of a year earlier than do men ($\beta = -0.77$, $p<0.05$). Health also shapes retirement planning. Respondents who report better health tend to delay retirement planning ($\beta = -0.44$, $p<0.01$), while spouses' positive health ratings delay expected age of retirement ($\beta = -0.27$, $p<0.01$).

Turning now to whether organizational context matters for their retirement planning, a first question is whether or not the outcomes in question actually vary significantly across workplaces. This is commonly assessed with unconditional (i.e., no covariates) intra-class correlations (Bryk & Raudenbush, 1992). We find significant variation across employing organizations for both the age when workers begin to plan (2% of variation is between organizations), and workers' expected age of retirement (8%).

Recall we propose that the age structure of an organization might influence the retirement planning efforts of its workers, producing between-employer variation. Organizations with a higher proportion of older workers or a wider age range of employees may be more likely to have a workplace culture supportive of retirement planning. Largely consistent with this, respondents working in organizations with higher average ages of employees begin retirement planning at earlier ages ($\beta = -0.30$, $p < 0.05$), regardless of respondents' own ages. The presence of older workers within an organization may spur their younger colleagues to begin planning earlier.

Organizational context also matters for anticipated age of retirement. Workers in organizations with an older average age of workers ($\beta = 0.27$, $p < 0.05$) and a wider age range ($\beta = 1.06$, $p < 0.001$) expect to retire later. While older workers may create a culture in which retirement planning is "in the air," they may also infuse that culture with expectations about the value of older workers. Specifically, those working within an older workforce tend to expect to retire closer to the traditional age of 65. Workers in organizations offering a greater degree of schedule flexibility and control expect to retire earlier ($\beta = -2.08$, $p < 0.05$), suggesting the possibility that such latitude extends to options for gradual or phased retirement. Overall, our model accounts for 19.4% of total variation in workers' expected ages of retirement.

Spousal Effects on Retirement Decisions

We also assess the degree to which partners in dual-earner couples perceive their two retirements as "tied" transitions, as well as factors that predict how their spouses' planned retirement timing influences the intended timing of retirement of respondents. We do this is by analyzing responses to the question "Do you expect your spouse/partner to retire earlier than you, about the same time as you, never retire, or is your spouse already retired?" Responses indicate that many couples plan on retiring in tandem. Nearly half of the men (49.4%) and women (46.3%) in our sample intend to retire about the same time as their spouses. Over one in four men (28.3%) and women (28.7%) expect to retire later than their spouses, with slightly fewer men (22.3%) and women (25.0%) expecting to retire earlier than their spouse.

Given the tendency to retire together, we further examine the issue of timing. Returning to simple descriptive statistics (see Table 1), 8 in 10 men (78.2%) report that their wives' retirement plans will have no effect on their own retirement timing. By contrast, over two out of five women (44.6%) report that their husband's plans will matter, causing them to either delay (16.4%) or speed up (28.2%) their own retirement.

Prima facie, these findings are compelling, but a full interpretation requires refinement. As these relationships may be more indicative of couple-level disparities between spouses rather than the gendered construction of retirement exits, associations should disappear when measures of spousal difference are taken into account. We assess this by examining the impact of gender, along with respondent-spouse differences in age, health, salary, and job prestige, through a non-hierarchical, multinomial logistic regression.

Coefficients in Table 3 show that even after measures of spousal disparity are included in the model, women are over one and a half times ($e^{0.524} = 1.69$) as likely as men to report that their spouse will influence them to speed up retirement. Women are also over two and a half times as likely ($e^{0.955} = 2.60$) to feel that their retirement will be delayed by their spouses' influence. Because these are gendered processes, we turn to models estimated separately for men and women. Men who earn more than their wives are less likely to see their spouses as influencing them to speed-up their retirement ($\beta = -0.029$, $p<0.01$). No other factors are significant predictors of men retiring earlier because of their wives' retirement. In contrast, a number of factors predict that wives will speed up their retirement timing because of their husbands. Women who are older than their husbands are more likely to do so ($\beta = 0.121$, $p<0.001$), while women who earn more than their husbands are less likely to see their spouses as influencing them to retire earlier that they would like ($\beta = -0.010$, $p<0.10$). No factors in these models were significant in predicting delays in retirement among either men or women.

Taken together, these findings show some tendency for women to tailor their later career clocks to those of their spouses, gauging retirement decisions both in their own and their partner's career stages. Women are more likely than men to plan their retirement transitions to correspond with that of their spouses, either speeding up or delaying their own retirement date, and such differences remain even after controlling for common measures of differences between spouses in age and resource inequalities.

Table 3. Multinominal Logistic Regression of Influence of Spouse on Referents' Anticipated Retirement Timing[a] by Gender and Spousal Differences.

Independent Variable	Speed Up[b]		Delay[b]	
	b	(SE)	B	(SE)
Entire sample ($N = 720$)				
Gender (1 = women)	0.524*	(0.237)	0.955***	(0.292)
r–s age difference	0.078***	(0.023)	0.038	(0.028)
r–s health difference	0.001	(0.06)	0.018	(0.077)
r–s salary difference ($1,000)	−0.011***	(0.003)	−0.001	(0.003)
r–s job prestige difference	−0.005	(0.007)	−0.011	(0.009)
Constant	−1.474***	(0.182)	−2.269***	(0.239)
Men ($N = 440$)				
r–s age difference	0.043	(0.032)	0.039	(0.04)
r–s health difference	0.042	(0.033)	−0.003	(0.113)
r–s salary difference ($1,000)	−0.029**	(0.09)	−0.008	(0.005)
r–s job prestige difference	−0.011	(0.004)	0.006	(0.013)
Constant	−1.362***	(0.209)	−2.093***	(0.28)
Women ($N = 280$)				
r–s age difference	0.121***	(0.034)	0.050+	(0.038)
r–s health difference	0.045	(0.085)	0.037	(0.104)
r–s salary difference ($1,000)	−0.010*	(0.004)	0.004	(0.005)
r–s job prestige difference	−0.001	(0.011)	−0.021	(0.013)
Constant	−1.054***	(0.178)	−1.372***	(0.195)

Notes: r–s indicates referent minus spouse differences, referents work at one of 10 organizations.
[a]Asked only for those over age 40 and not retired.
[b]Versus no effect.
***$p<0.001$.
**$p<0.01$.
*$p<0.05$.
+$p<0.10$.

Types and Intensity of Retirement Planning

"Retirement planning" often connotes preparing for future financial security (e.g., pensions, investments). However, given trends in health and longevity, "lifestyle" planning also matters. What is not clear is whether the same factors predicting financial planning serve to predict lifestyle planning as well. To answer this question, we estimate two hierarchical linear regression models of financial and lifestyle retirement planning (see Table 4).

Somewhat surprisingly, workers' ages are not significantly associated with their degree of financial planning for retirement. Though the reason is not clear, this may reflect the fact that financial planning by workers is increasingly a near life-long process, institutionalized through 401K and other defined contribution pension regimes. The absence of any age effect also underscores that younger cohorts of workers are beginning to plan earlier, perhaps reflecting their low confidence that established Social Security programs will be there for them when they are ready to retire. Lifestyle planning for retirement, by contrast, is significantly and positively related to age ($\beta = 0.69$, $p < 0.001$). Planning how one will spend time in retirement is far less institutionalized and normative than is financial planning. Such lifestyle concerns are more salient for older workers, who are closer to retirement.

Consistent with our previous findings, parenthood predicts lower levels of both financial and lifestyle planning for retirement. There is also evidence that entering one's first job later is associated with less lifestyle planning ($\beta = -0.47$, $p < 0.10$). Lower mastery (in terms of higher perceived constraints) is also associated with lower levels of retirement planfulness. By contrast, those with greater financial resources ($\beta = 0.372$, $p < 0.05$) and more autonomy at work ($\beta = 2.26$, $p < 0.05$) report higher levels of financial planning. Those in jobs with higher workloads also report higher levels of lifestyle planning ($\beta = 3.24$, $p < 0.05$). This may indicate a view of retirement as a release from undesirable working conditions.

Consistent with expectations and our earlier findings, women engage in less financial planning than men ($\beta = -3.67$, $p < 0.05$). But spousal health considerations are related to a greater intensity of financial planning, undoubtedly because of the perception that additional financial resources may be necessary to care for them.

At the organizational level, we find small but significant variation in financial (6%) and lifestyle (4%) planning for retirement across the employing organizations in our study. Specifically, both older workers and workers (of all ages) in organizations with older workforces are more likely to engage in lifestyle planning. We also speculated that working in organizations offering little job security would foster greater financial planning. Our models provide suggestive evidence. Higher average job security within organizations is associated with lower levels of financial planning, regardless of individual workers' own sense of job security. Taken together,

Table 4. Hierarchial Linear Regression Models of Financial and Lifestyle Planning for Retirement.

Variables	Financial Planning		Lifestyle Planning	
	Coefficient	(SE)	Coefficient	(SE)
Individual level (fixed effects)				
Intercept	68.27***	(0.68)	47.50***	(0.72)
Age	0.01	(0.11)	0.69***	(0.11)
Age at first job	−0.38	(0.27)	−0.47+	(0.27)
Age at first marriage	0.16	(0.19)	−0.20	(0.19)
Age at first child (< 27)	−5.18*	(2.26)	−7.28***	(2.18)
Age at first child (27+)	−2.91	(2.10)	−4.51*	(2.03)
Years of higher education	0.23	(0.33)	−0.07	(0.33)
Gap in schooling	−0.74	(1.45)	2.34+	(1.40)
Gender (women = 1)	−3.67*	(1.71)	−0.25	(1.64)
Age gap with spouse	0.17	(0.26)	−0.04	(0.25)
Age gap * gender	−0.45	(0.35)	0.05	(0.25)
Health rating	1.10*	(0.55)	−0.80	(0.54)
Caregiving (for infirm relative)	3.33+	(1.72)	4.89**	(1.68)
Spouse's health rating	0.11	(0.54)	0.33	(0.53)
Perceived constraints	−7.99***	(1.55)	−5.08***	(1.51)
Family income (logged)	3.72*	(1.91)	−1.72	(1.86)
Income adequacy	0.38***	(0.04)	0.09*	(0.04)
Job tenure	0.01	(0.12)	−0.08	(0.12)
Job security	0.05	(0.03)	−0.04	(0.03)
Schedule autonomy	2.26*	(1.09)	0.72	(1.03)
Work load	0.43	(1.39)	3.24*	(1.42)
Organizational context 2nd level (random intercept)				
Mean age	0.53	(0.40)	1.79*	(0.46)
Age spread	2.18	(1.57)	1.25	(1.37)
Job security	−0.40*	(0.14)	−0.19	(0.12)
Percent variation explained				
Individual level	10.4		7.4	
Organizational level	85.6		93.0	
Total variation	15.0		10.8	

Notes: *N* is 1,063 respondents in 10 organizations in both models.
***$p<0.001$.
**$p<0.01$.
*$p<0.05$.
+$p<0.10$.

our models explain 10.8–15.0% of the total variation in lifestyle and financial planning, respectively.

CONCLUSION

The life course regime established in the United States in the 20th century provided (for white, middle-class men at least) a lock-step template culminating in the leisure of retirement. Institutionalized role entry and exit portals, as well as qualifications for advancement, job security, pensions, and government supports such as Unemployment Insurance, Social Security, welfare, and educational loans created a life course regime marginalizing those out of step with its specifications. Today, as a result of a confluence of forces including increasing longevity, technological advances, shifting gender values, a changing workforce, and a globalizing economy, much of the existing infrastructure around the life course, especially retirement, is now obsolete. The growing age and gender heterogeneity of the workforce, along with the changing social contract that no longer linking job security with seniority, further challenge the traditional career/retirement template rendering retirement an increasingly incomplete institution (Moen & Altobelli, 2005). In light of this moving platform of social change, we are interested in whether, when, and what contemporary workers plan for their uncertain futures. Accordingly, we have examined the incidence, timing, and level of retirement planning as well as expectations about age of retirement among members of dual-earner households, most of whom are part of the large baby boom cohort wending its way toward later adulthood.

Our evidence suggests several biographical and contextual factors shaping retirement planning in this contemporary climate of risk and ambiguity. Consider first the importance of gender, as a shaper of both orientations and resources. Women in the two-earner households in this study are less apt than men to engage in any planning for retirement. Moreover, women make fewer financial plans and tend to hinge their own retirement planning activities on those of their husbands. Women are also more likely to expect to retire earlier. Gender may thus constitute a key frame of reference and meaning, given that retirement has historically been a male transition, part of men's but not women's taken-for-granted expectations. While most factors predicting the planning process operate similarly for both men and women *in similar circumstances*, men and women are seldom in similar circumstances. Thus, financial factors play a role in the level of planning and in the decision to retire early for both men and women, suggesting that retirement plans turn on having the economic resources to retire, something more likely for men than for women. Perceived income adequacy promotes the financial and lifestyle planning of all workers, regardless of gender, but women are less apt to feel their incomes are adequate to their needs. Both age and biographical pacing also influence the instigation, level, and timing of retirement planning and can shape workers' expectations about an early retirement. But again biographical pacing is connected with gendered life course scripts.

Second, our findings indicate that retirement planning occurs in *relational* environments, both at home and at work, with spouses and coworkers serving as important frames of reference. Our evidence, consistent with prior research, suggests that couples synchronize retirement exits (cf., Blau, 1998). But this too is a gendered process. Women tend to accommodate to their husband's plans, while men make their own plans without taking their spouses' plans into consideration.

We expected workers to shape their plans for retirement timing in accordance with the timing proclivities of their coworkers in particular workplace environments. Demographic characteristics and degree of employment security within an organization help explain employee planning and expectations, regardless of respondents' own personal characteristics. These findings lead us to suspect that employing organizations provide both a structural and cultural environment in which workers make decisions, including plans for retirement. The experiences of fellow employees become a backdrop against which employees' structure their own expectations about retirement.

This suggests the embeddedness of other career decisions within an organizational context, irrespective of workers' biographical circumstances or the large-scale blurring of retirement norms within society as a whole. Coworkers constitute a built-in reference group, shaped by, and shaping, organizational cultures, policies and practices. Additional evidence hints that reference groups experiencing prior downsizing and job insecurity influence how much planning employees engage in. Retirement and other career planning is likely bounded by the types of information workers receive from their social networks, including informally observing and interacting with coworkers and discussions between spouses around the kitchen table, as much as in formal "planning" meetings. Understanding such workplace and family dynamics requires further exploration. Of particular need is better information about the social processes and cultural climates of organizations in which workers spend so much of their adulthood.

Third, we have demonstrated real differences in various components of the planning process. Financial planning is far more common than lifestyle planning. Policy incentives and constraints (e.g., 401K plans and Keogh IRAs), Social Security eligibility, and cultural templates (exemplified by investment corporations' advertising) all encourage financial planning. By contrast, there are few institutionalized encouragements or templates for lifestyle planning. Adulthood in these post-retirement years remains unscripted, with few institutionalized opportunities for meaningful, productive engagement in a society that marginalizes those outside the conventional working-age years of adulthood. As the baby boom generation moves toward first 60 (2006) and then 65 (2011), it will be interesting to observe the ways in which planning processes play out in their actual retirement exits and life following retirement.

Our findings suggest that understanding career planning and expectations and ultimately the overall structuring of the life course can best be achieved by locating them in a complex matrix of biographical, historical, and institutional contexts, as well as within more proximate workplace and family circumstances. New life course insights can be gleaned by locating lives in the gendered and institutional contexts in which they are embedded, and by attending to the impacts of prior biographical pacing on subsequent expectations. Equally important are the multilayered social changes shaping decision-making. Given the scope and intensity of societal transformations, we anticipate that baby boomers' actual exits and retirement experiences, like their planning, will be only loosely connected to the taken-for-granted norms or experiences of previous generations. Members of this large cohort dominating today's workforce may well reinvent retirement, even as they are reinventing retirement planning, and even as they have individualized every prior phase of the contemporary life course.

NOTES

1. See, for example, results from a 2001 survey by the John J. Heldrich Center for Workforce Development (2001).
2. The two most frequent explanations for saving are for precautionary reasons related to life course risks (such as unemployment, illness, security – 28.8%) and for retirement (23.0%).
3. See, for example, the writings on feminist theory by Bem (1998), Anderson (1999), Thorne and Yalom (1982).
4. These data are supplemented with additional respondents from participating companies ($N = 85$) in a sample drawn from census block groups rather than companies. Analyses with a dummy code for these respondents revealed no difference.

5. For this estimation only, those few respondents who reported planning on "never retiring" ($N = 22$) were coded as planning to retire at age 70. Omitting them made no difference in the analyses.

6. Though it would be ideal to have separate sources of data about organizational characteristics, this information is simply not available. Contributing to the reliability of our aggregate measures is the large sample sizes within each of our 10 organizations (Raudenbush & Sampson, 1999) – average is 106, with a range of 30–346. The relatively small number of organizations, on the other hand, limits the number of organizational variables we can consider simultaneously. To guard against the possibility that contextual effects are simply due to the composition of organizations, we include individual-level measures of each organizational variable in our models (e.g., both mean organizational age and respondents' own ages are included).

7. For ease of interpretation, all individual and organizational variables are grand mean centered (Bryk & Raudenbush, 1992).

8. It may also reflect, however, a retrospective recall effect, with older workers having to recall further in the past than younger workers, thus partly inflating their reports of the age they began to plan.

ACKNOWLEDGEMENTS

Support was provided by the Alfred P. Sloan Foundation (Grant 2002-6-8) and the National Institute on Aging (P50 AG11711).

REFERENCES

Anderson, M. (1999). *Thinking about women: Sociological perspectives on sex and gender*. New York: Macmillan.

Becker, G. (1981). *A treatise on the family* (enlarged ed.). Cambridge, MA: Harvard University Press.

Bem, S. (1998). *An unconventional family*. New Haven, CT: Yale University Press.

Blau, D. (1998). Labor force dynamics of older married couples. *Journal of Labor Economics*, *16*, 595–629.

Blossfeld, H., & Drobnic, S. (2001). *Careers of couples in contemporary societies: From male breadwinner to dual earner families*. Oxford, UK: Oxford University Press.

Breiger, R. (1995). Social structure and the phenomenology of attainment. *Annual Review of Sociology*, *21*, 115–136.

Brim, O., & Ryff, C. (1980). On the properties of life events. In: P. B. Baltes, & O. G. Brim (Eds), *Life-span development and behavior* (Vol. 3, pp. 367–388). New York: Academic Press.

Bryk, A., & Raudenbush, S. (1992). *Hierarchical linear models for social and behavioral research: Applications and data analysis methods*. Newbury Park, CA: Sage.

Burtless, G., & Quinn, J. (2001). Living longer, living better: The policy challenges of an aging workforce. *The Public Policy and Aging Report*, *11*, 5–11.

Clausen, J. (1991). Adolescent competence and the shaping of the life course. *American Journal of Sociology*, *96*, 805–842.

Coleman, J. (1990). *Foundations of social theory*. Cambridge, MA: Harvard University Press.

Cook, K., & Levi, M. (1990). *The limits of rationality*. Chicago: University of Chicago Press.

Costa, D. (1998). *The evolution of retirement: An American economic history, 1880–1999*. Chicago: University of Chicago Press.

Elder, G. (1995). The life course paradigm: Social change and individual development. In: P. Moen, G. Elder & K. Lüscher (Eds), *Examining lives in context: Perspectives on the ecology of human development* (pp. 101–140). Washington, DC: American Psychological Association.

Elder, G. (1998). The life course as developmental theory. *Child Development*, *69*, 1–12.

Fullerton, H., & Toossi, M. (2001). The labor force: Steady growth, changing composition. *Monthly Labor Review*, *124*, 21–38.

Gendell, M. (2001). Retirement age declines again in 1990s. *Monthly Labor Review*, *124*, 12–21.

Giddens, A. (1984). *The constitution of society*. Berkeley, CA: University of California Press.

Giele, J., & Elder, G. (Eds) (1998). *Methods of life course research: Qualitative and quantitative approaches*. Thousand Oaks, CA: Sage.

Gist, J., Wu, K., & Ford, C. (1999). *Do baby boomers save and, if so, what for?* Washington, DC: Public Policy Institute.

Graebner, W. (1980). *A history of retirement*. New Haven, CT: Yale University Press.

Guo, G. (2000). Multilevel modeling for binary data. *Annual Review of Sociology*, *26*, 441–462.

Han, S. K., & Moen, P. (1999). Clocking out: Temporal patterning of retirement. *American Journal of Sociology*, *105*, 191–236.

Han, S. K., & Moen, P. (2001). Coupled careers: Pathways through work and marriage in the United States. In: H. Blossfeld & S. Drobnic (Eds), *Careers of couples in contemporary societies: From male breadwinner to dual earner families* (pp. 201–231). Oxford, UK: Oxford University Press.

Hardy, M., Hazelrigg, L., & Quadagno, J. (1996). *Ending a career in the auto industry: "30 and out"*. New York: Plenum.

Hayward, M., & Hardy, M. (1985). Early retirement processes among older men. *Research on Aging*, *7*, 491–515.

Hayward, M., Hardy, M., & Grady, W. (1990). Work and retirement among a cohort of older men in the United States, 1966–83. *Demography*, *27*, 337–356.

Henretta, J. (1994). Social structure and age based careers. In: M. Riley, R. Kahn & A. Foner (Eds), *Age and structural lag: The mismatch between people's lives and opportunities in work, family, and leisure* (pp. 57–79). New York: Wiley.

Henretta, J., O'Rand, A., & Chan, C. (1993a). Gender differences in employment after spouse's retirement. *Research on Aging*, *15*, 148–169.

Henretta, J., O'Rand, A., & Chan, C. (1993b). Joint role investments and synchronization of retirement: A sequential approach to couples' retirement timing. *Social Forces*, *71*, 981–1000.

Herzog, A., Kahn, R., Morgan, J., Jackson, J., & Antonucci, T. (1989). Age differences in productive activities. *Journal of Gerontology: Social Sciences*, *44*, 129–138.

Hogan, D., & Astone, N. (1986). The transition to adulthood. *Annual Review of Sociology*, *12*, 109–130.

Homans, G. (1950). *The human group*. New York: Harcourt Brace.

John J. Heldrich Center for Workforce Development. (2001). *Work trends VIII – holding on: Americans assess a changing economic landscape*. New Jersey: Rutgers, The State University of New Jersey.

Kahneman, D., & Tversky, A. (1983). Choices, values, and frames. *American Psychologist*, *39*, 341–350.

Kotter, J. (1995). *The new rules: How to succeed in today's post-corporate world*. New York: Free Press.

Lachman, M., & Burack, O. (1993). Planning and control processes across the life-span – An overview. *International Journal of Behavioral Development*, *16*, 131–143.

Liao, T. (1994). *Interpreting probability models: Logit, probit, and other generalized linear models*. Thousand Oaks, CA: Sage Publications.

Lindenberg, S. (1985). An assessment of the new political economy: Its potential for the social sciences and for sociology in particular. *Sociological Theory*, *3*, 99–114.

Marshall, V., Heinz, W., Krüger, H., & Verma, A. (Eds), (2001). *Restructuring work and the life course*. Toronto: University of Toronto Press.

Merton, R. (1968). *Social theory and social structure*. New York: The Free Press.

Moen, P. (1985). Continuities and discontinuities in women's labor force activity. In: G. H. Elder Jr. (Ed.), *Life course dynamics: Trajectories and transitions, 1968–1980* (pp. 113–155). Ithaca, NY: Cornell University Press.

Moen, P. (Ed.), (2003). *It's about time: Couples and careers*. Ithaca, NY: Cornell University Press.

Moen, P., & Altobelli, J. (in press). Strategic selection in a retirement project: Will Americans develop hybrid arrangements. In: J. James & P. Wink (Eds), *The crown of life: Dynamics of the early post-retirement period*. New York, NY: Springer Publishing Company.

Moen, P., Dempster-McClain, D., & Williams, R. (1989). Social integration and longevity: An event history analysis of women's roles and resilience. *American Sociological Review*, *54*, 635–647.

Moen, P., Dempster-McClain, D., & Williams, R. (1992). Successful aging: A life course perspective on women's roles and health. *American Journal of Sociology*, *97*, 1612–1638.

Moen, P., & Han, S. (2001). Reframing careers: Work, family, and gender. In: V. Marshall, W. Heinz, H. Krüger & A. Verma (Eds), *Restructuring work and the life course* (pp. 424–445). Toronto: University of Toronto Press.

Moen, P., Kim, J. E., & Hofmeister, H. (2001). Couples' work/retirement transitions, gender, and marital quality. *Social Psychology Quarterly*, *64*(1), 55–71.

Moen, P., & Roehling, P. (2005). *The career mystique: Cracks in the American dream*. Boulder, CO: Rowman and Littlefield.

Mortimer, J. (2003). *Working and growing up in America*. Cambridge, MA: Harvard University Press.

Musick, M. A., Regula Herzog, A., & House, J. S. (1999). Volunteering and mortality among older adults: Findings from a national sample. *Journal of Gerontology*, *54B*, S173–S180.

Nakao, K., & Treas, J. (1990). *Computing 1989 occupational prestige scores*. GSS Methodological Report No. 70. Chicago, IL: National Opinion Research Center.

Neugarten, B., & Hagestad, G. (1976). Age and the life course. In: R. Binstock & E. Shanas (Eds), *Handbook of aging and the social sciences* (pp. 35–55). New York: Van Nostrand Reinhold.

O'Rand, A. (1996). Structuration and individualization: The life course as a continuous, multilevel process. In: A. Kerckhoff (Ed.), *Generating social stratification: Toward a new research agenda* (pp. 3–16). Boulder, CO: Westview Press.

O'Rand, A., & Henretta, J. (1999). *Age and inequality: Diverse pathways through later life.* Boulder: Westview.

Pearlin, L., Menaghan, E., Lieberman, M., & Mullan, J. (1981). The stress process. *Journal of Health and Social Behavior*, *22*, 337–356.

Prenda, K., & Lachman, M. (2001). Planning for the future: A life management strategy for increasing control and life satisfaction in adulthood. *Psychology and Aging*, *16*, 206–216.

Quadagno, J., & Quinn, J. (1997). Does social security discourage work? In: E. Kingson & J. Schulz (Eds), *Social security in the 21st century* (pp. 127–146). New York: Oxford University Press.

Quinn, J. (1998). *New paths to retirement.* Philadelphia: University of Pennsylvania Pension Research Council.

Quinn, J. (2002). Retirement trends and patterns among older American workers. In: S. Altman & D. Shactman (Eds), *Policies for an aging society* (pp. 293–315). Baltimore: Johns Hopkins University Press.

Raudenbush, S., Bryk, A., Cheong, Y. F., & Congdon, R. (2000). *HLM5:Hierarchical linear and nonlinear modeling.* Lincolnwood, IL: Scientific Software International.

Raudenbush, S., & Sampson, R. (1999). 'Ecometrics': Toward a science of assessing ecological settings, with application to the systematic social observation of neighborhoods. *Sociological Methodology*, *29*, 1–41.

Reitzes, D., Mutran, E., & Fernandez, M. (1998). The decision to retire: A career perspective. *Social Science Quarterly*, *79*, 607–619.

Riley, M., Kahn, R., & Foner, A. (Eds), (1994). *Age and structural lag: The mismatch between people's lives and opportunities in work, family, and leisure.* New York: Wiley.

Rodin, J. (1990). Control by any other name: Definitions, concepts, and processes. In: J. Rodin, C. Schooler & K. Schaie (Eds), *Self-directedness: Cause and effects throughout the life course* (pp. 1–17). Hillsdale, NJ: Lawrence Erlbaum.

Rosenfeld, R. (1980). Race and sex differences in career dynamics. *American Sociological Review*, *42*, 210–217.

Rosenfeld, R. (1992). Job mobility and career processes. *Annual Review of Sociology*, *18*, 39–61.

Settersten, R. (1999). *Lives in time and place: The problems and promises of developmental science.* Amityville, NY: Baywood Publishing Company.

Settersten, R., & Hagestad, G. (1996). What's the latest? II. Cultural age deadlines for educational and work transitions. *The Gerontologist*, *5*, 602–613.

Settersten, R., & Mayer, K. (1997). The measurement of age, age structuring, and the life course. *Annual Review of Sociology*, *23*, 233–261.

Shanahan, M. J. (2000). Pathways to adulthood: Variability and mechanisms in life course perspective. *Annual Review of Sociology*, *2*, 667–692.

Shanahan, M., Hofer, S., & Miech, R. (2002). Planful competence, the life course, and aging: Retrospect and prospect. In: S. Zarit, L. Pearlin & K. Schaie (Eds), *Societal impacts on personal control in the elderly.* New York: Springer.

Shibutani, T. (1961). *Society and personality: An interactionist approach to social psychology.* Englewood Cliffs, NJ: Prentice-Hall.

Siegel, J. (1993). *A generation of change: A profile of America's older population.* New York: Russell Sage.

Sorensen, A. (1983). Women's employment patterns after marriage. *Journal of Marriage and the Family*, *45*, 311–321.

Spitze, G., Logan, J. R., Joseph, G., & Lee, E. (1994). Middle generation roles and the well-being of men and women. *Journal of Gerontology*, *49*, S107–S116.

Streib, G., & Schneider, C. (1971). *Retirement in American society*. Ithaca, NY: Cornell University Press.

Stryker, S., & Serpe, R. (1994). Identity salience and psychological centrality: Equivalent, overlapping, or complementary concepts? *Social Psychology Quarterly*, *57*, 16–35.

Thorne, B., & Yalom, M. (1982). *Rethinking the family: Some feminist questions*. Boston: Longman.

Uhlenberg, P., & Miner, S. (1996). Life course and aging: A cohort perspective. In: R. Binstock & L. George (Eds), *Handbook of aging and social sciences* (4th ed.). New York: Academic Press.

U.S. Bureau of Census. (2001). *Statistical abstract of the United States: 2001* (121st ed.). Washington, DC: U.S. Department of Commerce.

Weiss, R. (1990). Losses associated with mobility. In: S. Fisher & C. L. Cooper (Eds), *On the move: The psychology of change and transition* (pp. 3–12). Chichester, England: Wiley.

Weiss, R. (1997). Adaptation to retirement. In: I. Gotlib & B. Wheaton (Eds), *Stress and adversity over the life course* (pp. 232–248). New York: Cambridge University Press.

Wharton, A., Rotolo, T., & Bird, S. (2000). Social context at work: A multilevel analysis of job satisfaction. *Sociological Forum*, *15*, 60–65.

Williams, R. M., Jr. (1975). Relative deprivation. In: L. A. Caser (Ed.), *The idea of social structure* (pp. 355–378). New York: Harcourt Brace and Jovanovich.

PART IV: CONSEQUENCES

MAPPING SOCIAL CONTEXT ON MENTAL HEALTH TRAJECTORIES THROUGH ADULTHOOD

Philippa Clarke and Blair Wheaton

ABSTRACT

Current research raises the possibility that adult life course transitions are a function of higher meso-level social structural processes. The purpose of this paper is to investigate how the mental health consequences of life course transitions are a function of neighborhood context and how this relationship varies over stages of the life course. Using growth curve models with national data on American adults, we find that a considerable part of the mental health effect of neighborhood operates through life course transitions. Role transitions themselves are also a function of social context, with strongest effects seen during stages of the life course typified by gains and losses of statuses and roles.

There is now a large body of work on the mental health consequences of life course transitions at the individual level (e.g., Barrett, 2000; Elder, 1998; Elder, George & Shanahan, 1996; Marks & Lambert, 1998; Simon & Marcussen, 1999; Wheaton, 1990; Williams, 2003). There are multiple messages in this research, including the variability of mental health impacts

The Structure of the Life Course: Standardized? Individualized? Differentiated?
Advances in Life Course Research, Volume 9, 269–301

ISSN: 1040-2608/doi:10.1016/S1040-2608(04)09010-0

following transitions (Wheaton, 1990). However, it is clear that, once the meaning of transitions is better specified through individual appraisals, biography, or contextual circumstances in the role, some transitions seriously compromise mental health and its trajectory over ensuing stages of the life course (Brown & Harris, 1978; George, 1999; Menaghan, 1989; Wheaton, 1990). There is also an emerging literature, focusing on higher ecological levels, suggesting that disadvantaged states, communities, or neighborhoods are inversely related to mental health either concurrently or later in the life course, and that this impact often operates through the precipitation of stressors at the individual level (Aneshensel & Sucoff, 1996; Boardman, Finch, Ellison, Williams & Jackson, 2001; Ross, 2000; Ross & Mirowsky, 2001; Ross, Mirowsky, & Pribesh, 2001; Wheaton & Clarke, 2003; Yen & Kaplan, 1999). Neighborhood research at pre-adult stages of the life course has also demonstrated that *untimely* life course transitions, such as dropping out of high school and teenage parenting, are more prevalent in disadvantaged communities (Brooks-Gunn, Duncan, Klebanov & Sealand, 1993; South & Baumer, 2000), largely due to peer attitudes and behavior and more community tolerance of these transitions.

The overlapping themes in these literatures raises the possibility that, including the sequencing and timing of life course transitions, the general structure of the life course are themselves a function of higher level social structural processes operating through neighborhood context. This is a question that links the micro and macro levels of social scientific inquiry (Alexander, Giesen, Munch, & Smelser, 1987) and has the potential to provide a more complete understanding of the multilevel processes shaping mental health over the adult life course. Our paper begins from the premise that the forces shaping adult life course transitions not only operate at the individual level. Macro-social processes such as recessions, corporate downsizing and relocation, and changes in the structure of an economy can affect the *distribution* of resources and either promote or weaken the concentration of poverty and unemployment at the neighborhood level (Massey & Eggers, 1990). Increases in poverty concentration have ensuing consequences for the social and economic characteristics of neighborhoods, precipitating a cascade of mutually reinforcing conditions, such as elevated crime rates, poor school quality, increased welfare dependence, and housing deterioration (Massey, 1990). In turn, we expect that highly concentrated social disadvantage will foster individual life course transitions out of marriage and employment, primarily because of the increased stress on, and reduced support resources for, these kinds of roles in disadvantaged neighborhoods (Buka, Brennan, Rich-Edwards, Raudenbush & Earls, 2003). In essence, we argue that the structure of the life course and its implicit ties to life course stressors is embedded within a dynamic socio-structural context.

STAGES OF ADULTHOOD AND THE SALIENCE OF SOCIAL CONTEXT

Adulthood is often broadly divided into early, middle, and late stages. Mirowsky and Ross (1992) defined adulthood less in terms of stages and more as a continuous ebb and flow of social forces that result in unique configurations of demands and benefits at these three broad stages of adulthood. Stage distinctions are primarily age-based and the issues that arise at each stage are defined in terms of dominant age-based tasks. If we invert this priority to consider the stages of adulthood as defined by typical problems and goals, we may find new life course taxonomies for adulthood.

Given the wide variation in the end of initial schooling, it is advisable to define the first stage of adulthood in terms of the tasks following the end of schooling. This stage could start at 16 or at 30, but the stage tasks which follow are similar. We call this stage the *sorting period* of adulthood, defined by a period of uncertainty, transience, choice, and turnover in relationships, roles, and jobs. This period could typically last from 5 to 10 years after the end of education. The 20s have become a focal decade for this stage, with the delay in major role commitments such as marriage, parenting, and final career choice occurring now more typically in the late rather than early 20s (Booth, Crouter, & Shanahan, 1999). Thus, we can think of this sorting period in the 20s as the "new adolescence" (Arnett & Taber, 1994).

After role situations stabilize, with or without major commitments, we define the next stage as the d*evelopmental period.* This period is characterized by the assumed expansion of responsibilities and commitments within roles and the push for achievement of fundamental life goals (Arnett & Taber, 1994). It is in this stage that most of the differences in work and family trajectories emerge most clearly. At the beginning of this stage, the population is less differentiated with respect to employment, marriage, and parenting than at the end. It is at this point in life that paths typically diverge. By the end of this stage there are stable differences in role commitments, achievements, and long-term life chances.

Somewhere after 40, we enter the *midlife period* (Brim, Ryff, & Kessler, 2004). This period of life has emerged as "the third age" of adulthood, with specific and enhanced choices and options concerning work (past 65), thus mixing stable components of the fruition of earlier investments with the possibility of new direction. Finally, we reach a fourth stage, a winding down period of *later life* commitments and responsibilities and a progressive experience of disengagement, precipitated, in part, by losses in physical functioning and social roles (Mirowsky & Ross, 1992; Schieman, Van Gundy, & Taylor, 2001). It is at this last stage that a clear change in direction occurs relative to the adult plateau of the midlife stage.

We predict that social contexts are most important during the developmental period of adulthood, and, secondarily, during the later life period. The developmental period is the primary phase of adulthood in which variability in experience is developed and maintained, and in later life the dependence on local environments and their amenities increases dramatically (Glass & Balfour, 2003; Klinenberg, 2003). It is also true that during the early and late stages of adulthood, environments are more homogeneous across individuals. At the beginning of adulthood, differentiation in achievements and role commitments have not reached full fruition. At the end of adulthood, there is a progressive segregation of the aged and fixed or declining income. If this is the case, increasing homogenization of social life implies less variability in environments and thus less potential for impact. If there are two opposing forces in later life, one promoting and the other demoting the importance of context, we would expect at least diminished importance relative to the crucial developmental period.

LIFE TRANSITIONS AND SOCIAL CONTEXTS

The history of research on life transitions includes the accumulated work on life events and mental health (Thoits, 1983) and the divergent threads of hypotheses needed to explain relatively modest findings on the relationship between life events and mental health through the 1970s. The search for resources or coping styles which ameliorate the impact of stress, or more generally, sources of differential vulnerability to stress, followed from this set of findings. However, another strain of work on life events and transitions has considered the meaning and threat potential of life transitions based on social circumstances and the context of the transition (Brown, 1981; Wheaton, 1990). Most of this work conceptualizes context at the individual level, either as life history or as a constellation of current and recent role circumstances.

In this chapter, we look beyond the individual level to consider the importance of the aggregate social contexts that define the settings we live in, work in, and move through. Although there has been considerable attention to the effect of macro historical and economic contexts for the consequences of life course transitions (e.g., Elder, 1974; Elder, Shanahan, & Clipp, 1994; Moen, 1998), we focus on the more proximal meso level contexts (neighborhoods) in which individuals reside in day-to-day life. Taking residence as inextricably bound to personal identity, we focus on neighborhoods as a source of availability for the meaning, and therefore, consequences of major life transitions. Specifically, the purpose of this chapter is to investigate how the mental health consequences of role transitions and the concomitant general structure of the life course are a function of the neighborhood context in which one lives and how this relationship varies over stages of the adult life course.

There is good reason to believe that the effects of neighborhoods may have differential salience at different points in the life course. Massey (1998, p. 571) asserts that "neighborhood effects vary in their nature and intensity at different stages of the life cycle," suggesting that we must pay attention to age variations in the structural context of the life course. In a recent work, we found that neighborhood disadvantage in early adulthood had no effect on concurrent mental health over and above neighborhood disadvantage in childhood (Wheaton & Clarke, 2003), indicating the significance of the lingering impact of past neighborhoods at formative stages of life. We hypothesize that neighborhood context is likely to gain greater import again between early adulthood and the midlife stage, during the developmental adult role stage of the life course when economic and family responsibilities are more vulnerable to change. At this stage, individuals draw on resources and peers in their local communities for the exchange of support and resources. But as individuals gain economic security and develop broader, established, and stable social networks in midlife, the significance of one's immediate neighborhood for the meaning and consequence of life course transitions may be minimized. And coming full circle in later life, one's social context is likely to take on greater meaning as older adults spend a greater amount of time in their immediate neighborhoods and are more reliant on the services and resources in that community (Glass & Balfour, 2003; Klinenberg, 2003).

We focus on the effect of socioeconomic disadvantage in neighborhoods and examine measures of neighborhood-related stress that could precipitate life course transitions. We also attempt to explain the processes by which role transitions, and neighborhood disadvantage affect mental health by including measures of economic hardship (Mirowsky & Ross, 2001; Pearlin, Menaghan, Lieberman, & Mullan, 1981), personal mastery (Pearlin &

Schooler, 1978), self esteem (Rosenberg, 1981), and social interaction. Economic hardship may surface from repeated role exits over time (debts and child support would accumulate, becoming a greater proportion of total income), with a demoralizing effect on mental health (Mirowsky & Ross, 2001). Protracted role instability could also constrain "the extent to which one regards one's life chances as being under one's own control" (Pearlin & Schooler, 1978, p. 5), and this sense of mastery is also likely to be compromised in disadvantaged neighborhoods characterized by powerlessness, disorder, fear and mistrust (Ross et al., 2001). Conversely, cohesive neighborhoods and successful role acquisition over the life course are likely to lead to more positive self evaluations, with salubrious effects on mental health (Pearlin & Lieberman, 1979; Turner & Lloyd, 1999). Finally, we hypothesize that social networks are likely to be developed following repeated role entries, as individuals become more socially integrated and experience increased social interaction. On the other hand, we expect role instability and role loss to decrease the opportunities for social interaction, as divorce, widowhood, and unemployment create a more constricted social network that has adverse consequences for mental health (House, Umberson & Landis, 1988).

DATA AND METHODS

Data

Data are drawn from the first two waves of the National Survey of Families and Households (NSFH) (Sweet & Bumpass, 1996). The first wave, conducted in 1987–1988, was a national multistage area probability sample of 13,007 persons aged 17 to 92 years. Follow-up interviews were conducted five years later (1992–1994) with 10,005 persons. The detailed data collected on life course transitions in a national sample of American adults make this an ideal database with which to explore our research question.

Neighborhood data were obtained for each wave of the survey using the census tract as a proxy for neighborhood. Census tracts have on average about 4,000 people and are designed to capture homogenous areas that roughly map to neighborhoods. Each respondent's address at each wave was coded to the 1990 census tract and linked to the US Decennial Census in 1990. The merge was conducted by the NSFH investigators in order to maintain the confidentiality of respondents and only a proxy geographic identifier was made available for our analyses.

Measures

Mental health is measured at both waves with 12 items from the Center for Epidemiologic Studies Depression Scale (Radloff, 1977). Respondents were asked "How many days during the past week have you...": (1) felt bothered by things that usually don't bother you; (2) not felt like eating; (3) felt that you could not shake the blues; (4) had trouble keeping your mind on what you were doing; (5) felt depressed; (6) felt that everything was an effort; (7) felt fearful; (8) slept restlessly; (9) talked less than usual; (10) felt lonely; (11) felt sad; (12) felt you could not get going. Responses are averaged to produce an index of depression scored from 0 to 7. Alpha reliability is 0.929 in the first wave and 0.927 in the second wave. *Age* at each wave is measured by subtracting the reported date of birth from the date of each interview. Descriptive statistics for these and other variables used in the analyses are presented in Appendix A.

We focus on two key life transitions that have salience over the adult life course: employment transitions and marital transitions.[1] The number of *role entries* at each wave is the cumulative sum of marriages and job entries (full-time or part-time[2]). *Role exits* sums the number of marital separations, widowhood events, and job losses (full-time or part-time jobs). We use the sum of the number of exits (and entries) to capture the cumulative burden (and cumulative advantage) of the history of one's role transitions across the entire life course. Although both marital and employment transitions are aggregated, we also conduct analyses to examine the effects of each type of transition separately and report these findings in the results. However, we use the aggregated measure in order to capture the increasing interdependence and overlap in role transitions that may be increasingly characteristics of structures of the life course in the modern era (George, 1993; Rindfuss, Swicegood & Rosenfeld, 1987).

Neighborhood socioeconomic disadvantage is an average of six census indicators: percent female headed families, percent households on public assistance income, percent families in poverty, male unemployment rate, family income (normed on $210,000 and reverse coded), and percent adults with less than a primary school education. Factor loadings range from 0.623–0.940 ($\alpha = 0.908$).

We control for four background influences in our analyses: (1) individual social class; (2) individual sociodemographic characteristics (gender and race); (3) functional status; and (4) residential stability. All could account for the relationship between depression and either neighborhood disadvantage or role transitions if these mental health risk factors predispose individuals to live in more disadvantaged neighborhoods or experience more role instability over the adult life course.

Social class is measured by years of completed education at each wave. *Female* is a dummy variable that is coded 1 for females and 0 for males. *Race* is coded 1 for African Americans and 0 for others.[3] *Functional status* at both waves is assessed by a series of six questions that ask respondents whether they have a physical or mental condition that (1) limits their ability to work for pay, (2) care for personal needs, (3) move about inside the house, (4) perform day-to-day household tasks, (5) climb a flight of stairs, and (6) walk six blocks. The count of reported limitations forms an index with values that range from 0 to 6. The index has an alpha reliability of 0.850 in wave 1, and 0.875 in wave 2. We also include a control for *residential stability* and assess the extent to which neighborhood effects are modified for long-term residents. This is measured at each wave by the number of years in current neighborhood.

Mediators

Mastery is measured by a single question common to both waves of the survey. Respondents were asked to what extent they agree with the statement: "I always felt my life would work out the way I wanted it". Responses are coded on a 5-point scale ranging from strongly disagree to strongly agree. *Self esteem* is a mean index of responses to three items from the Rosenberg (1981) scale: (1) "I feel that I am a person of worth, on equal plane with others"; (2) "On the whole, I am satisfied with myself"; (3) "I am able to do things as well as other people". Responses range from strongly disagree (coded 1) to strongly agree (coded 5), and alpha reliability is 0.668 in wave 1 and 0.654 in wave 2.

Measures of *economic hardship* typically incorporate subjective difficulty in paying bills, and buying food, clothing or medical care (Mirowsky & Ross, 2001; Pearlin et al., 1981). A similar measure is available at the second wave of NSFH but not in wave 1, so we derived a measure of economic strain based on the ratio of total debt (long-standing credit card bills, installment loans, personal loans from banks and friends/relatives) plus family support payments, over total annual household income, in dollars, expressed as a percentage.

We also examine the frequency of *social interaction*.[4] At both waves respondents were asked how often they attend social events at church; participate in recreational activities, professional societies, sports or hobby organizations; spend social evenings with friends, neighbors, and work colleagues. Responses were scaled from 0 to 4, indicating "never" to "several times per week". The social interaction index averages the responses; alpha reliability is 0.688 in wave 1 and 0.692 in wave 2.

Stress at each wave is a discrete count of recent stressful events that could be implicitly cultivated by neighborhood environment (Boardman et al., 2001; Sampson, Morenoff & Gannon-Rowley, 2002), including delinquency in respondents' children (trouble with the police, suspension from school, repeating a grade, teenage pregnancy), spousal drug and alcohol use, marital violence, and providing care to someone in the household with a disability or chronic illness.

Statistical Analysis

We use growth curve models to examine the effects of role transitions and neighborhood disadvantage on mental health trajectories over the adult life course (Singer & Willett, 2003). Age is used as the indicator of time, generating a synthetic cohort through adulthood. In order to facilitate parameter interpretation, we center age at the youngest value in this sample (age 17). We analyze a three-level model, with multiple observations nested within persons over time, and then nested again within neighborhoods. The structure of the model can be expressed by equations at three levels. At level 1 (within-person model) depression scores at time t are nested within individuals (i), and then embedded within neighborhood (j) contexts:

$$CESD_{tij} = \pi_{0ij} + \pi_{1ij}(age - 17)_{tij} + e_{tij} \quad (1)$$

where π_{0ij} is the expected depression score for person i in neighborhood j at age 17 (since age is centered), and π_{1ij} captures the rate of change in depression over the life course.

These parameters are then modeled as a function of individual characteristics (at level 2) and neighborhood characteristics (at level 3).[5] The level 2 (between person) submodel assumes that mental health trajectories vary across individuals. We explicitly model these differences as follows:

$$\pi_{0ij} = \beta_{00j} + \beta_{01j}(role\_transitions)_{ij} + r_{0ij} \quad (2a)$$

$$\pi_{1ij} = \beta_{10j} + \beta_{11j}(role\_transitions)_{ij} \quad (2b)$$

Here, the intercept and slope from Eq. (1) are modeled as a function of individual role transitions, where β_{01j} represents the difference in the initial depression score (at age 17) for someone who has experienced a role transition, within neighborhood j, and β_{11j} captures the difference in the rate of change in depression over the life course in neighborhood j for someone with a one unit difference in the number of role transitions. At level 3 the differences between individual trajectories are modeled as a function of neighborhood characteristics. For example:

$$\beta_{00j} = \gamma_{000} + \gamma_{001}(ND)_j + u_{00j} \quad (3a)$$

$$\beta_{10j} = \gamma_{100} + \gamma_{101}(ND)_j \quad (3b)$$

where γ_{001} captures the difference in the initial depression scores at age 17 for persons living in disadvantaged neighborhoods (ND), and γ_{101} represents the difference in the mental health trajectories for those in disadvantaged neighborhoods. Substituting the equations across the three levels gives us the full composite model.

We use the MIXED procedure in SAS to estimate all models. The distribution of the residuals at all three levels shows a good approximation to normality with little deviation from the diagonal in the normal probability plots. Because only two waves of data are currently available in the NSFH,[6] it is not possible to estimate random slopes for time in our models (Singer & Willett, 2003, pp. 151–156). However, we do estimate fixed interactions between all independent variables and time (age) to examine whether their effects vary over stage of the life course.

Although the attrition rate over the NSFH survey is substantial and nonrandom (Mirowsky & Reynolds, 2000), the advantage of the growth curve model is that all observations are used even if persons are only followed at one time point. The final sample size for these analyses is 22,475 (23,012 person period observations, less 497 with missing CESD score, 8 missing age, and 32 missing tract data). Neighborhood disadvantage, as well as respondent's education, social interaction, sense of mastery, self esteem, and residential stability, are grand mean centered in order to give meaningful interpretation to the intercept (Singer & Willett, 2003).

Analyses begin by estimating an unconditional growth model and plotting the trajectories of depression over the adult life course. We then examine how mental health trajectories differ by current neighborhood disadvantage, as well as by cumulative role transitions at the individual level. A key step here focuses on the extent to which accumulated life course transitions explain the effects of current neighborhood context. Final models include additional mediators to account for the direct effects of role transitions and neighborhood disadvantage on mental health.

There is a problem due to the limited residential histories in our data. While role transitions during each stage of the life course may be influenced by current context, the existing neighborhood may not necessarily reflect the context in which *earlier* adult roles were entered or exited and this is more likely to be true at later stages of the life course. However, the literature would suggest that even though people move, they rarely escape from a socioeconomic class of neighborhood (South & Crowder, 1997; Quillian, 2002). In this case, current neighborhood disadvantage acts as a proxy for past neighborhood socioeconomic context. This is consistent with the modestly strong correlation (0.65) found in earlier work between neighborhood disadvantage across ten years in a national sample (Wheaton & Clarke, 2003). The correlation between neighborhood disadvantage across five years in the NSFH (0.79) is of course much higher. Nonetheless, the imperfect correlation also raises the possibility that repeated role loss increases the likelihood of moving from a neighborhood of low disadvantage to one of high disadvantage or conversely that role entries may lead to upward neighborhood mobility.

We do two things in our analyses to address this limitation. First, we include the number of years in current neighborhood as a control in our models and test whether the effects of neighborhood disadvantage vary according to residential stability. Second, we account for role selection into neighborhoods by explicitly testing the effects of prior transitions that occurred in past neighborhoods *separately* from role transitions that occurred in the current neighborhoods.

Finally, we use spline regression analysis (Marsh & Cormier, 2002) to specify distinct age segments over the adult life course where mental health slopes are constant rather than the more widely used function fitting strategy (Mirowsky & Ross, 1992). Following a stepwise strategy to identify significant points in the life course where depression trajectories shift abruptly, the spline model captures the various slope changes smoothly over adulthood by joining the regression lines without a break. We use the spline approach for two reasons. First, this allows us to test whether there is some validity to our assumptions about stages of adulthood and where those stages typically occur – at least with respect to changes in depression. Second, this approach also allows us to test specific interactions of stage of adulthood with role transitions and neighborhood context rather than the general effects across all stages of adulthood. Thus, we can test for the sensitivity of depression to role transitions and neighborhood context across specific stages of adulthood.

RESULTS

Results from the unconditional growth model are presented in Table 1 (Model A). The spline regression identified four distinct age segments in the trajectories of mental health over the life course,[7] and the coefficients

Table 1. Results for Unconditional Growth Model, Plus Neighborhood Disadvantage and Role Transitions, in Multilevel Models for Change in Adult Life Course Depression.

	Model A	Model B	Model C
Fixed effects			
Initial status (at age 17)			
Intercept	1.9060***	1.9347***	1.8853***
ND		−0.0333	
Role entries			0.0884
Role exits			−0.2220
Role entries*exits			−0.0068*
Rate of change			
Age 17 to 22	−0.0999***	−0.1091***	−0.0703
Age 23 to 38	−0.0149***	−0.0138***	0.0039
Age 39 to 68	−0.0012	−0.0018	−0.0156***
Age 69+	0.0145***	0.0154***	0.0263***
ND*age 17 to 22		0.0264	
ND*age 23 to 38		0.0085***	
ND*age 39 to 68		−0.0043***	
ND*age 69+		−0.0008	
Role entries*age 17 to 22			−0.0570
Role exits*age 17 to 22			0.1116*
Role entries*age 23 to 38			−0.0121***
Role exits*age 23 to 38			0.0115***
Role entries*age 39 to 68			0.0070***
Role exits*age 39 to 68			−0.0080***
Role entries*age 69+			−0.0064
Role exits*age 69+			−0.0009
Variance components			
Level 2 initial status	0.8136***	0.8126***	0.7792***
Level 3 initial status	0.0433***	0.0202**	0.0258***

Note: ND = neighborhood disadvantage.
*$p<0.05$.
**$p<0.01$.
***$p<0.001$ (two-tailed tests).

approximate the typical "U"-shaped depression curve (Kessler, Foster, Webster & House, 1992; Mirowsky & Ross, 1992; Miech & Shanahan, 2000; Schieman et al., 2001). The level 1 model captures these splines by replacing the single time variable in Eq. (1) with four separate variables for each age segment:

$$CESD_{tij} = \pi_{0ij} + \pi_{1ij}(age17 - 22)_{tij} + \pi_{2ij}(age23 - 38)_{tij} + \pi_{3ij}(age39 - 68)_{tij} + \pi_{4ij}(age69+)_{tij} + e_{tij}$$

where the coefficients capture the rate of change in depression over distinct phases of the adult life course. For example, π_{1ij} is the depression slope for person i in neighborhood j who is between the ages of 17 and 22 at time t.

As can be seen from Table 1 (Model A), depression over the first spline segment from age 17 to 22 (inclusive) drops precipitously as young adults emerge from adolescence. This life stage, a stage of emerging adulthood (Arnett, 2000) that we call the *sorting period*, is characterized by experimentation and the exploration of life's possibilities. The next spline segment, which we term the *developmental period* of adulthood (age 23 to 38), is typified by gains in statuses and roles including early career path, marriage, and asset acquisitions (Arnett & Taber, 1994), and mental health continues to improve albeit at a slower rate. This is followed by the *midlife period* (age 39 to 68) where depression levels off to its lowest point in the adult life course. This segment represents the emotional advantage of midlife, where the stability of marital and employment roles, and the social and economic status they confer, lead to optimal mental health (Mirowsky & Ross, 1992; Mirowsky, 1996). Finally the later stages of the life course (over age 68) are marked by increased depression, perhaps as role exits, declines in function, and decreased sense of control pose a challenge for mental health (Mirowsky & Ross, 1992; Schieman et al., 2001). These results are plotted in Fig. 1 and all subsequent models are run with time specified by these four spline age segments.

The variance components for Model A (Table 1) indicate that after specifying the effects of time, significant variation remains in initial depression scores (at age 17) both between individuals and between neighborhoods. The next set of models seeks to explain this variation through individual and contextual factors. Model B examines how mental health trajectories vary by neighborhood context. The significant interaction term between neighborhood disadvantage and the developmental age segment (age 23 to 38) indicates that for those adults living in disadvantaged neighborhoods, depression trajectories decline less steeply in this stage of adulthood. This

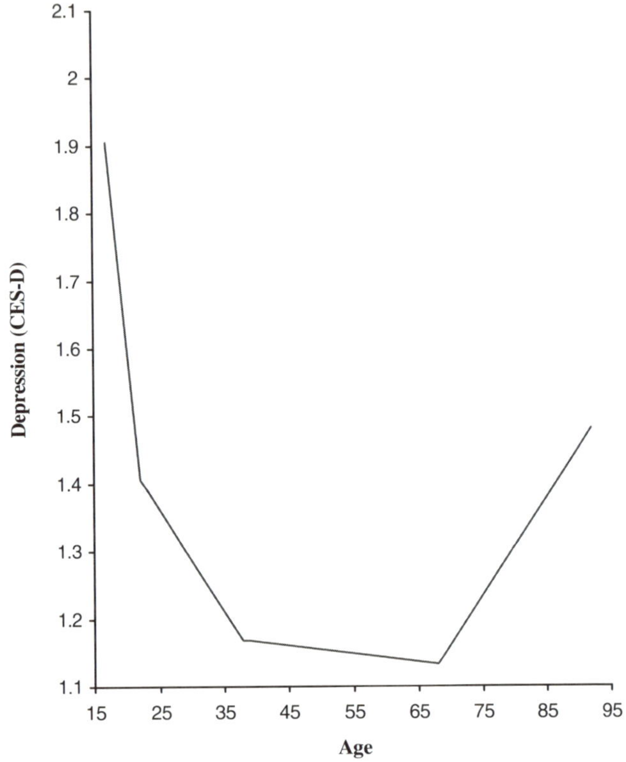

Fig. 1. Trajectories of Depression over the Adult Life Course.

difference is plotted in Fig. 2 at one standard deviation above and below the average level of neighborhood disadvantage. Between the ages of 23 and 38 depression trajectories decline at a rate of 0.022 per year for those living in low disadvantage neighborhoods. In contrast, this rate of decline is reduced to only 0.005 for those living in neighborhoods characterized by high disadvantage, as indicated by the much flatter slope through this stage of the life course.

Although depression trajectories in disadvantaged neighborhoods follow a steady decline through midlife, the rate of decline is not sufficient to match the depression scores of those living in less disadvantaged contexts, where depression slopes are essentially flat over this stage of the life course (the slopes for midlife adults living in average or low disadvantage neighborhoods are not significantly different from zero). As a result, adults in

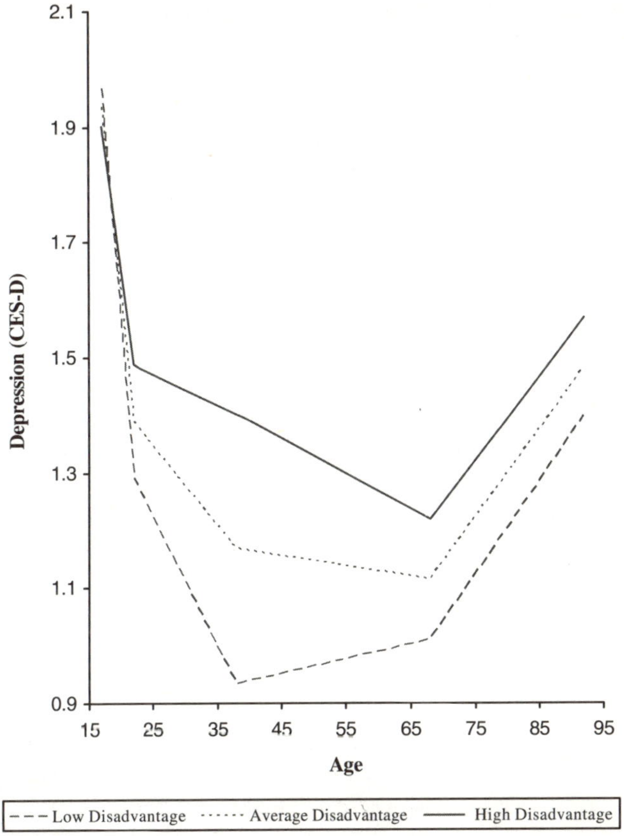

Fig. 2. Trajectories of Depression by Neighborhood Disadvantage over the Adult Life Course.

disadvantaged communities enter the later stage of the life course with worse mental health, even though their trajectories of depression are parallel to those in more advantaged neighborhoods at that stage.

The flatter mental health trajectory over the developmental period of adulthood among those in disadvantaged neighborhoods leads us to question whether the establishment and maintenance of statuses and roles over this stage of the life course is somehow impeded in disadvantaged neighborhoods. The final column in Table 1 (Model C) examines how mental health trajectories vary by role transitions over the life course. Significant interactions between role transitions and time (age) indicate that depression trajectories over the life course vary by combinations of role accumulation and loss. There is also a significant interaction *between* role entries and exits, suggesting that the mental health effects of role entries is conditional upon the accumulated experience of role exits, and correspondingly, that the effect of role exits varies by the number of previous role entries. In effect, the interaction captures disorder and instability in the progression of adult roles over the life course (Hogan, 1978; Marini, 1984; Rindfuss et al., 1987), expressed as the relative balance of entries and exits in a sequence of transitions within a life trajectory. Thus, it is not just the experience of a role transition that has implications for mental health, but the relative *excess* or *imbalance* of role exits to role entries (and of role entries to role exits) experienced cumulatively within the course of long-term trajectories (Elder, 1985; George, 1993) that is consequential for adult mental health.

A visual representation of this interaction is displayed in Fig. 3, which plots the trajectories of depression when role entries exceed role exits (heavy dashed line), when role exits are more frequent (solid line), and for those who experience the average number of role entries and exits at each stage of the life course (lighter dashed line). In the former cases, role transitions are calculated at 0.5 standard deviations above and below the mean value for each age segment.[8] (see Appendix B for age-specific values.) This is a very similar picture to what we saw when plotting the effects of neighborhood disadvantage (Fig. 2). Trajectories of depression decline less steeply (and in fact do not decline at all) through the developmental period of adulthood when persons experience more role exits. Conversely, when role acquisition exceeds role loss, trajectories of depression decline rapidly through this stage of adulthood, consistent with the decline seen for those living in neighborhoods with low disadvantage. Mental health trajectories begin to converge again through midlife, although depression scores remain higher for those who have experienced more role instability just as they remained higher for those in disadvantaged neighborhoods. Trajectories begin to diverge in later life as the net excess of role losses drives depression scores higher in comparison to those whose rate of role acquisition remains in a positive balance. This divergence in later life is less pronounced in Fig. 2, perhaps because current contexts do not fully capture the life-long roster of role accumulation and loss.

The similarity of the plots in Figs. 2 and 3 raises the distinct possibility that part of the mental health effect of neighborhood disadvantage may be operating through its influence on life course transitions. The next set of growth curve models (Table 2) investigates this possibility by testing the extent to which the effects of neighborhood disadvantage are mediated by

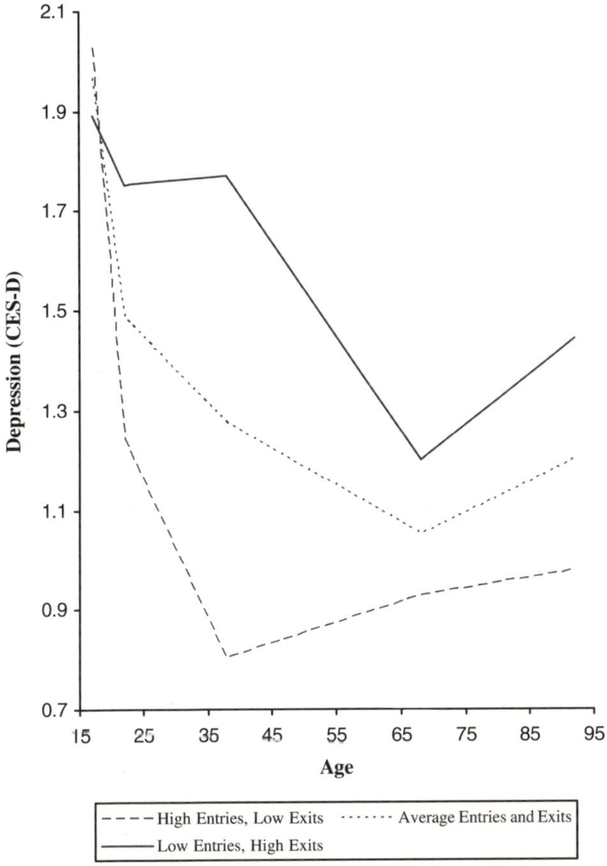

Fig. 3. Trajectories of Depression by Role Transitions over the Adult Life Course.

the relative balance of role entries and role exits over the adult life course. Model A in Table 2 replicates Model B from Table 1 with the effects of neighborhood disadvantage calculated at the midpoint of each spline age segment. The effects represent the numerical equivalent to the visual plot in Fig. 2. There is no significant difference in the mental health trajectories of young adults living across neighborhoods, but throughout the developmental and midlife stages of adulthood, individuals living in disadvantaged neighborhoods have depression scores that are about 0.17 higher than those in neighborhoods with average disadvantage. This difference is halved in later life, but still remains significant.

Table 2. Effects of Neighborhood Disadvantage and Role Transitions, Plus Controls and Mediators, in Multilevel Models for Change in Adult Life Course Depression.

	Model A	Model B	Model C	Model D[a]	Model E[a]	Model F[a]	Model G[a]	Model H[a]
Fixed effects								
Effect of ND								
At age 20	0.0458	−0.0172		−0.0226	−0.0032	−0.0403	−0.0089	−0.0107
At age 30	0.1662***	0.0785***		0.0517***	0.0459**	0.0374**	0.0453**	0.0262
At age 54	0.1694***	0.0336**		0.0154	0.0139	0.0203	0.0151	0.0182
At age 81	0.0952**	−0.0597		−0.0556	−0.0640	−0.0474	−0.0603	−0.0583
Effect of role entries at low exits[b]								
At age 20			−0.0660	−0.0701	−0.0784	−0.0557	−0.0703	−0.0588
At age 30			−0.2187***	−0.2048***	−0.2076***	−0.1792***	−0.2155***	−0.1833***
At age 54			−0.1561***	−0.1500***	−0.1529***	−0.1413***	−0.1489***	−0.1407***
At age 81			−0.0929	−0.1067	−0.1150	−0.1134	−0.0984	−0.1156
Effect of role exits at low entries[b]								
At age 20			0.0808	0.0856	0.1041	0.1181	0.1044	0.1562
At age 30			0.2977***	0.2829***	0.2851***	0.2465***	0.2938***	0.2534***
At age 54			0.2164***	0.2091***	0.2060***	0.1886***	0.2084***	0.1833***
At age 81			0.1196*	0.1270*	0.1286*	0.1245*	0.1283*	0.1272*
Control variables								
Education (years)		−0.1388***	−0.1337***	−0.1273***	−0.1342***	−0.0908***	−0.1222***	−0.0971***
Female		0.2662***	0.1981***	0.1996***	0.1959***	0.1914***	0.2018***	0.1920***
African American		0.1416***	0.1467***	0.1176***	0.1199***	0.1438***	0.1116***	0.1386***
Functional status		0.2534***	0.2369***	0.2373***	0.2361***	0.2015***	0.2361***	0.2001***
Residential stability		−0.0382***	−0.0269*	−0.0293*	−0.0294*	−0.0334**	−0.0271*	−0.0324**

Mediating variables								
Economic hardship					0.0018***			0.0019***
Mastery						−0.1210***		−0.1200***
Self esteem						−0.2203***		−0.2182***
Social interaction							−0.0311***	−0.0069
Variance components								
Level 2 initial status	0.8126***	0.6716***	0.6662***	0.6661***	0.6796***	0.6150***	0.6592***	0.6171***
Level 3 initial status	0.0202**	0.0153**	0.0138**	0.0125*	0.0123*	0.0084	0.0144**	0.0093
Pseudo-R^2 statistic								
% Variation in CESD explained	2.22	11.80	12.97	13.11	13.27	17.99	13.33	18.24

Note: ND = neighborhood disadvantage; CESD = Center for Epidemiologic Studies Depression Scale.

[a]The effect of neighborhood disadvantage is calculated at the (age-specific) average number of role entries and exits at the midpoint of each age segment. The effects of role entries and exits are calculated for the midpoint of each age segment, at the overall average level of neighborhood disadvantage.

*$p<0.05$.

**$p<0.01$.

***$p<0.001$ (two-tailed tests).

[b]Low exits and entries are calculated at 0.5 standard deviations below the mean.

We then add controls to account for individual socioeconomic and sociodemographic selection into neighborhoods (Model B, Table 2) to rule out the possibility that the association between neighborhood disadvantage and depression is spurious due to common individual level characteristics.[9] Here we see that social selection into neighborhoods accounts for a large portion of the relationship between neighborhood disadvantage and adult depression. Fully 80% of the neighborhood effect in midlife is confounded by individual socioeconomic and sociodemographic characteristics. However, there is still a small remaining net effect of neighborhood disadvantage on depression ($\beta = 0.034$, $p < 0.01$) over and above the individual characteristics that predispose socioeconomically and demographically underprivileged groups to live in disadvantaged communities by this stage of the life course. By later life, compositional effects completely account for any neighborhood effect ($\beta = -0.06$, ns). The strongest net effect of neighborhood context remains for the developmental period of adulthood. Although roughly half of the unadjusted effect of neighborhood disadvantage is explained by the controls, neighborhood context has the greatest consequences for the mental health of adults who are in this more vulnerable life course stage of status attainment ($\beta = 0.076$, $p < 0.001$). There were no significant interactions between individual controls and neighborhood disadvantage, and the mental health effects of contextual disadvantage did not vary by length of time in neighborhood.

Model C in Table 2 presents the adjusted effects of role transitions on mental health trajectories.[10] The effect of role exits is calculated at 0.5 standard deviations below the average number of role entries for each age segment, while the effect of role entries is calculated at 0.5 standard deviations below the average number of role exits for each age segment (see Appendix B for age-specific values and endnote 8). Thus, in the developmental period of adulthood, the mental health effect of the number of role exits is calculated for adults with 1.85 role entries, and in midlife the effect of role exits is calculated at 2.40 role entries. As expected, transitions into employment and marriage have a negative effect on depression across the life course, while an imbalance of role exits compromises mental health. The greatest mental health effects of role transitions occur during the developmental and midlife stages of adulthood, when the emotional benefits of social and economic status are tied to stable role acquisition. Thus, for example, each increase in the number of transitions out of marriage or employment through the developmental period of adulthood results in a 0.300 increase in depression score for someone with a history of just under two role entries. The effects of role transitions on depression do not vary by gender, but 25% of the elevated mental health risk for women in Model B is mediated by role transitions. This is consistent with Mirowsky's (1996) finding that the gender gap in depression is largely a function of differences in status attainment over the adult life course.

The effects of neighborhood disadvantage and role transitions are included simultaneously in Model D of Table 2. Here we assess the mediating role of life course transitions for the effects of contextual disadvantage and find that in the midlife period over half the adjusted mental health effect of neighborhood disadvantage is explained by role transitions ($\beta = 0.015$, ns). Role transitions also explain 34% of the neighborhood effect in the developmental period of adulthood, but a significant direct effect of neighborhood disadvantage remains ($\beta = 0.052$, $p < 0.001$). As argued earlier, this could either be a mediating effect or a function of role selection into neighborhoods. In order to distinguish between these two processes we separated the effects of role transitions that occurred in previous neighborhoods from those that occurred in the current context (results not shown) and found that possible role selection into neighborhoods accounts for very little of the contextual effect on mental health (less than 8% in the developmental and midlife stages). Current and prior transitions are therefore aggregated for all subsequent models under the assumption that current neighborhood socioeconomic characteristics act as a proxy for past neighborhood context.

The combined micro and meso level factors in Model D explain over 13% of the total variation in depression, but significant mental health variation between persons and between neighborhoods remain. The final four models in Table 2 add each mediator in turn followed by all mediators at once to explain the mechanisms by which role transitions and neighborhood disadvantage affect depression over the life course. Model E adds the measure of economic hardship, which was hypothesized to explain the mental health effect of role transitions. Economic hardship is associated with increased depression ($\beta = 0.002$, $p < 0.001$), but plays no substantial mediating role for life course transitions. The coefficients for role entries and exits change little from Model D, perhaps because our derived measure inadequately taps subjective economic hardship. However, part of the direct effect of neighborhood disadvantage is explained by economic hardship, suggesting that living in a disadvantaged community during the developmental period of adulthood compromises mental health in part through financial difficulties.

While economic hardship plays little mediating role for life course transitions, Model F indicates that 10–13% of the mental health effect of role entries and exits operates through personal mastery and self esteem. These mediating effects are strongest for role exits in the developmental and midlife stages of adulthood where transitions out of marriage and employment lead to increased depression through a reduced sense of mastery over one's life and more negative self evaluations. Mastery and self esteem are also compromised in adults living in disadvantaged neighborhoods, and account for over 27% ([0.0517–0.0374]/0.0517 = 0.277) of the direct neighborhood effect on depression in this life stage.

We also hypothesized that role transitions would affect mental health through the social ties and social interaction that are linked to role acquisitions. However, the measure of social interaction plays little mediating role for life course transitions (Model G in Table 2), but it does account for 12% ([0.0517–0.0453]/0.0517 = 0.124) of the neighborhood effect in the developmental stage. Opportunities for diverse social interaction through recreational, professional, and other interest groups may be constrained in disadvantaged communities, with subsequent implications for mental health. This is especially consequential for adults in this stage of early marriage, parenting, initial career path, and asset acquisition.

The final model in Table 2 (Model H) adds all mediators simultaneously, which account for the significant neighborhood effect in the developmental period of adulthood ($\beta = 0.026$, ns). Thus, in this model we have explained the total mental health effect of social context over this stage of the life course through a combination of role transitions, coupled with financial strain, reduced opportunities for social interaction, and psychological spillover in the form of negative self evaluations and loss of personal control. Social interaction no longer has a significant direct effect on mental health, mediated perhaps through psychological mastery and self esteem. We have explained over 18% of the total variation in depression and reduced the intercept variation in depression between neighborhoods to almost zero.

As a final step in our analyses, we estimate a set of models with life course transitions as the dependent variable in order to paint a more complete picture of the mechanisms by which neighborhood disadvantage influences role entries and exits over different stages of the life course. These models include the core set of control variables used in the previous models. We also include stress as a potential mediator.[11]

Results for these models are presented in Table 3.[12] Neighborhood disadvantage has the expected effect on role transitions, serving to reduce the number of entries into marriage and employment in stages beyond the sorting period of emerging adulthood. Contextual disadvantage also predictably operates to increase the number of role exits through early adulthood and midlife. But the effect is reversed in later life where living in a disadvantaged neighborhood is actually associated with a reduction in role

Table 3. Effect of Neighborhood Disadvantage on Role Transitions over the Adult Life Course.

	Models for Role Exits			Models for Role Entries		
Fixed effects						
Effect of ND						
At age 20	0.0968**	0.0853*	0.0779*	−0.0264	0.0108	0.0028
At age 30	0.0.1012***	0.0899***	0.0738***	−0.0451**	−0.0059	−0.0224
At age 54	0.0903***	0.0589***	0.0507***	−0.0530***	−0.0244	−0.0332*
At age 81	−0.0769*	−0.1143***	−0.1051***	−0.1615***	−0.1333***	−0.1240***
Control Variables						
Education (yrs)		−0.0690***	−0.0549***		−0.0110	0.0029
Female		0.5901***	0.5547***		0.3521***	0.3162***
African American		−0.1605***	−0.1662***		−0.3177***	−0.3236***
Functional status		0.0880***	0.0801***		0.0440***	0.0361***
Residential stability		−0.2093***	−0.2041***		−0.2009***	−0.1952***
Mediating Variable						
Stress			0.1669***			0.1673***
Variance components						
Level 2 initial Status	1.5638***	1.4394***	1.4175***	1.4603***	1.393***	1.3743***
Level 3 initial Status	0.1009***	0.0793***	0.0697***	0.1015***	0.0884***	0.0789***

Note: ND = neighborhood disadvantage.
*$p<0.05$.
**$p<0.01$.
***$p<0.001$ (two-tailed tests).

exits. Older adults in socioeconomically disadvantaged neighborhoods may experience financial strains that prevent them from retiring from their jobs. It is also possible that transitions into widowhood occur much earlier for adults in disadvantaged communities, perhaps at the tail end of midlife, reducing the number of these role exits that occur after age 69 in disadvantaged neighborhoods.

The addition of socioeconomic and sociodemographic controls somewhat weakens the effect of neighborhood disadvantage on role exits, but the effects remain statistically significant.[13] In contrast, adjustment for controls accounts for the neighborhood effect on role entries through the middle stages of adulthood. The only remaining effect exists for adults in later life and even this is reduced. Thus, the apparent benefits of living in advantaged neighborhoods for increased opportunities for entry into statuses and roles is largely a function of a common underlying process that selects individuals with more personal opportunities for role entries into advantaged communities. Opportunities for role entry are only vulnerable to contextual characteristics in later life. Note also that education is protective for role exits but has no effect on role entry, and that women and disabled adults experience more overall role instability, with an elevated risk of both exits and entries. African Americans experience fewer role transitions of either type.

If disadvantaged neighborhoods lead to more transitions out of employment and marital roles, stress is a potential mechanism by which this process occurs. When stress is added to the final column of the role exits model (Table 3), it explains 18% of the effect of neighborhood disadvantage on role exits in the developmental period of adulthood. The mediating effects of stress are somewhat weaker in the other stages of adulthood. Adults during the developmental period of adulthood, when they are in the vulnerable stages of embarking on careers and settling into married life, are more likely to be susceptible to the disruption caused by stressful events, including delinquency in one's children, spousal drug and alcohol use, and marital violence – behavior that may be aggravated in disadvantaged neighborhood environments (Boardman et al., 2001; Sampson et al., 2002). Stress plays less of a role in the link between neighborhood disadvantage and role entries in older adults.

CONCLUSION

This chapter demonstrates three important findings. First, up to half of the adjusted neighborhood effect on adult mental health operates through life course transitions, specifically the balance of role entries and exits across stages of adulthood. This principally involves an imbalance of role exits in disadvantaged neighborhoods, which have the strongest mediating effect in the developmental and midlife stages of adulthood. Role transitions completely account for any adjusted neighborhood effect in midlife and account for over a third of the mental health effect of neighborhoods in the developmental period of adulthood. These findings integrate the existing literature on the mental health effects of role transitions with research on the mental health effects of neighborhood by demonstrating that life course transitions, as a potential stressor, mediate the effects of neighborhood context on adult mental health. In addition, by examining role transitions over the entire adult life course, we specifically test the strength of their mediating relationship for the mental health effects of neighborhoods at distinct stages of adulthood. We identify life course transitions in the middle periods of adult life as accounting for much of the mental health risk of neighborhoods. These midlife marital and job transitions affect mental health largely through their psychological consequences for personal mastery and positive self evaluations.

Second, our results emphasize that role transitions are themselves a function of higher meso level social contexts. Research on the determinants of adult life course transitions tends to exist only at the micro level of social inquiry (e.g., Barrett, 2000; Marks & Lambert, 1998; Simon & Marcussen, 1999; Williams, 2003). Yet, our results prompt us to look beyond the individual to the characteristics of the surrounding neighborhood, which may create an environment that precipitates role entry or exit. Our results suggest that the role transitions of older adults are particularly susceptible to the adjusted effects of neighborhood disadvantage, since it is only in later life that both the number of role entries and role exits are influenced by social context. But the deleterious effect of neighborhood on role exits is strongest during the developmental period of adulthood (age 23 to 38) when a local environment of poverty and unemployment exposes individuals to pernicious influences that create tensions in their lives at a vulnerable point during the early stages of marriage and career.

This finding highlights the importance of focusing on social ecological conditions and their consequences for individual role transitions, particularly in view of rapid social and economic change. As a case in point, consider the city of Roanoke, Virginia, where 25% of the city's manufacturing jobs disappeared between 1993 and 2001, leaving a median family income that is 30% below the state median and a family poverty rate that is double the state rate (US Census Bureau, 2000, as reported in the *New York Times*, May 2, 2004, p. 20). At the same time, figures from the 2000 Census indicate that the rate of separation or divorce in Roanoke is one of the highest in the United States; 50% higher than the national average and having grown at three times the national average (30%) since 1990 (US Census Bureau, 2000). Our paper suggests links between these statistics in arguing that socioeconomically disadvantaged neighborhoods breed environments that contribute to role exits, in part through the precipitation of individual-level stressors that stem from exposure to factors such as poor school quality, crime, and inadequate health and social services (Massey, 1990; Sampson et al., 2002).

Finally, our results demonstrate that neighborhoods have the strongest net effect on mental health and on the consequences of life transitions for mental health during the developmental period of adulthood. As a consequence, the mental health benefits of midlife (Mirowsky & Ross, 1992) are not realized to the same magnitude for individuals living in disadvantaged environments. The vulnerability during the developmental period of adulthood, as individuals begin to embrace work and family responsibilities, may make one more susceptible to social contexts at this stage of the life course. We find that the decreased the opportunities for social interaction, lower personal mastery and self esteem, and greater economic hardship that stem from life in a disadvantaged community have adverse mental health consequences for individuals embarking on their adult life.

Despite our prediction that later life would also be a period of mental health sensitivity to neighborhood differences, our results suggest otherwise. We noted that there are multiple forces in later life that may work in opposing directions, some potentially enhancing the importance of context, via increased dependence on residential environments and increased stability in these environments, and some potentially weakening the importance of context, via the increased social and economic homogeneity of social life in old age. Our findings suggest the latter set of influences predominate. In effect, we see no evidence of the importance of increased dependence, but we do see indirect evidence that the increased homogeneity of neighborhood environments results in a convergence in mental health.

At the same time, the midlife period is a stage of increasing convergence in experience, which does conform to our expectations. The diversity in mental health trajectories, and in the balance of role entries to exits introduced during the developmental period, is notably reduced during this period. A careful look at Figs. 2 and 3 shows that the lines representing trajectories at different levels of neighborhood disadvantage and for different combinations of role transitions all tend toward convergence at this point. Thus, midlife is the signal for increased independence from neighborhood as adulthood proceeds. This is also a time of life with increased stability and similarity of experience. Children leave home, the later career is stabilized in terms of position and type of work, and the returns on long-term investments in relationships or marriage come to fruition (Wethington, Kessler, & Pixley, 2004).

This chapter advocates a framework for interpreting the mental health effects of life course transitions that consider both individual and contextual factors in tandem, as well as the various mediating processes that operate as explanatory mechanisms at both levels of social reality. Rather than focusing only on individual level predictors, we emphasize the importance of specifying social ecological forces that influence both individual role transitions and mental health. The marked differentiation of experiences evident across adulthood also demands that researchers pay attention to differential contextual salience during distinct stages of the life course. We expect that this fundamental fact emerging from our findings will have implications for the study of individual life course processes through adulthood.

Changes in the structure of the modern life course, coupled with the effects of rapid social and economic change at the ecological level, give added weight to our results, particularly in view of the mental health consequences for younger adults in the process of status attainment. As the standardized life course is replaced by increasing variability and disorder in the sequencing and progression of adult roles (Buchmann, 1989; Hogan, 1978; Marini, 1984; Moen & Han, 2001; Rindfuss et al., 1987; Shanahan, 2000), the mental health effects of this role instability are likely to be compounded by corporate restructuring and downsizing. These in turn have the potential to transform the distribution of resources and opportunities at the neighborhood level (Massey, 1990). Global restructuring at the macro level has ensuing consequences for local environments, placing a structured set of opportunities and constraints on the potential for ordered, progressive role transitions (Shanahan, 2000). Such transformation have important implications for adult mental health. Our goal in this chapter was to draw attention to the structural context of the life course, at the neighborhood level, and its implications for the experience of role transitions. A broader approach integrating both local socio-ecological characteristics and individual life course elements across adulthood will help specify and clarify the meaning of time in lives.

NOTES

1. Parenting roles are not included, since entry into parenting does not confer the same social and economic status as marriage and employment. Exit out of parenting roles could arguably never occur. We also exclude cohabitation transitions since it is unclear whether they represent the same social and economic status as marriage.

2. We are unable to distinguish between full-time and part-time employment due to a data collection procedure for the job history data in wave 2.

3. There were no statistically significant differences in the education-adjusted depression scores for non-Hispanic Whites, Hispanics and others (Native American, Asian, others). But African Americans differed significantly from all other racial groups in their mental health.

4. Although there are measures of emotional and instrumental support in the NSFH, received support is less related and consequential for mental health than the perception of available support (Wethington & Kessler, 1986), which may be captured better through increased frequency of social interaction within a diverse social network.

5. Note that role transitions and neighborhood disadvantage are not, strictly speaking, at levels 2 and 3, respectively, since we measure them as time-varying variables. However, they are incorporated in these equations as time-invariant for illustrative purposes. In any case, the composite model is equivalent to what we specify here because we do not estimate random components apart from the intercept.

6. A third wave of data has been collected for 2000, but was not available at the time of this writing.

7. In comparison to a typical quadratic model, the information criteria are smaller in the spline model, suggesting a better fitting model. The AIC is considerably smaller (21.1 points), and the numerical change in the BIC statistic (8 points) is "strong" evidence of a better fitting model, according to Raftery (1995).

8. We used the age-specific mean ± 0.5 standard deviations (rather than the typical mean ± 1 standard deviation) in order to constrain the values of role exits and entries to real ranges of combinations through each age segment. Thus, for example, the effect of low entries at high exits during the developmental period of adulthood is plotted at 1.85 entries and 1.70 exits, thereby preventing an impossible scenario of role exits exceeding role entries.

9. We tested whether the effects of the controls varied over stages of the life course, but interactions with time did not result in any improvement in model fit. All controls are therefore included in the models as main effects only.

10. We also disaggregated role transitions into employment and marital transitions (results not shown). Although we found somewhat stronger mental health effects of employment transitions in later life, and of marital transitions in midlife, both transition types remained significant over stages of adulthood. We therefore aggregated employment and marital transitions for all subsequent analyses.

11. We also examined the effects of social interaction as a potential mediator of neighborhood disadvantage on role transitions, since the proportion of unemployed men in the neighborhood may represent isolation from job opportunities and networks, but social interaction did not have any mediating effect in models for entries or exits.

12. Again, in these models, we separated the effects of transitions that occurred in the current neighborhood from transitions that occurred in previous neighborhoods by looking only at the effect of current neighborhood on current transitions. Past transitions were then added after the other control variables to account for role selection into neighborhoods. We found a similar pattern of effects to the results presented in Table 3 (although somewhat weaker). For example, the positive effects of current neighborhood disadvantage on concurrent role exits are strongest during the developmental period of adulthood, and the association remains negative for older adults. However, the positive effect of current neighborhood on role exits during the developmental period actually becomes stronger when controlling for past role selection into current context because the confounding/role selection process is an overall negative effect. This is largely because past transitions are tied up with current transitions (e.g., a history of past role entries increases the number of current exits, while previous exits reduces the number of current exits). We therefore combine past and current transitions as the dependent variables in Table 3, again following the assumption that current neighborhoods act as a general proxy for past neighborhood disadvantage.

13. The effect of current neighborhood disadvantage on lifetime role exits did not vary by length of time in current neighborhood.

ACKNOWLEDGMENT

This work was supported, in part, by a post-doctoral fellowship to Philippa Clarke, funded by the Social Sciences and Humanities Research Council of Canada.

REFERENCES

Alexander, J., Giesen, B., Munch, R., & Smelser, N. (1987). *The micro-macro link*. Berkeley: University of California Press.

Aneshensel, C., & Sucoff, C. (1996). The neighborhood context of adolescent mental health. *Journal of Health and Social Behavior*, *37*, 293–310.

Arnett, J. (2000). High hopes in a grim world: Emerging adults' views of their futures and "Generation X". *Youth and Society*, *31*, 267–286.

Arnett, J., & Taber, S. (1994). Adolescence terminable and interminable: When does adolescence end? *Journal of Youth and Adolescence*, *23*, 517–537.

Barrett, A. (2000). Marital trajectories and mental health. *Journal of Health and Social Behavior*, *41*, 451–464.

Boardman, J., Finch, B., Ellison, C., Williams, D., & Jackson, J. (2001). Neighborhood disadvantage, stress and drug use among adults. *Journal of Health and Social Behavior*, *42*, 151–165.

Booth, A., Crouter, A., & Shanahan, M. (1999). *Transitions to adulthood in a changing economy: No work, no family, no future?* Westport, Connecticut: Praeger.

Brim, O., Ryff, C., & Kessler, R. (2004). *How healthy are we? A national study of well-being at midlife*. Chicago: University of Chicago Press.

Brooks-Gunn, J., Duncan, G., Klebanov, P., & Sealand, N. (1993). Do neighborhoods influence child and adolescent development? *American Journal of Sociology*, *99*, 353–395.

Brown, G. (1981). Contextual measures of life events. In: B. Dohrenwend & P. Dohrenwend (Eds), *Stressful life events and their contexts* (pp. 187–201). New Brunswick, NJ: Rutgers University Press.

Brown, G., & Harris, T. (1978). *Social origins of depression: A study of psychiatric disorder in women*. London: Tavistock.

Buchmann, M. (1989). *The script of life in modern society: Entry into adulthood in a changing world*. Chicago: University of Chicago Press.

Buka, S., Brennan, R., Rich-Edwards, J., Raudenbush, S., & Earls, F. (2003). Neighborhood support and the birth weight of urban infants. *American Journal of Epidemiology*, *157*, 1–8.

Elder, G. (1974). *Children of the great depression: Social change in life experience*. Chicago: University of Chicago Press.

Elder, G. (1985). Perspectives on the life course. In: G. H. Elder Jr (Ed.), *Life course dynamics: Trajectories and transitions* (pp. 23–49). Ithaca, NY: Cornell University Press.

Elder, G. (1998). The life course and human development. In: R.M. Lerner (Ed.), *Handbook of child psychology theoretical models of human development*, (5th ed., Vol. 1, pp., 939–991). New York: Wiley.

Elder, G., George, L., & Shanahan, M. (1996). Psychosocial stress over the life course. In: H. B. Kaplan (Ed.), *Psychosocial stress perspectives on structure, theory, life-course, and methods* (pp. 247–292). San Diego, CA: Academic.

Elder, G., Shanahan, M., & Clipp, E. (1994). When war comes to men's lives: Life course patterns in family, work and health. *Psychology and Aging*, *9*(Special Issue), 5–16.

George, L. (1993). Sociological perspectives on life transitions. *Annual Review of Sociology*, *19*, 353–373.

George, L. (1999). Life course perspectives on mental health. In: C. Aneshensel & J. Phelan (Eds), *Handbook of the sociology of mental health* (pp. 565–583). New York: Kluwer Academic/Plenum.

Glass, T., & Balfour, J. (2003). Neighborhoods, aging, and functional limitations. In: I. Kawachi & L. Berkman (Eds), *Neighborhoods and health* (pp. 303–334). New York: Oxford University Press.

Hogan, D. (1978). The variable order of events in the life course. *American Sociological Review*, *43*, 573–586.

House, J., Umberson, D., & Landis, K. (1988). Structures and processes of social support. *Annual Review of Sociology*, *14*, 293–318.

Kessler, R., Foster, C., Webster, P., & House, J. (1992). The relationship between age and depressive symptoms in two national surveys. *Psychology and Aging*, *7*, 119–126.

Klinenberg, E. (2003). *Heat wave: A social autopsy of disaster in Chicago (Illinois)*. Chicago: University of Chicago Press.

Marini, M. (1984). The order of events in the transition to adulthood. *Sociology of Education*, *57*, 63–84.

Marks, N., & Lambert, J. (1998). Marital status continuity and change among young and midlife adults: Longitudinal effects on psychological well-being. *Journal of Family Issues*, *19*, 652–687.

Marsh, L., Cormier, D. (2002). *Spline regression models.* Sage University Papers Series on Quantitative Applications in the Social Sciences, 07–137. Thousand Oaks, CA: Sage.

Massey, D. (1990). American apartheid: Segregation and the making of the underclass. *American Journal of Sociology*, *96*, 329–357.

Massey, D. (1998). Back to the future: The rediscovery of neighborhood context. *Contemporary Sociology*, *27*, 570–572.

Massey, D., & Eggers, M. (1990). The ecology of inequality: Minorities and the concentration of poverty, 1970–1980. *American Journal of Sociology*, *95*, 1153–1188.

Menaghan, E. (1989). Role changes and psychological well-being: Variations in effects by gender and role repertoire. *Social Forces*, *67*, 693–714.

Miech, R., & Shanahan, M. (2000). Socioeconomic status and depression over the life course. *Journal of Health and Social Behavior*, *41*, 162–176.

Mirowsky, J. (1996). Age and the gender gap in depression. *Journal of Health and Social Behavior*, *37*, 362–380.

Mirowsky, J., & Reynolds, J. (2000). Age, depression and attrition in the National Survey of Families and Households. *Sociological Methods and Research*, *28*, 476–504.

Mirowsky, J., & Ross, C. (1992). Age and depression. *Journal of Health and Social Behavior*, *33*, 187–205.

Mirowsky, J., & Ross, C. (2001). Age and the effect of economic hardship on depression. *Journal of Health and Social Behavior*, *42*, 132–150.

Moen, P. (1998). Recasting careers: Changing reference groups, risks and realities. *Generations*, *22*, 40–45.

Moen, P., & Han, S. (2001). Reframing careers: Work, family and gender. In: V. Marshall, W. Heinz, H. Kruger & A. Verma (Eds), *Restructuring work and the life course* (pp. 424–445). Toronto: University of Toronto Press.

Pearlin, L., & Lieberman, M. (1979). Social sources of emotional distress. In: R. Simmons (Ed.), *Research in community and mental health* (pp. 217–248). Greenwich, Connecticut: JAI Press.

Pearlin, L., & Schooler, C. (1978). The structure of coping. *Journal of Health and Social Behavior*, *19*, 2–21.

Pearlin, L., Menaghan, E., Lieberman, M., & Mullan, J. (1981). The stress process. *Journal of Health and Social Behavior*, *22*, 337–356.

Quillian, L. (2002). Why is black-white residential segregation so persistent? Evidence on three theories from migration data. *Social Science Research*, *31*, 197–229.

Radloff, L. (1977). The CES-D scale: A self report major depressive disorder scale for research in the general population. *Applied Psychological Measurement*, *1*, 385–401.

Raftery, A. (1995). Bayesian model selection in social research. *Sociological Methodology*, *25*, 111–163.

Rindfuss, R., Swicegood, C., & Rosenfeld, R. (1987). Disorder in the life course: How common and does it matter? *American Sociological Review*, *52*, 785–801.

Rosenberg, M. (1981). The self concept: Source, product and social force. In: M. Rosenberg & R. H. Turner (Eds), *Social psychology: Sociological perspectives* (pp. 593–624). New York: Basic Books.

Ross, C. (2000). Neighborhood disadvantage and adult depression. *Journal of Health and Social Behavior*, *41*, 177–187.

Ross, C., & Mirowsky, J. (2001). Neighborhood disadvantage, disorder, and health. *Journal of Health and Social Behavior*, *42*, 258–276.

Ross, C., Mirowsky, J., & Pribesh, S. (2001). Powerlessness and the amplification of threat: Neighborhood disadvantage, disorder and mistrust. *American Sociological Review*, *66*, 568–591.

Sampson, R., Morenoff, J., & Gannon-Rowley, T. (2002). Assessing "neighborhood effects": Social processes and new directions in research. *Annual Review of Sociology*, *28*, 443–478.

Schieman, S., Van Gundy, K., & Taylor, J. (2001). Status, role and resource explanations for age patterns in psychological distress. *Journal of Health and Social Behavior*, *42*, 80–96.

Shanahan, M. (2000). Pathways to adulthood in changing societies: Variability and mechanisms in life course perspective. *Annual Review of Sociology*, *26*, 667–692.

Simon, R., & Marcussen, K. (1999). Marital transitions, marital beliefs and mental health. *Journal of Health and Social Behavior*, *40*, 115–125.

Singer, J., & Willett, J. (2003). *Applied longitudinal data analysis*. New York: Oxford University Press.

South, S., & Baumer, E. (2000). Deciphering community and race effects on adolescent premarital childbearing. *Social Forces*, *78*, 1379–1408.

South, S., & Crowder, K. (1997). Escaping distressed neighborhoods: Individual, community, and metropolitan influences. *American Journal of Sociology*, *102*, 1040–1084.

Sweet, J., & Bumpass, L., (1996). *The National Survey of Families and Households – Waves 1 and 2: Data Description and Documentation*. Center for Demography and Ecology, University of Wisconsin-Madison (http://www.ssc.wisc.edu/nsfh/home/htm).

Thoits, P. (1983). Dimensions of life events that influence psychological distress. In: H. B. Kaplan (Ed.), *Psychosocial stress: Trends in theory and research* (pp. 33–103). New York: Academic Press.

Turner, R., & Lloyd, D. (1999). The stress process and the social distribution of depression. *Journal of Health and Social Behavior*, *40*, 374–404.

US Census Bureau. (2000). Census 2000 Summary File 3: Highlights from the Census 2000 Demographic Profiles. http://factfinder.census.gov.

Wethington, E., & Kessler, R. (1986). Perceived support, received support, and adjustment to stressful life events. *Journal of Health and Social Behavior*, *27*, 78–89.

Wethington, E., Kessler, R., & Pixley, J. (2004). Turning points in adulthood. In: O. Brim, C. Ryff & R. Kessler (Eds), *How healthy are we? A national study of well-being at midlife* (pp. 586–613). Chicago: University of Chicago Press.

Wheaton, B. (1990). Life transitions, role histories and mental health. *American Sociological Review*, *55*, 209–223.

Wheaton, B., & Clarke, P. (2003). Space meets time: Integrating temporal and contextual influences on mental health in early adulthood. *American Sociological Review*, *68*, 680–706.

Williams, K. (2003). Has the future of marriage arrived? A contemporary examination of gender, marriage, and psychological well-being. *Journal of Health and Social Behavior*, *44*, 470–487.

Yen, I., & Kaplan, G. (1999). Poverty area residence and changes in depression and perceived health status: Evidence from the Alameda County Study. *International Journal of Epidemiology*, *28*, 90–94.

APPENDIX A. DESCRIPTIVE STATISTICS OF MEASURES USED IN ANALYSES: NATIONAL SURVEY OF FAMILIES AND HOUSEHOLDS, 1987–1994

	Wave 1		Wave 2	
Variable	Mean	Std Dev	Mean	Std Dev
Depression	1.22	1.39	1.13	1.28
Age	43.40	17.65	47.72	16.39
Female	0.52	0.50	0.52	0.49
African American	0.11	0.31	0.11	0.31
Education	12.56	3.12	12.82	2.97
Role entries	2.51	1.37	3.02	1.51
Role exits	1.27	1.36	1.74	1.60
Residential stability	10.02	11.55	13.29	11.80
Functional status	0.32	1.0	0.77	1.52
Neighborhood disadvantage	22.47	7.76	21.76	7.33
Economic hardship	18.15	46.17	10.03	26.02
Mastery	3.59	0.96	3.66	0.95
Self esteem	4.06	0.63	4.04	0.66
Social interaction	0.63	0.45	1.11	0.59
Stress	0.29	0.72	0.54	1.05

APPENDIX B. AVERAGE NUMBER OF ROLE ENTRIES AND EXITS (± STANDARD DEVIATION) BY AGE

	Role Entries	Role Exits
Age 17–22	1.027 ± 0.983	0.291 ± 0.727
Age 23–38	2.455 ± 1.209	1.101 ± 1.208
Age 39–68	3.139 ± 1.488	1.730 ± 1.603
Age 69+	2.853 ± 1.470	2.194 ± 1.495

WORK AND ACTIVITY CHARACTERISTICS ACROSS THE LIFE COURSE

Patricia Drentea

ABSTRACT

Those who study aging have a long-standing interest in the age-related patterns of work/activity characteristics. Such questions have become increasingly important in recent years in light of perceived changes in the nature and timing of social roles and the increasing 'standardization' and 'individualization' of the life course. Using data from two surveys, the 1995 Aging Status and Sense of Control, and from the 1987–1988 National Survey of Families and Households, bivariate and multivariate statistics were used to examine the work/activity characteristics of both the employed and non-employed over the life course. Regression models examine to what extent are the observed age patterns a function of roles and statuses. The characteristics of one's daily pursuits are age-linked, and also are in part structured by employment status, family status, health, and education. The age-linked patterns of our daily pursuits are important for understanding benefits and disadvantages to aging and employment. As such, we can contextualize the characteristics we experience in our main activities beyond our individual lives, and into age-structured phenomena.

The Structure of the Life Course: Standardized? Individualized? Differentiated?
Advances in Life Course Research, Volume 9, 303–329

ISSN: 1040-2608/doi:10.1016/S1040-2608(04)09011-2

How do the characteristics of our daily pursuits vary over the life course? Is the age-linked pattern mainly a function of employment status or of other social roles and personal circumstances as well? Previous research from both sociology and gerontology has examined the importance of activities across the life course (Baltes, Wahl, & Ulrich, 1990; Frankish, Milligan, & Reid, 1998; Havighurst, 1963; Kohn & Schooler, 1982; Larson, Zuzanek, & Mannel, 1985; Mannell, 1993; McIntosh & Danigelis, 1995; Zimmer, Hickey, & Searle, 1995). Types of activities and subsequent psychosocial outcomes change across the life course for many reasons. These include health, changing opportunities and preferences, changes in family patterns, decreased social activity, and finally financial changes (Kelly, 1993; Mirowsky & Ross, 1999b; Parnes & Less, 1985; Ross & Drentea, 1998; Schieman, Gundy, & Taylor, 2001; Shanahan, 2000). As types of activities change (Herzog, Kahn, Morgan, Jackson, & Antonucci, 1989; Verbrugge, Gruber-Baldini, & Fozard, 1996), we would expect *the characteristics* of our activities to change as well.

This chapter draws upon Kohn and colleagues' framework of examining job characteristics, but applies these characteristics to age. It examines how age and social roles structure our daily pursuits in terms of characteristics such as complexity, autonomy, and fulfillment. Age is an ascribed status that comes with a set of normative roles. This research uses the *five views of aging* to hypothesize the age-linked patterns of characteristics expected to emerge. Of interest here is how the *characteristics* of our activities vary by age. This research models how age-linked phenomena such as education, health, family status, and employment status structure activity characteristics. As such, we can contextualize the characteristics we experience in our main activities beyond our individual lives and as age-structured phenomena that is connected to the broader structure of the life course.

THE IMPORTANCE OF ACTIVITY CHARACTERISTICS

Theory and research regarding how characteristics of one's work and activities affect both personality and well-being have been a prolific area of study in sociology (Kohn 1983; Link, Lennon, & Dohrenwend, 1993; Spenner, 1988). It has determined that it is not just work or activities, but characteristics regarding work and activities that affect personality and other social psychological outcomes (Kohn, 1983). Specifically, substantive complexity, non-routine work and autonomy have been shown to be predictors of psychological well-being (Hughes, Galinsky, & Morris, 1992; Kohn, 1994; Kohn, 1983; Moore & Hayward, 1990; Zimmer et al., 1995). These characteristics enhance intellectual flexibility, which helps individuals to solve their own problems creatively, alleviate distress, and enhance emotional well-being (Bird & Ross, 1993; Kohn, 1994; Kohn & Schooler, 1982; Lennon & Rosenfield, 1992; Menaghan, 1991; Mirowsky & Ross, 1992). Manageability, or the perception that one has the resources to handle the challenges with which they are faced, has been linked to lower depression and anxiety (Antonovsky, 1987; Flannery, Perry, Penk, & Flannery, 1994). While Kohn and his colleagues have shown that activity characteristics are important for mental health, we know little about how age-related experiences are linked to activity characteristics. Most research that examines age patterns of activities study the activities themselves. For instance, some research concentrates on discretionary versus obligatory activities and finds that we spend more time as we age in obligatory and discretionary activities such as self-care and leisure (Baltes et al., 1990; Verbrugge et al., 1996) and less time in committed activities such as paid work. Others show the gradual decline in most productive activities over time (Antonucci, Jackson, Gibson, & Herzog, 1994; Herzog et al., 1989).

AGE STRATIFICATION

The concept of age stratification suggests that society structures our daily pursuits by our chronological age. Many scholars have found support for this "standardization" across the life course (Shanahan, 2000). For instance, the young go to school, middle age marks a time of work and family, and old age is a time of leisure and retirement. This has traditionally been called the "three boxes of life" (Riley et al., 1999).

Admittedly scholars, including the Rileys, have found much variation in our lives, and concluded that age stratification is loosening its grip on our aging society. Many have found increasing variation or individualization across the life course (Riley et al., 1999; Rindfuss, Swicegood, & Rosenfeld, 1987; Rindfuss, 1991; Shanahan, 2000). More age integration is taking place in society while the age norms of social roles continue to loosen. While useful theoretically in that age certainly *does* structure some events, activities, norms, roles, and statuses, it is only one way to examine how activity characteristics may be structured. Age stratification is however closely connected with different views of what aging is and means for individuals and society. Thus, different views of aging provide a foundation for thinking about age stratification, its relation to changing structures of the life course, and its connection to other dimensions of stratification that are the basis of social differentiation in aging, the life course, and its consequences.

FIVE VIEWS OF AGING

This chapter uses the *five views of aging* developed by Mirowsky and Ross (1992) to examine the meaning of age. I use these five views to examine what is meant by age, and then discuss how each perspective would predict the age-linked patterns of work and activity characteristics. This framework allows for biological, social, and psychological reasons for some characteristics to peak over time. It is important to note here that predicting any type of age-linked pattern for a population is extremely complex. While the "three boxes of life" roughly carve out three main phases of life, it is inadequate to predict the potential complexity and constellations of events and circumstances that can occur in people's lives. Still, it is a sociological axiom that there are social patterns of characteristics common to a population that are based primarily on life-cycle properties. Armed with the disclaimer that human lives are messy (Rindfuss et al., 1987) and may be getting messier (Shanahan, 2000), we proceed to examine the five views. They are *Age as Stage*, *Age as Maturity*, *Age as Decline*, *Age as Survival*, and *Age as Historical Trend.*

AGE AS STAGE

This view is informed by age stratification theory, showing that human life goes through a life cycle which includes phases that are primarily structured by changes in marital, job, and economic statuses. It follows role theory where early adulthood involves role acquisition in family formation, and first jobs, middle life includes role enactment of the myriad of the roles we acquire, and finally later life involves roll loss as we retire, launch children and lose loved ones (Elder, 1985). From this perspective, aging is characterized by the broad patterning of social roles that define the structure of the life course.

Both roles and role-related activity patterns unfold in predictable ways. Early on, family lives become more complicated with marriages, children, divorces, remarriages, and even further children. The middle adult years of the life cycle are a challenging time in which most have many job and family responsibilities and transitions (Belsky & Kelly, 1994; Pearlin & Skaff, 1996; Rindfuss, 1991). Mid-life is then a dynamic time where individuals gain problem-solving skills, intellectual flexibility, and a better understanding of the complexity of the world (Mirowsky & Ross, 1999b). The later middle years, between ages 40 and 60, lives may become more manageable as much of the upheaval of the early years such as having children, family formation, and job changes will settle down. Children later become more independent and less reliant upon us (Herzog et al., 1989) and typically we have greater financial resources (Mirowsky & Ross, 1999a). In the later years many of our daily pursuits change from highly structured, obligatory activities such as paid work and child care, to leisure and more discretionary activities (Mirowsky & Ross, 1999b; Verbrugge et al., 1996). Thus, provided good health and reasonable financial resources, retirement brings more flexibility and freedom. Retirement may change our leisure activities (Parnes & Less, 1985) and eliminates some schedules and obligations (Kelly, 1993).

Age as life-cycle stage then predicts a rise and fall of activity characteristics. Variety, complexity, problem solving, and possibly even fulfillment should peak in the middle years. Moreover, manageability and autonomy will assume the opposite pattern with the middle years being the least manageable and autonomous. Alternatively, fulfillment may rise in later life as we have more discretionary time. Controlling for work and family life, education and health should then explain any age pattern, since it is these roles and circumstances that create the activity characteristic.

AGE AS MATURITY

The age as maturity perspective considers age as a one-way progression of improvement – a summation of experience (Mirowsky & Ross, 1992). This perspective states that people get better at living as they grow older. Born out of Freud's theory of development and embraced by Erikson (1964) and Levinson (1978), it shows that we move through life solving competing conflicts within ourselves. After each transition we graduate to the next phase or stage. The early and middle adulthood is characterized by concerns over the self, of being an individual versus forming a family, of leading a productive life and so on. The later stages are more involved with being reflective, where we have become comfortable with ourselves and can be wise and concerned with others. Even within the sociological realm, there is support for this perspective (see Gove, Ortega, & Style, 1989). In this perspective, individuals learn how to handle both their disposition and situation, leading to predicted increased manageability, fulfillment, and autonomy (especially with retirement) over time. It is difficult to hypothesize whether they would have more or less complexity, problem solving, and variety over time. Alternatively, we may see problem solving increase with age. Yet, this would depend primarily on what *they* value as important characteristics. Controlling for employment and marital status should have little effect on most characteristics, because no matter what our statuses and roles, we will become better at living our lives. Similarly, health and socioeconomic status, within reason, should not dramatically affect the main upward trajectory – since the maturational perspective stresses that we solve our problems, adjust to hardship, and we adapt to our limitations.

Age as Decline

The age as decline perspective brings into play the biological realities of aging, age is about physical decline. This perspective is also reminiscent of disengagement theory where we gradually disengage from our roles and statuses in society and turn inward. As part of the physical decline, our strength and stamina diminish. Aging is about accumulated biological decline, the senescence of cells and organisms, and a decline in mental abilities as well (Hayflick, 1994). Accordingly, *age as decline* predicts decreased manageability, fulfillment, autonomy, and likely variety (as our functional status and mobility decrease). As we age, we may experience more problem solving and more complexity as handling the decline gets challenging. Controlling for roles and statuses should also not affect this decline much. However, health and socioeconomic status are important contingencies and should explain much of the expected downward slope.

Age as Survival

The *age as survival* view states that age structures the traits associated with survival so that those traits that have a selective advantage become more common with an elderly population and those with a selective disadvantage become scarcer. Thus the older one is, the more likely they will have the traits associated with survival. From this perspective, we may assume that those who feel life is not manageable (i.e., overwhelming), and who have little autonomy may be underrepresented in the oldest age groups and concomitantly most probability samples. As such, this view would predict an increase in fulfillment, manageability, and autonomy across the age categories. This view does not lend itself to predictions about variety, problem solving, and complexity. Controlling for health, education, marital, and employment status should explain much of the relationship with age because these controls are all associated with morbidity and mortality.

Age as Historical Trend

Finally, the *age as historical trend* view situates each cohort within their historical era. Life course literature shows how early experiences, such as coming of age in the Great Depression, have affected later life outcomes (Elder & Liker, 1982). This view states that activity characteristics would be dependent on one's cohort, situated in history. Thus, predictions would have to examine each cohort within each time period for each outcome, yielding hundreds of hypotheses that are clearly beyond the scope of this paper. Still, we return to this perspective in the discussion.

Increasing Heterogeneity

Another phenomenon to consider when examining age-linked patterns is that there is even more variation in later life (Dannefer & Uhlenberg, 1999). Indeed, the accumulation of advantage or disadvantage compounds over time, affecting later life outcomes. Thus in all cases, I expect more variation or heterogeneity in the oldest age groups.

In this chapter, the age-related patterns of work and activity characteristics for both the employed and the non-employed are examined. First, the age-dependent patterns are reported by the bivariate relationship between age and the different work/activity characteristics. Second, multiple regression analyses examine whether employment and marital status, health, gender, minority status, and education account for the age patterns observed. Here, gender and minority status are control variables. Results are included in Tables 2 and 3 and are graphed in Figs. 1 and 2. When the age coefficients are weakened or no longer significant after adding a group of variables, we can say part or all of the age effect is explained by those variables. Put differently, the variation in the activity characteristic is not directly associated with age.

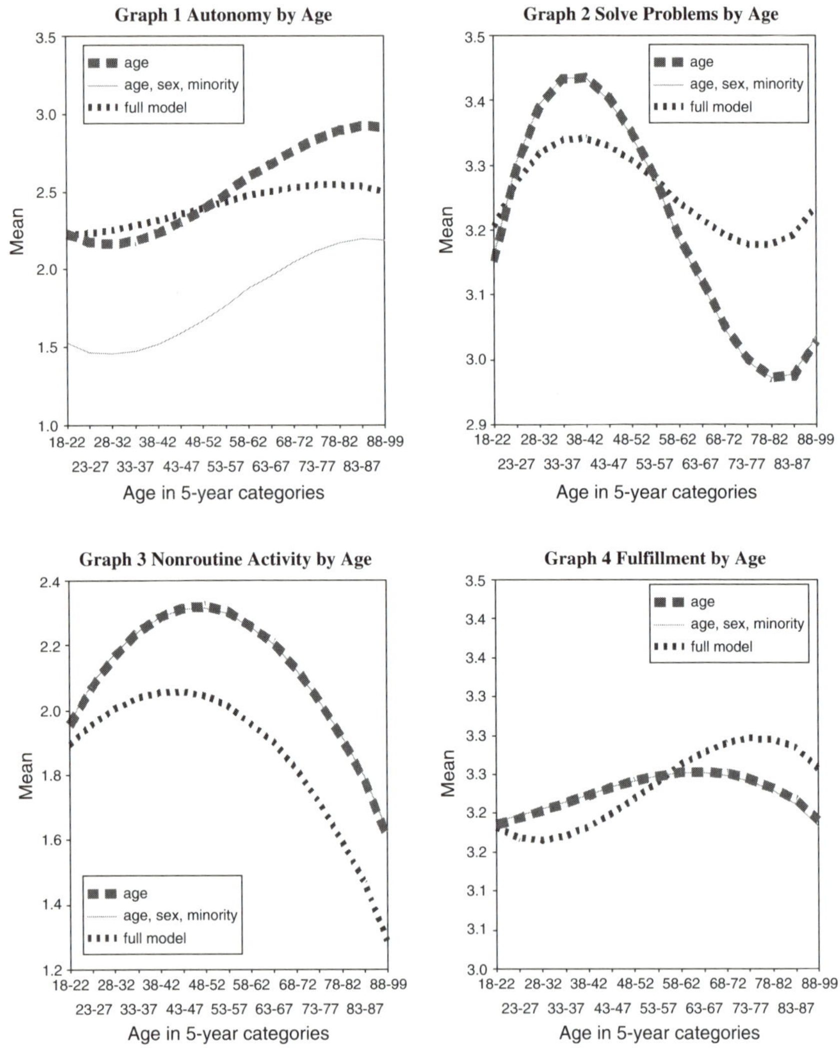

Fig. 1. ASOC Data.

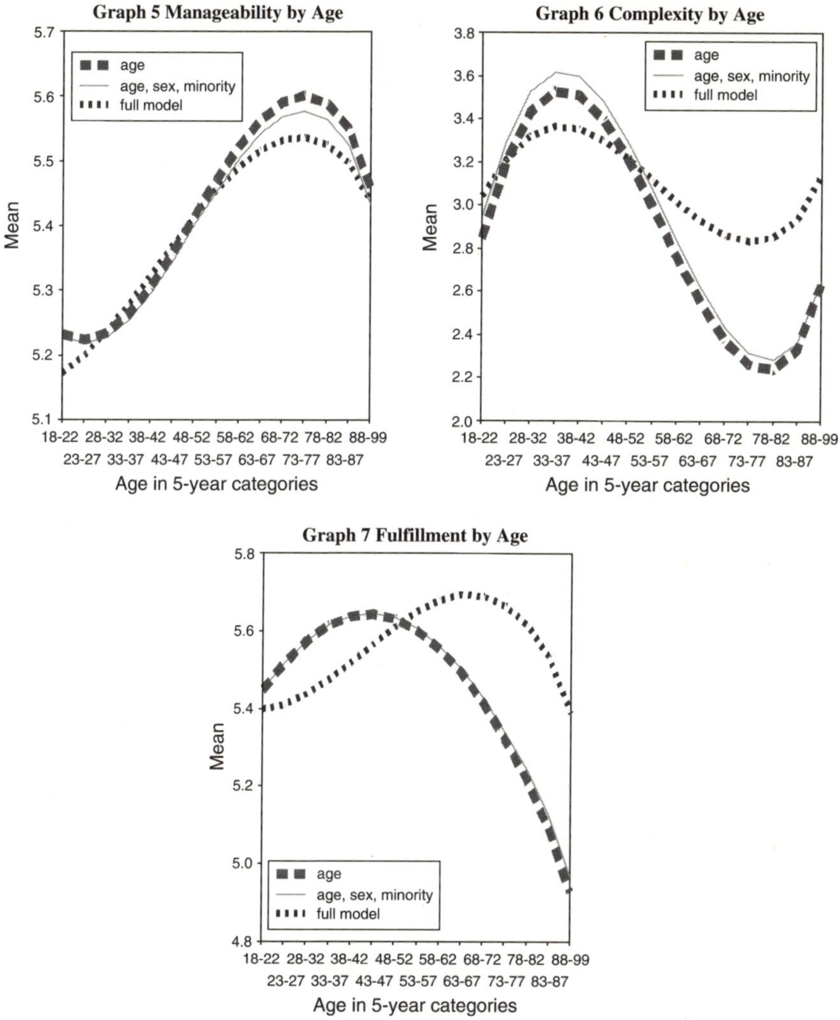

Fig. 2. NSFH Data.

DATA AND VARIABLES

This research uses two nationally representative samples of the U.S. population, the Aging, Status and Sense of Control (hereafter ASOC) survey and the National Survey of Families and Households (hereafter NSFH). A major strength of this research is that both surveys had questions regarding characteristics for respondents – not just those working for pay (Spenner, 1988). Kohn and his colleagues' research has also been criticized for not including dimensions of satisfaction (Spenner, 1988). This chapter also extends their work by examining the fulfillment one experiences in their daily pursuits, along with examining autonomy, complexity, problem solving, variety, and manageability.

The ASOC data were collected in 1995 as part of a project examining the decline in the sense of control after about age 50 and subsequent declines in health and well-being. It is a national telephone probability sample of U.S. households. The response rate for this survey was 71.6%, and there are 2,592 respondents ranging in age from 18 to 95. The survey was limited to English-speaking adults (see Ross & Drentea, 1998 for more information on the survey). The NSFH includes interviews with a probability sample of 13,005 respondents during 1987–1988 and includes an oversample of several groups including blacks, Chicanos and Puerto Ricans, single parents, people with step-children, cohabiting couples, and those who have recently married (see Sweet, Bumpass & Call, 1988 for more information). It is a multistage area probability sample in the coterminous U.S. of the non-institutionalized population age 19 and older. The response rate was 74% (Sweet et al., 1988).

Both data sets use the same sets of sociodemographic measures. In both ASOC and NSFH, *age* is number of years old, I computed *age squared* and *age cubed* in order to describe the functional form that age takes with each dependent variable. While we typically use polynomials to the squared term, I also test age cubed to examine late life shifts in the characteristics. *Sex* is coded 1 for female, 0 for male. *Minority Status* is coded 1 for non-whites and 0 for whites. *Education* is highest grade completed. Below are the measures that differ slightly across data sets.

ASOC Measures

The ASOC includes several measures of life stage, including marital status, which differentiates *married* from *divorced, single* and *widowed, number of children* (total number of children under age 18 living in household), and employment status (which differentiates *employed full-time, employed part-time, keeping house full-time, retired, unable to work because of disability, temporarily unemployed* or *laid off,* or *going to school*). Each of these employment statuses were coded as a dummy variable where 1 is their employment status and 0 is not. These categories are mutually exclusive and represent self-reports of one's main employment status. We also include *self-reported health* based on a five-point scale ranging from very poor to very good.

The main outcomes of interest are *work and activity characteristics*. These are based on a series of questions about the characteristics of the activities of their main employment status. The word "work" was used for those who worked full or part-time, the words "your daily activities" were used for all other respondents. First, *autonomy* is measured by a series of questions asking whether the respondent does "not [have] anyone supervise [their] daily activities or work or "[has] someone who supervises [them]," "how free [they] feel to disagree with the person who supervises [their] work/activities" (ranging from 'not at all free' to 'don't have anyone supervising them,') "who usually decides *how* [they] will do [their] work/activities" (ranging from 'someone else' to '[them]'), and "who usually decides *what* [they] will do in [their] work/activities" (ranging from 'someone else' to 'them'). Responses were summed and averaged, a higher number indicates high autonomy ($\alpha = 0.776$).

A second activity variable is *solving problems*. Respondents were asked a single item: "In my work/activities I have to figure out how to solve problems." This variable ranges from strongly disagree to strongly agree. *Non-routine work/activities* are also measured by a single variable: "does your work(activities) usually involve doing. . ." (1) the same thing in the same way repeatedly, (2) the same kind of thing in a number of different ways, or (3) a number of different kinds of things. Finally, *fulfillment* is measured by the questions "my work(activities) gives me a chance to do things I enjoy" (ranging from strongly disagree to strongly agree), "my work/activities gives me a chance to develop and to learn new things" (ranging from strongly disagree to strongly agree), "my work/activities gives me a chance to interact with people I like (ranging from strongly disagree to strongly agree), and "if a good friend told you he or she was interested in doing what you do (having the same job as you), would you…" (1) advise against it, (2) have doubts about recommending it, (3) depends on the person (friend), and (4) strongly recommend it. Responses were averaged and the index ranges from 1 to 4. A high score indicates more fulfillment ($\alpha = 0.667$).

NSFH Measures

As before, the NSFH data contain several measures of life stage. Marital status differentiates respondents who are *married* from those who are *separated, widowed, divorced* and *never married.* Having *children* is measured by the total number of kids in the house upto age 18. For employment statuses, I created seven employment statuses comparable to the ASOC analysis through a process of deduction. First, those who were in the military were omitted ($n = 81$) for comparability. Second, respondents who were in school full-time were coded as *students.* Those who do not have a job and reported having a limitation that renders them unable to work were coded as *unable to work*. Those who worked less than 35 h on average each week were coded as *employed part-time*, and those who worked 35 or more hours on average each week were coded as *employed full-time.* Those who were not working but were looking for work in the past 4 weeks were coded as *unemployed.* Men who did not fall into any of these above categories and who were over age 50 were coded as *retired.* Women who were 65 or older and who reported being retired at wave 2, were coded as retired at wave 1. Women who were under age 65, who did not fit in any of the above categories were coded as *homemakers.* Comparisons of the two data sets showed that the percentage of individuals in each of the employment statuses were remarkably similar. *Self-reported health* is again measured on a five-point scale ranging from very poor to excellent.

The main *work and activity characteristics* were again based on a series of items that asked respondents about the work they do around the house and in their paid jobs. Respondents were asked "how would you describe the work you do around the house?" and "how would you describe the work you do at your paid job?" In order to have work and activity characteristics comparable to ASOC, I computed work/activity variables for each question. The variables were created as follows: if the respondent worked, the score on each response reflects their work, and if they do not work (e.g. are retired, homemakers etc.), the responses reflect the work they do around the house. For each question, the respondent circled a number between 1 and 7 to reflect where they fall on the continuum.

The NSFH analyses consider three dimensions of activities. First, *manageability* is measured in terms of a ranking of respondent's daily pursuits. This ranged from 1 to 7, where 1 was "overwhelming" and 7 was "manageable." Second, *complexity* was measured in terms of a ranking of daily pursuits, where 1 was "simple" and 7 was "complicated." Finally, *fulfillment* is an index using the following sets of work/activity characteristics: "boring" to "interesting"; "unappreciated" to "appreciated"; "lonely" to "sociable"; and finally, "poorly done" to "well done." Each response ranged from 1 to 7 and the summary measure was an average over the four items ($\alpha = 0.680$). Means and standard deviations for all variables are shown in Table 1.

Table 1. Means and Standard Deviations for both Data Sets (Weighted Samples).

Survey Data	ASOC		NSFH	
	Mean	Standard Deviation	Mean	Standard Deviation
Age	47.6	17.68	43.55	17.66
Sociodemographic background				
Female	0.57	—	0.53	—
Minority	0.18	—	0.2	—
Education	13.4	2.65	12.56	3.12
Kids	0.71	1.05	0.75	1.15
Married	0.58	—	0.61	—
Separated	—	—	0.03	
Widowed	0.11	—	0.07	—
Divorced	0.14	—	0.08	—
Single/never married	0.16	—	0.2	—
Health	4.07	0.94	4.01	0.83
Work statuses				
Retired	0.18	—	0.13	—
Full-time	0.49	—	0.49	—
Part-time	0.09	—	0.09	—
Homemaker	0.11	—	0.15	—
Unable to work	0.03	—	0.06	—
In school	0.04	—	0.04	—
Unemployed	0.03	—	0.03	—
Work and activity characteristics				
Autonomy	2.41	0.59	—	—
Manageability	—	—	5.37	1.82
Solve problems	3.27	0.68	—	—
Complexity	—	—	3.15	2.08
Non-routine	2.2	0.86	—	—
Fulfillment	3.22	0.55	5.57	1.17
N	2592		12897	

ANALYTIC STRATEGY

In both data sets, descriptive statistics were weighted for generalizability, but multivariate analyses use unweighted data. Unweighted data were used because the Ordinary Least Squares (OLS) estimates will be unbiased, consistent, and have smaller standard errors than weighted OLS estimates (see Winship & Radbill, 1994 for handling sample weights in regression analysis using complex data). First, in a bivariate representation, the mean level of each work and activity characteristic by age groups using standard error plots with means and 95% confidence intervals (not shown due to space limitations). The variation both within each age category and across the age categories was analyzed. Respondents were grouped by 5-year age groups across the life course to detect nuances in life-stage differences in each characteristic. In each case, there is increasing variation of activity characteristics in later life. However, due to relatively few respondents in these late life categories, the results should be interpreted with caution. Regression analyses of each work and activity characteristic is then shown. The models first regress autonomy, solving problems, non-routine work and fulfillment (ASOC) and manageability, complexity, and fulfillment (NSFH) on age to establish the association between age and each characteristic. In the second and third set of equations, sociodemographic background characteristics, health, family, status, and employment statuses are added to examine whether they account for the age pattern, net of control variables. Visual representations of the equations are graphed, solved at mean levels of the other variables (Graphs 1–7). In the graphs, the heavy broken line represents Eq. (1), the bivariate association with age (Eqs. (1), (4), (7), and (10)), the light, finely broken line shows the associations of age with each activity characteristic after sex and minority status are controlled (Eqs. (2), (5), (8), and (11)), and the full model (Eqs. (3), (6), (9), and (12)) with all controls is shown by the light broken line.

DOES AGE OR AGE-LINKED ROLES STRUCTURE ACTIVITY CHARACTERISTICS?

ASOC Data

Table 2 shows autonomy regressed on age, sociodemographic background variables, health, family status, and employment status. All three age

Table 2. OLS Coefficients: Work/Activity Characteristics Regressed on Age, Sociodemographic Background Characteristics, Family and Work Status (ASOC Unweighted Sample).

	Autonomy			Solving Problems			Non-routine			Fulfillment		
	Eq. (1)	Eq. (2)	Eq. (3)	Eq. (1)	Eq. (2)	Eq. (3)	Eq. (1)	Eq. (2)	Eq. (3)	Eq. (1)	Eq. (2)	Eq. (3)
Age	0.016**	0.016**	0.008**	−0.009**	−0.009**	−0.004*	0.003*	0.003*	−0.001	0.002	0.002	0.004**
	(15.089)	(14.881)	(6.334)	(−6.626)	(−6.915)	(−2.070)	(1.955)	(1.717)	(−0.453)	(1.575)	(1.496)	(2.517)
Age^2	3.171E-4**	3.174E-4**	3.6E-05	−4.895E-4**	−4.806E-4**	−2.110E-4**	−4.390E-4**	−4.279E-4**	−3.243E-4**	−2.4E-05	−2.2E-05	5.8E-05
	(6.412)	(6.425)	(0.671)	(−7.849)	(−7.735)	(−2.846)	(−5.413)	(−5.298)	(−3.274)	(−0.449)	(−0.425)	(0.949)
Age^3	−7.991E-6**	−8.103E-6**	−3.126E-6*	1.106E-5**	1.118E-5**	5.457E-6**	3.1E-07	5.3E-07	−7.5E-07	−8.4E-07	−9.1E-07	−2.5E-06
	(−4.890)	(−4.966)	(−1.976)	(5.29[illegible])	(5.368)	(2.459)	(0.116)	(0.198)	(−0.258)	(−0.456)	(−0.491)	(−1.314)
Sociodemographics												
Female		0.062*	−0.024		−0.089**	−0.003		−0.142**	−0.05		0.023	0.082**
		(2.97)	(−1.195)		(−3.346)	(−0.123)		(−4.104)	(−1.356)		(1.037)	(3.579)
Minority		−0.04	−0.03		−0.126**	−0.096**		−0.133**	−0.079*		−0.022	0.033
		(−1.417)	(−1.224)		(−3.522)	(−2.802)		(−2.879)	(−1.736)		(−0.733)	(1.143)
Education			0.013**			0.043**			0.042**			0.019**
			(3.731)			(8.729)			(6.489)			(4.605)
Health			0.028**			0.035**			0.065**			0.127**
			(2.689)			(2.38)			(3.319)			(10.378)
Marital status (reference category = married) and children												
Divorced			0.049*			0.011			−0.122**			−0.063*
			(1.785)			(0.281)			(−2.396)			(−1.986)
Single			−0.031			−0.035			−0.082			−0.111**
			(−0.970)			(−0.771)			(−1.369)			(−3.008)
Widow			0.060*			0.015			−0.065			−0.045
			(1.945)			(0.340)			(−1.112)			(−1.215)
Kids			0.014			0.018			−0.037*			−0.001
			(1.299)			(1.194)			(−1.792)			(−0.115)

Table 2. (*Continued*)

	Autonomy			Solving Problems			Non-routine			Fulfillment		
	Eq. (1)	Eq. (2)	Eq. (3)	Eq. (1)	Eq. (2)	Eq. (3)	Eq. (1)	Eq. (2)	Eq. (3)	Eq. (1)	Eq. (2)	Eq. (3)
Work statuses (reference category = full time)												
Retired			0.579**			−0.311**			0.154*			−0.064
			(15.567)			(−5.944)			(2.218)			(−1.459)
			−0.058			−0.318**			−0.133*			−0.03
Part-time			(−1.631)			(−6.479)			(−2.030)			(−0.736)
			0.699**			−0.396**			−0.196**			−0.300**
Homemaker			(20.955)			(−8.450)			(−3.142)			(−7.721)
			0.669**			−0.373**			−0.248**			−0.573**
Unable to work			(12.311)			(−4.853)			(−2.406)			(−8.919)
			0.517**			0.032			−0.145			0.125*
In school			(8.976)			(0.399)			(−1.352)			(1.879)
			0.743**			−0.278**			0.106			−0.210**
Unemployment			(12.295)			(−3.255)			(0.937)			(−2.852)
Constant	2.3	2.272	1.836	3.406	3.476	2.716	2.315	2.417	1.491	3.233	3.224	2.424
Adjusted R^2	0.209	0.212	0.42	0.059	0.067	0.161	0.027	0.036	0.085	0	0	0.147
F-statistics	224.849	137.5	97.68	53.719	37.341	26.726	24.025	19.668	13.405	1.234	1.06	22.971

Note: T ratios are in parentheses.
$^{*}p < 0.05$.
$^{**}p < 0.01$.

variables are significantly associated with autonomy, indicating that the cubic function best describes the bivariate relationship between age and autonomy. In Eq. (3), I added controls, marital and employment status variables. The divorced and widowed report more autonomy than those who are married. Retirees, homemakers, the disabled, students, and the unemployed, all have more autonomy compared to full-time workers – the comparison category. This model predicts 42% of the variance in autonomy. In all these regression analyses, the constant indicates the mean level of each characteristic for white men at age 45.

Graph 1 shows the s-shaped pattern with age in which autonomy increases in the latter stages of life (see heavy broken line, Graph 1). In the bivariate relationship, autonomy is lower in younger to middle ages, with a slight decrease in autonomy around the ages of 28–32 (see Eq. (1), Table 2). Around age 28, many people finish with school or experience their first entry-level positions – which typically have little autonomy. As age increases, autonomy increases as well. Around ages 50–70 – typical years for retirement – autonomy increases sharply, and then gradually increases even more until the late 80s. With sex and minority status controlled, the same pattern is maintained, only with slightly lower levels of autonomy. After controls, one's employment status largely explains the upswing of autonomy later in life. All employment statuses except for part-time work are associated with having a high degree of autonomy. Put differently, the age-dependent pattern of autonomy has more to do with what one is doing during particular stages of the life course rather than with age (see light dotted line, Graph 1).

Graph 2 shows a strong association of age with solving problems. In Eq. (4) (Table 2), the linear, second- and third-order polynomial functions of age are significantly associated with solving problems, indicating two bends (see heavy broken line, Graph 2). Problem solving peaks about age 35 and then declines through age 80. In Eq. (3), minority status and all of the work statuses are associated with solving fewer problems than full-time workers (except for being in school). Education, better health, being in school, and working full-time are also associated with a high level of problem solving. In addition, the age effect is slightly attenuated, indicating that employment status explains some of the effect of age on solving problems (see light dotted line, Graph 2). Interestingly, retirees solve the fewest number of problems in their daily pursuits. The age-dependent pattern of solving problems appears to reflect both age and work statuses.

Both the linear and square function of age are significantly associated with non-routine work, indicating that non-routine work peaks in middle age, and then becomes more routine again (see heavy broken line, Graph 3). The relationship resembles an inverted U-shaped pattern over the life course. One's daily pursuits are more varied through the late 20 s, 30 s, and 40, peak about age 50, and decrease thereafter. Non-minority status, education, and health are significantly associated with more non-routine activities, while having more children and those who are divorced have less non-routine activities (see Eq. (7), Table 2). Compared to full-time workers, retirees have more variety than full-time workers, while all other employment statuses have significantly less non-routine activities. The quadratic age pattern remains in the full model, but there is less overall variety after controls for sociodemographic, health, and life stage variables (see light dotted line, Graph 3).

There is no bivariate association of age with fulfillment as shown in Eq. (1) (see Graph 4). On average, individuals report a fairly high level of fulfillment. Once I add all variables to the equation, there is a small positive linear effect of age on fulfillment (see Eq. (12), Table 2). The divorced and single have lower levels of fulfillment as compared to the married. Homemakers, the disabled, and the unemployed have less fulfillment in their daily pursuits than full-time workers. Being in school is more fulfilling than full-time work. Overall, fulfillment is more a product of education, marital, and employment statuses, and has little association with age.

NSFH Data

Regarding manageability, there is a slight s-form with an upswing and then downswing of manageability (see heavy broken line, Graph 5). Manageability is lowest around age 30 and steadily increases until about age 72. Only in the two greatest age categories (ages 88–99) does manageability decrease slightly. There is also greater variance among the oldest groups. This may be due to more persons being overwhelmed because of functional impairment in the oldest age groups, or the transition or hardships associated with the loss of loved ones, or financial hardship for some.[1] In Eq. (3) (Table 3), higher-order functions of age are no longer significant, indicating that the later life downswing of manageability with age is due to the positive effect of minority status, health, and employment status (see light dotted line, Graph 5). Better health is associated with greater manageability. Retirees, homemakers, part-time workers, students, and the unemployed all have more manageability compared to full-time workers – the comparison category.

Table 3. OLS Coefficients: Work/Activity Characteristics Regressed on Age, Sociodemographic Background Characteristics, Family and Work Status.

	Manageability			Complexity			Fulfillment		
	Eq. (1)	Eq. (2)	Eq. (3)	Eq. (1)	Eq. (2)	Eq. (3)	Eq. (1)	Eq. (2)	Eq. (3)
Age	0.011**	0.011**	0.009**	−0.030**	−0.031**	−0.014**	−0.00075	−0.0006	0.009**
	(5.703)	(5.755)	(3.752)	(−14.861)	(−15.457)	(−5.539)	(−0.631)	(−0.503)	(6.189)
Age^2	8.8E-05	8.9E-05	−3E-05	−0.01**	−0.001**	−5.778E-4**	−3.456E-4**	−3.383E-4**	3.7E-05
	(1.24)	(1.256)	(−0.321)	(−16.536)	(−16.950)	(−5.912)	(−7.562)	(−7.416)	(0.650)
Age^3	−6.115E-6*	−6.077E-6*	−3E-06	3.480E-5**	3.454E-5**	1.775E-5**	4.3E-07	6E-07	−5.478E-6**
	(−2.197)	(−2.184)	(−0.964)	(11.302)	(11.435)	(5.387)	(0.237)	(0.333)	(−2.842)
Sociodemographics									
Female		0.073*	0.009		−0.463**	−0.193**		−0.150**	0.040*
		(2.205)	(0.242)		(−12.935)	(−4.933)		(−7.059)	(1.757)
Minority status		0.108**	0.109**		−0.709**	−0.485**		0.068**	0.168**
		(2.973)	(2.804)		(−17.944)	(−11.886)		(2.913)	(7.027)
Education			−0.009			−0.103**			−0.009**
			(−1.492)			(16.408)			(−2.584)
Health			0.133**			−0.006			0.198**
			(6.263)			(−0.248)			(15.087)
Marital status and number of kids (reference category = married)									
Separated			0.11			−0.124			−0.154**
			(1.405)			(−1.510)			(−3.200)
Widowed			0.011			0.107			−0.160**
			(0.175)			(1.555)			(−3.958)
Divorced			0.073			−0.008			−0.109**
			(1.443)			(−0.156)			(−3.498)
Never married			0.013			−0.189**			−0.102**
			(0.252)			(−3.501)			(−3.234)
Kids			−0.007			−0.037*			−0.027**
			(−0.417)			(−2.087)			(−2.587)
Work statuses (reference category = full time)									
Retired			0.193**			−0.968**			−0.592**
			(2.480)			(−11.793)			(−12.336)
Part-time			0.244**			−0.510**			0.025

Table 3. *(Continued)*

	Manageability			Complexity			Fulfillment		
	Eq. (1)	Eq. (2)	Eq. (3)	Eq. (1)	Eq. (2)	Eq. (3)	Eq. (1)	Eq. (2)	Eq. (3)
			(3.991)			(−7.928)			(0.6777)
Homemaker			0.212**			−0.997**			−0.644**
			(3.890)			(−17.371)			(−19.202)
Unable to work			0.09			−0.609**			−0.767**
			(1.112)			(−7.111)			(−15.323)
In school			0.394**			−0.829**			−0.349**
			(3.932)			(−7.852)			(−5.651)
Unemployment			0.294**			−1.208**			0.722**
			(3.091)			(−12.056)			(−12.329)
Constant	5.358	5.285	4.816	3.395	2.703	2.703	5.643	5.712	5.16
Adjusted R^2	0.005	0.006	0.011	0.04	0.076	0.135	0.009	0.013	0.099
F-statistic	20.74	15.296	8.707	180.765	214.087	107.208	41.082	36.132	75.911

Note: *T* ratios are in parentheses.
$^{*}p<0.05$.
$^{**}p<0.01$, one-tailed tests.

Those unable to work (disabled) do not experience significantly less manageability than full-time workers, after controlling for other factors.

All three age variables are significantly associated with complexity (see Eq. (1), Table 3), meaning that the third-order polynomial best describes the bivariate relationship between age and complexity. There are two bends in the curve, one occurring in mid-life and another in later middle-age. Complexity peaks about age 35 and then declines until the late 70s (see heavy broken line, Graph 6). There is a slight upturn in the later cohorts. Women and those with more education indicate higher levels of complexity, whereas minorities report lower levels of complexity in their lives (see Eq. (3), Table 3). Never having been married is associated with less complexity than those who are married, as is having more children. The age effects are slightly attenuated, indicating that complexity is in some part due to sociodemographics, work, and family status (see light dotted line, Graph 6). Retirees, homemakers, part-time workers, the disabled, students, and the unemployed all indicate less complexity in their activities compared to full-time workers – the comparison category. Retirement is associated with a low degree of complexity, rendering one's daily pursuits even less complex than any other employment status except that of homemakers.

Finally in the fulfillment regressions, Eq. (1) in Table 3 shows an inverted U-shaped association between age and fulfillment. It peaks in middle-age, and then decreases (see heavy dashed line, Graph 7). All marital statuses are associated with lower fulfillment compared to being married (see Eq. (3), Table 3). More children in the home is associated with less fulfillment as well. Once status variables are added, the first and cubic terms for age become significant (see light dotted line, Graph 7). Here, fulfillment peaks in the late 60 s, once other variables are controlled. Retirees report lower levels of fulfillment than full-time workers. Similarly, homemakers, the disabled and the unemployed have less fulfillment in their daily pursuits than full-time workers. Part-time work is as fulfilling as full-time work, indicating that those who have jobs experience the highest fulfillment. Overall, there is still a significant effect of age on fulfillment that is not completely diminished by SES, health, sociodemographics, marital, or employment status.

Summary

Two nationally representative surveys were used to examine the relationship of age with work/activity characteristics. To sum up the hypotheses, *age as stage* predicted a rise and fall of variety, complexity, problem solving, and possibly fulfillment. Variety, complexity, problem solving all show a rise and fall, with mid-life being the most active period of the life course. Fulfillment showed little variation across the age categories, especially after controls were implemented. Manageability and autonomy were predicted to show a U-shaped pattern, yet they were more accurately represented as both simply increasing with age. The controls for work and family life, SES and health explained much of the age pattern. Thus the *age as stage* hypothesis is strongly supported.

The *maturational perspective* predicted increased manageability, fulfillment, and autonomy over time. This was found for autonomy and manageability and to some extent with fulfillment in NSFH. This perspective also predicted problem solving to rise with older age, but was only partially supported. It was also expected that other variables should have little affect on most characteristics. This was not the case as controlling for SES, gender, family, and employment status explained much of the relationships. Overall, there is little support for the maturational perspective.

The *age as decline* perspective predicted decreased manageability, fulfillment, autonomy, and likely variety. After about age 50, non-routine activities did decline. In contrast, fulfillment rose slightly, while autonomy and manageability increased substantially. This hypothesis also predicted that we may experience more problem solving and complexity as we age. Complexity did increase over time, yet problem solving declined after age 35. Finally, it predicted that roles and statuses should not affect the expected declines. In the cases where there were declines, this is the case. In sum, *age as decline* had mixed results. Autonomy and manageability showed an age distribution that was exactly the opposite of that predicted. Problem solving and variety however did decline.

The *age as survival* perspective only predicted an increase in fulfillment, manageability, and autonomy across the age categories. Indeed, fulfillment, manageability, and autonomy did increase as age categories grew higher. Control variables explained some of the relationship with age. Age as survival has some support as those with great autonomy and fulfillment likely stayed in the sample and lived longer. There was also support for the hypothesis of increasing variation in later life of activity characteristics.

CONCLUSION

This research shows that while age certainly structures the characteristics of daily pursuits (Kahn, 1994; Riley et al., 1994; Riley & Riley, 1994;

Verbrugge et al., 1996), age-related patterns are clearly shaped by the structure of the life course. This is seen by both the strong support for the age as life stage perspective, as well as the general significance of life stage variables indexing marital and employment status. It is also worth noting the comparative strength of the work status variables, indicating that our main work and activities structure much of our lives. This lends support to the notion of standardization across the life course (Shanahan, 2000), particularly with respect to work and work roles. Still, this shows that a pattern emerges, not due to roles, but rather to the characteristics of the roles.

Gerontologists have long been interested in activity, while sociologists are interested in work. This study merges these interests, viewing the activities of the employed and those who are not employed as comparable. The characteristics of individuals' daily pursuits are important for understanding benefits and disadvantages of aging. Examining how age-affiliated life circumstances, such as family and employment status, affect activity characteristics allow us to contextualize the characteristics we feel. For instance, a 35-year-old mother of two may find comfort to see it is a common experience for problem solving to peak at that time. We may all find solace in the finding that autonomy increases with age.

Keeping more people active and employed longer may aid both society and individuals (Riley et al., 1994). As a society, we benefit by keeping more people employed longer for the social security system to stay solvent (Kohli, 1994). Thus, it is important for both government and business to understand what it is the people gain from employment in order to understand how best to keep them in the labor force beyond retirement age. This study suggests that the elderly gain a sense of autonomy in later life. Thus, altering employment opportunities for the elderly to have a high degree of autonomy may persuade some to either remain in or seek those positions.

There are some limitations to this research. First, non-response bias is a problem in most survey research. Regarding this project, non-response patterns from certain employment statuses, family structures, or age ranges would be especially problematic for these findings. There is no way to know whether there are patterns in who did not respond to these surveys. However, given the large representative samples and analyses of two data sets, we can be more confident.

A second limitation is how comparable are activity characteristics on the work and non-work roles. Kohn and colleagues constructed separate questionnaires to assess the applicability of their measures for non-work roles such as homemaker and found that, while most measures were applicable, it was helpful to change the content of the questions for the role (Schooler,

Kohn, Miller, & Miller, 1983). In this project, most activity characteristics are inherently transferable to all the employment statuses, for instance, everyone can answer how much they solve problems. Related, workers report only on their work characteristics, even though their work may not be the most salient characteristic in their lives. This may reflect the assumption that work is the most salient for workers and reports characteristics only about work. It is however plausible that for most workers, their work *is* the main activity of the day and therefore is highly salient.

Third, these results are cross-sectional and cohort interpretations are always a possibility. Certainly, longitudinal data would appropriately address the aging-period-cohort effects. One main cohort difference that may affect these findings is that younger cohorts have more education than ever before (NRC, 1999). Thus we may expect that as those young cohorts age, they may behave differently than the older cohorts currently do – as we know that early life experience greatly affects later life outcomes (Elder & Liker, 1982).

Still, there is room for further consideration of differentiation of activity across the life span. Future research should examine the race and sex variations in each activity characteristic. It is likely that they will vary significantly by these ascribed statuses, as well as achieved statuses like education, and health status. We expect a myriad number of combinations of work and family scenarios that would change the outcomes of the activity characteristics. We should also examine whether on/off time roles would change the activity characteristics. For instance, being a younger retiree – a trend in the U.S. today, may be very different than being a normatively aged retiree. The younger retiree may be wealthy and healthy, leading to better outcomes; or she/he may have experienced forced retirement, thus yielding more negative outcomes. Future research could also examine how age patterns and activity characteristics ultimately affect psychological well-being outcomes such as depression and so on.

NOTE

1. We hesitate to emphasize this too strongly as manageability had especially wide confidence intervals associated with the later years in life.

ACKNOWLEDGMENTS

I am indebted to John Mirowsky, Ross Macmillan and anonymous reviewers for their comments on a previous draft. I thank John Mirowsky (PI) and

Catherine E. Ross (co-PI) for use of their 1995 Aging, Status, and Sense of Control data set, funded by the National Institute on Aging (RO1 AG12393).

REFERENCES

Antonovsky, A. (1987). *Unraveling the mystery of health: How people manage stress and stay well.* San Francisco: Jossey-Bass Publishers.

Antonucci, T. C., Jackson, J. S., Gibson, R. C., & Herzog, A. R. (1994). Sex differences in age and racial influences on involvement in productive activities. In: M. S. Muncie (Ed.), *Gender roles through the life span: A multidisciplinary perspective* (pp. 259–282). IN: Ball State University Press.

Baltes, M. M., Wahl, H.-W., & Ulrich, S.-F. (1990). The daily life of elderly Germans: Activity patterns, personal control, and functional health. *Journal of Gerontology: Psychological Sciences, 454*, 173–179.

Belsky, J., & Kelly, J. (1994). *The transition to Parenthood.* New York: Dell.

Bird, C. E., & Ross, C. E. (1993). Houseworkers and paid workers: Qualities of the work and effects on personal control. *Journal of Marriage and the Family, 55*, 913–925.

Dannefer, D., & Uhlenberg, P. (1999). Paths of the life course: A typology. In: V. Bengston & K. W. Schaie (Eds), *Handbook of theories of aging* (pp. 306–326). NY: Springer Publishing Company.

Elder, G. H., & Liker, J. K. (1982). Hard times in women's lives: Historical influences across forty years. *American Journal of Sociology, 88*, 241–269.

Erikson, E. (1964). *Childhood and society*. NY: Norton.

Flannery, R. B., Perry, J. C., Penk, W. E., & Flannery, G. J. (1994). Validating Antonovsky's sense of coherence scale. *Journal of Clinical Psychology, 504*, 575–577.

Frankish, J. C., Milligan, C. D., & Reid, C. (1998). A review of relationships between active living and determinants of health. *Social Science and Medicine, 473*, 287–301.

Gove, W. R., Ortega, S. T., & Style, C. B. (1989). The maturational and role perspectives on aging and self through the adult years: An empirical evaluation. *American Journal of Sociology, 94*, 1117–1145.

Havighurst, R. A. (1963). Successful aging. In: R. Williams, C. Tibbetts & W. Donahue (Eds), *Processes of aging* (pp. 299–320). New York: Altherton.

Hayflick, L. (1994). *How and why we age.* New York: Ballantine Books.

Herzog, A. R., Kahn, R. L., Morgan, J. N., Jackson, J. S., & Antonucci, T. C. (1989). Age differences in productive activities. *Journal of Gerontology: Social Sciences, 444*, S129–S138.

Hughes, D., Galinsky, E., & Morris, A. (1992). The effects of job characteristics on marital quality: Specifying linking mechanisms. *Journal of Marriage and the Family, 54*, 31–42.

Kahn, R. L. (1994). Opportunities, aspirations, and goodness of fit. In: M. W. Riley, R. Kahn & A. Foner (Eds), *Age and structural lag: Society's failure to provide meaningful opportunities in work, family, and leisure* (pp. 37–56). New York: Wiley.

Kelly, J. (1993). *Activity and aging: Staying involved in later life.* Newbury Park: Sage Publications, Inc. Kohli, M. (1994). Work and retirement: A comparative perspective. In: M.W. Riley, Kahn, R., Foner, A. (Eds.), *Age and structural lag: Society's failure to provide meaningful opportunities in work, family, and leisure* (pp. 80–106). New York: Wiley.

Kohn, M. (1994). Job complexity and adult personality. In: D. Grusky (Ed.), *Social stratification: class, race, and gender in sociological perspective* (pp. 430–439). Boulder, CO: Westview.

Kohn, M. L. (1983). *Work and personality: An inquiry into the impact of social stratification.* Norwood, NJ: Ablex Publishing Corporation.

Kohn, M. L., & Schooler, C. (1982). Job conditions and personality: A longitudinal assessment of their reciprocal effects. *American Journal of Sociology*, *87*, 1257–1286.

Larson, R., Zuzanek, J., & Mannel, R. (1985). Being alone versus being with people: Disengagement in the daily experience of older adults. *Journal of Gerontology*, *40*, 375–381.

Lennon, M. C., & Rosenfield, S. (1992). Women and mental health: The interaction of job and family conditions. *Journal of Health and Social Behavior*, *334*, 316–327.

Levinson, D. (1978). *Seasons of a man's life*. New York: Knopf.

Link, B. G., Lennon, M. C., & Dohrenwend, B. P. (1993). Socioeconomic status and depression: The role of occupations involving direction, control, and planning. *American Journal of Sociology*, *98*, 1351–1387.

Mannell, R. (1993). High-investment activity and life satisfaction among older adults: Committed, serious leisure, and flow activities. In: J. Kelly (Ed.), *Activity and aging: staying involved in later life* (pp. 125–145). Newbury Park: Sage Publications, Inc.

McIntosh, B. R., & Danigelis, N. L. (1995). Race, gender, and the relevance of productive activity for elders' affect. *Journal of Gerontology: Social Sciences*, *50B4*, S229–S239.

Menaghan, E. G. (1991). Work experiences and family interaction processes: The long reach of the job. *Annual Review of Sociology*, *17*, 419–444.

Mirowsky, J., & Ross, C. E. (1992). Age and depression. *Journal of Health and Social Behavior*, *33*, 187–205.

Mirowsky, J., & Ross, C. E. (1999a). Economic hardship across the life course. *American Sociological Review*, *64*, 548–569.

Mirowsky, J., & Ross, C. E. (1999b). Well-being across the life course. In: A. V. Horwitz & T. L. Scheid (Eds), *A handbook for the study of mental health: Social contexts, theories and systems* (pp. 328–348). Cambridge, UK: Cambridge University Press.

Moore, D. E., & Hayward, M. D. (1990). Occupational careers and mortality of elderly men. *Demography*, *27*, 31–53.

National Research Council NRC. (1999). *The changing nature of work: Implications for occupational analysis.* Washington, DC: National Academy Press.

Parnes, H., & Less, L. (1985). Variation in selected forms of leisure activity among elderly males. In: Z. Blau (Ed.), *Current perspectives on aging and the life cycle: Work, retirement and social policy*, (Vol. 1, pp. 223–242). Greenwich CT: JAI Press Inc.

Pearlin, L. I., & Skaff, M. M. (1996). Stress and the life course: A paradigmatic alliance. *The Gerontologist*, *362*, 239–247.

Riley, M. W., Kahn, R., & Foner, A. (1994). *Age and structural lag: Society's failure to provide meaningful opportunities in work, family, and leisure*. New York: Wiley.

Riley, M. W., & Riley, J. W., Jr. (1994). Structural lag: Past and future. In: M. W. Riley, R. Kahn & A. Foner (Eds), *Age and structural lag: society's failure to provide meaningful opportunities in work, family, and leisure* (pp. 15–36). New York: Wiley.

Riley, M. W., Foner, A., & Riley, J. W. (1999). The aging and society paradigm. In: V. Bengtson & K. W. Schaie (Eds), *Handbook of theories of aging* (pp. 327–343). NY: Springer Publishing Company.

Rindfuss, R. R. (1991). The young adult years: Diversity, structural change, and fertility. *Demography*, *28*, 493–512.

Rindfuss, R. R., Swicegood, C. G., & Rosenfeld, R. A. (1987). Disorder in the life course, how common and does it matter? *American Sociological Review*, *52*, 785–801.

Ross, C. E., & Drentea, P. (1998). Consequences of retirement activities for distress and the sense of personal control. *The Journal of Health and Social Behavior*, *39*, 317–334.

Schieman, S., Gundy, K. V., & Taylor, J. (2001). Status role, and resource explanations for age patterns in psychological distress. *The Journal of Health and Social Behavior*, *42*, 80–96.

Schooler, C., Kohn, M. L., Miller, K. A., & Miller, J. (1983). Housework as work. In: M. L. Kohn & C. Schooler (Eds), *Work and personality: An inquiry into the impact of social stratification* (pp. 242–260). Norwood, NJ: Ablex Publishing Corporation.

Shanahan, M. J. (2000). Pathways to adulthood in changing societies: variability and mechanisms in life course perspective. *Annual Review of Sociology*, *26*, 667–692.

Spenner, K. I. (1988). Social stratification, work, and personality. *Annual Review of Sociology*, *14*, 69–97.

Sweet, J. A., Bumpass, L L., & Call, V. (1988). *The design and content of the national survey of families and households*. NSFH Working Paper no. 1. National Survey of Families and Households, University of Wisconsin.

Verbrugge, L. M., Gruber-Baldini, A. L., & Fozard, J. L. (1996). Age differences and age changes in activities: Baltimore longitudinal study of aging. *Journal of Gerontology: Social Sciences*, *51B*, S30–S41.

Winship, C., & Radbill, L. (1994). Sampling weights and regression analysis. *Sociological Methods and Research*, *232*, 230–257.

Zimmer, Z., Hickey, T., & Searle, M. S. (1995). Activity participation and well-being among older people with arthritis. *The Gerontologist*, *354*, 463–471.

LIFE COURSE OF ACADEMIC PROFESSIONALS: SUBSTANTIVE TASKS, FALSE ASSUMPTIONS, INSTITUTIONAL ACCOMMODATIONS, AND PERSONAL ADJUSTMENTS

Victor Shaw

ABSTRACT

This paper conceptualizes a general academic career pathway on the basis of common knowledge and collective experiences among academic professionals. Five major stages, including initiation, routinization, secularization, solidification, and graduation, are proposed and described with respect to their false assumptions and substantive tasks. Recognizing the importance of institution and professional context, the chapter proposes ways in which academic institutions may reform their existing evaluation and reward systems to the benefit of career-making academicians. Emphasizing the indispensability of human agency, it offers suggestions for individual scholars to frame and structure their academic endeavors in the perspective of success and goal attainment.

The Structure of the Life Course: Standardized? Individualized? Differentiated?
Advances in Life Course Research, Volume 9, 331–347

ISSN: 1040-2608/doi:10.1016/S1040-2608(04)09012-4

A career-making academician has to undergo education, locate an institution for affiliation and employment, and build a professional network around stakeholders in funding and publication. Although not every prospective academic professional is able to secure all necessary conditions in his or her career-making endeavor, a well-integrated combinational presence of all necessary conditions still does not guarantee any aspiring academic a successful career in the modern and postmodern world of scholarship (Lewis, 1975; Clark, 1987; Huer, 1991; Rajagopal & Lin, 1996; Tierney, 1997; Blaxter, Hughes, & Tight, 1998; Bianco-Mathis & Chalofsky, 1999; Alstete, 2000; Baez, 2002).

From a life course point of view, career-making and career pathway in academia are work, professional pursuit, and job trajectory. Jobs and work careers are critical events, patterns, and paths in life course (Marshall, Heinz, Kruger, & Verma, 2001). Life course, featuring the movement of individuals in time and space or the unfolding of social processes over individual life span, is shaped not only by micro individual factors, such as human agency and linked lives, but also by macro social forces, including cultural beliefs, institutional establishments, and social change (Mayer & Tuma, 1990; Elder, 1995). Heinz (1991) notes how life course as a sequence of status passages differentiates according to social and economic circumstances, specifically how biographical planning and stock-taking evolve in response to changes in education, labor market, employment, and culture (Heinz & Kruger, 2001). Zollinger (2002) delineates four life-course orientations and outcomes: innovation versus tradition, rebellion versus conformity, notable success versus ordinary attainment, and precocity versus delay. These distinctions highlight the interrelationships among life career paths, socialization, and social structure. When Hermanowicz (2002) studied three age cohorts of physical scientists employed in three types of institutions, elite, pluralist, and communitarian schools, he found that achievement ambition, to a large extent, is influenced by the profession or institution as an organized set of narratives, which further is "master" formed and possessed by the professional or organizational world in which professionals work.

This chapter conceptualizes a general academic career pathway on the basis of common knowledge and collective experiences among academic professionals. Major stages are proposed and described with respect to their false assumptions and substantive tasks. Recognizing the importance of institution and professional context, the chapter proposes ways in which academic institutions may reform their existing evaluation and reward systems to the benefit of career-making academicians. Emphasizing the indispensability of human agency, it further offers suggestions for individual scholars to frame and structure their academic endeavors in the perspective of success and goal attainment.

STAGES OF ACADEMIC CAREERS: SUBSTANTIVE TASKS AND FALSE ASSUMPTIONS

An academic career pathway reflects the general career process that academic professionals in a society or historical era move through in their lifelong scholarly pursuits (Gould, 1978; Levinson, 1978; Brown & Brooks, 1996; Shaw, 2002). It is specific to a society because scholars may follow different career paths under different social environments. It is specific to a historical era because academicians may take different career routes due to different historical forces. A typical academic career pathway in a particular society is not necessarily universalistic. Although it may hold true as a whole, some academicians may deviate from the general pathway by passing through its stages in different sequences or by dropping out in the beginning or the middle of the journey. The academic career pathway is also related to age but is not necessarily age-graded. While many academic professionals move from junior to senior levels as they age, some reach higher stages when they are young and some remain in lower stages when they are old (Finkelstein, 1984; Clark & Centra, 1985; Clark & Lewis, 1985; Piper, 1992; MacDonald, 1995; Rossides, 1998). Parallel to the general life course, an academic career is not just an ontogenetic development followed by a self-defined, self-motivated, and self-contained academician. It is a sociogenic process featuring the interplay of individual academician as organism or human agency and the knowledge enterprise as environment or systemic force in historical time and societal space (Dannefer, 1984; Elder, 2003).

In the context of modern and postmodern society, this paper identifies five general stages, *initiation*, *routinization*, *secularization*, *solidification*, and *graduation*, in the academic career pathway. Each stage features a unique developmental theme. There are not only major tasks to tackle but also common false assumptions to overcome. Epitomized in a self-statement, a false assumption represents a misperception about or a misstep by an academician in his or her career pathway. In more general terms, it symbolizes misguided human agency in life course.

Initiation

At the stage of initiation, individual academicians face five major tasks. The first involves proficiency and competency. They attend graduate school, learn the academic language, and command essential skills. The second deals with norms and normative behavior. They follow requirements, familiarize themselves with prevailing customs, and internalize basic rules. A third task emphasizes identification and identity. They meet insiders, worship leaders, and emulate their favorite role models in the field. The fourth concerns specialization and specialty. They build motivation, develop interests, and decide on a focused area of inquiry. The last involves adventure and experimentation. They conduct exercise in research, take practice of presentation and publication, and taste the larger professional waters through various academic media.

The false assumptions that academic beginners often make typically fall under five categories. The first is: "I can conquer the world." One is obsessed with grand ideas, but ignores technical details. One is overjoyed by acquaintance with masterly achievement in research, but underestimates the meticulous effort involved in developing a great idea and making a masterpiece. One focuses on substance, but fails to see the emotion and sentiment involved in academic undertakings. For example, one openly and defiantly criticizes a professor for misinterpreting a theory in class. Or without regard for normal style and conventional procedure, one sends a self-conceived draft to an editor in an attempt to overthrow a dominant paradigm.

The second is: "I am not fully responsible." One is reluctant to put the whole of one's learning or the truth of one's position into scholarly presentation and publication. One condones one's own mistakes. One is self-inhibited and fearful. One defers written examinations, postpones final defenses, or bypasses opportunities for professional presentation, competition, and publication. For example, when some professors ask one to contribute a chapter to a volume they are editing, one turns it down on the grounds that they perceive themselves as lacking expertise in the area. The third is: "I am not ready yet to enter the profession." One settles for the ease of student life on campus and idealizes the romance of student life within the walls of the university while dramatizing the brutality of survival in the academic market. One registers for classes one after another and participates in aimless discussion in classroom settings. One engages in entertaining yet unproductive intellectual exercises with university faculty and students but never graduates. While both institutional forces, notably the gendered and class-based context of higher education, and external factors typically associated with balancing school, work, and family in the life span play a role, individual orientation, personal adjustment, or human agency in general contributes to the widely observed phenomenon on university campus

(Etzkowitz, Kemelgor, Neuschatz, & Uzzi, 1992). That is, it takes more and more Ph.D. candidates longer and longer time to wrap up their training in graduate school. Lastly, "I don't think I can make it there." One admires and perhaps fears founding fathers, influential figures and even one's own advisors and mentors. One mystifies disciplinary theories and methods, overestimates the talent and effort required for quality work and significant achievements, and holds oneself in awe of academic establishments.

Routinization

Routinization is the stage when one settles into a tenure-track position at an institution. The first major task one has to deal with is to get to know the job, the institution, the profession, and the disciplinary establishment. Specifically, one needs to learn rules, familiarize oneself with existing conventions, empathize with prevailing sentiments, make contacts, and establish a network of interaction, reference, and support. One also builds a teaching portfolio by identifying a set of courses, preparing syllabi and course materials, setting up ground rules for conducting class, interacting with students in and out of class, grading, and handling complaints, and cultivating a teaching style or habit characteristic of one's own fluency, ease, and comfort. A third task is developing a research agenda.[I] Typically, one needs to retreat from grand ideas embraced during graduate school and overcome characteristic feelings of incompetence, unpreparedness, and unsureness one went through as a student. One then must identify one's own strengths and weaknesses, delve into an area of specialty, program oneself into a research way of life, and place oneself properly, in terms of scholarly uniqueness, quality, and productivity, in the knowledge enterprise. A fourth element is to open and maintain a track of service. One needs to make oneself known and available for service related to one's training and expertise. Depending upon one's needs, interests, and visibility, one may actively seek opportunities for service or firmly turn down various requests for service. Fifth, one puts the academic career in proper perspective with various commitments in life. Settling into a community, one confronts questions of marriage, parenthood, homeownership, investment, volunteering, and even hobbies. Along with academic concerns and commitments, such personal issues shape and reshape one's career in general and routinization in particular (Moen & Orrange, 2002; Glassner & Hertz, 2003).

A routinizing academician faces five common false assumptions. One is: "My advisors and classmates are out there for me." One calls on one's graduate advisors, asks them for advice, for substantive assistance, and may even pressure them to call a reviewer, an editor, or some committee members on one's behalf. One talks to former classmates about one's pains and sufferings. One who acts in this fashion under this assumption may unfairly drag someone in one's graduate training into one's routinization process. Another false assumption is: "I saw that done or I did that in graduate school, differently from what you guys do here." One tells one's students and colleagues, in classrooms and in department meetings, what one saw or did in graduate school, implying that what they do now and here is awkward, backward, or outright wrong. Such comparisons may unfortunately offend members of the old guard in one's college or university. Still another is: "This is not what I expected" where one struggles between the ideal and reality, feeling that one's students are under-prepared and ill-mannered, lamenting that one's colleagues are cold-blooded and hostile, and perhaps resenting that one's leaders are repressive and evil-minded. This may result in complaint and the search for an exit from one's situation. One then is likely to change jobs frequently.

The fourth assumption is: "I cannot do research because I am preoccupied with teaching and service." Out of an instinctual fear for research, one spends time in the laboratory but never turns out anything, collects data but never analyzes them, or juggles a lot of ideas but never puts anything on paper. One may do every little thing in teaching and service to avoid the laboratory or fieldwork. Ultimately, this may make it difficult to pursue research in the later phase of one's career. The fifth common assumption is: "I have to put a lot of things in life on hold so that I can get my career under control." One isolates oneself in one's office and calls around, answers calls, attends meetings, and chats with students and colleagues. One follows most of one's activities and days in academic settings and on scholarly matters.

Secularization

Secularization begins when one is tenured and becomes immersed in one's institutional as well as disciplinary establishments. At this stage, an academician identifies with prevailing norms and conventions, by practicing them, exemplifying them, teaching them to one's students, and defending them when they are breached. In teaching, one relies on one's own established methods, style, and reputation and may be known for teaching a set of content courses, for being casual, permissive, discursive, inaccessible, or for being boring or formal, demanding, organized, inspiring, and helpful to students. With colleagues, one complainingly or jokingly talks about students being lazy, unprepared, unintelligent, uncooperative, disrespectful, or not as good as they used to be some time ago. Sitting on committees, one may argue for or against tough standards. In research, one works year after year in an area, using familiar theoretical and methodological approaches and producing similar findings and publications. One reviews manuscripts in the area, sits on editorial boards, and may even rise to the association leadership in one's discipline. In service, one responds to calls for advice, consultation, and expert opinion from the community, the government, and the media. One may set up a consulting firm, organize a conference, run an association, edit a journal or book series, or engage in other academic undertakings. One makes money, gains respect, and extends influence. Finally, secularization makes one settled into a peculiar work routine and lifestyle. One may be known by janitors, security guards, secretaries, or neighbors for leaving the laboratory late every day, going into the office frequently in the middle of the night, guzzling several cups of coffee after lunch, or taking walks around the campus before midnight. Most important, one relates to people in other occupations in a way that typifies one's calling in academe. For instance, one may sound like a scholar even when talking about news and movies with next-door neighbors. Life activities and routines characteristic of academic efforts do not exist merely as side products. They serve as powerful reinforcements in scholarly endeavours. Developing a lifestyle compatible to academic endeavour, therefore, can be considered as one of the main tasks a secularizing academician works on in a scholarly career.

The most common false assumption held by a secularized academician is: "I know it all." One teaches classes off the top of one's head. One pages through new publications without serious reading. One writes papers and makes judgments following a set track of thought. One takes much for granted and does not question assumptions, even those that are irrational, unreasonable, and questionable. To newcomers, one tends to assume: "I am an insider." Under this assumption, one pours out stories, experiences, and versions of reality to the newer arrivals. One joins old colleagues and friends to monitor, gossip about, scare, manipulate, or even discriminate against the newcomers. One labels them "naive," "inexperienced," or "unrealistic" when the newer arrivals experiment with something new or something one simply dislikes. To students, one is habituated to assume: "I am always right." One lectures students, corrects them, scolds them, orders them to conduct different exercises, or even forces them to attempt something out of their reach. One fails to realize that one can be wrong and may learn from students. One may be quick to assume that they are more educated, informed, and rational and consequently look down upon common citizens as being mundane, ignorant, and gullible. One criticizes politicians as being wicked, opinionated, and manipulative. One laments the media as being biased, emotionally charged, and misleading. One is too presumptuous to appreciate the beauty and vividness of the larger social mosaic. The last false assumption an academician is likely to make regards future and change: "I am not going to be different." One counts and brags about one's years of service. One takes comfort in what one has accomplished in teaching, research, and service. One sticks to routine and accustomed ways of thinking and acting. One resists change, innovation, and reform. One defends traditional positions. One refuses to back down even in confronting mistakes made.

Solidification

Solidification does not necessarily follow the stage of secularization for all career-making individuals. It builds upon or emerges from secularization among a small number of academicians. Sitting in full professorships, a great many faculty on university campuses feel they have arrived at their career destination. At most, they keep doing what they are familiar with doing, becoming ever more secular along their career pathway. A few, however, attempt to rise above their secular experiences. They reach the uncommon stage of solidification when they are successful. There are three paths toward solidification. One is through scholarly endeavors and accomplishments. One makes extraordinary discoveries, puts forth revolutionary concepts and theories, develops unusual methods and techniques, produces masterpieces, and may even spearhead a new area of inquiry. One may become the president of one's disciplinary association or be awarded highest honors in his or her discipline or for the whole knowledge enterprise, such as the Nobel Prize. An alternative pathway is by way of management. One is fortunate to be elected or pushed, often through political maneuvering, into the chairmanship of one's department. The latter makes one eligible to apply for a managerial position at the dean's level. The experience as dean sets a stage for a further ascendance to leadership at the university level.

On the second pathway via management, one sharpens public speaking skills, strategizes human relations at different levels and in different settings, manipulates resources and opportunities, plays fund-raising tactics, and relates properly to the larger political environment surrounding his or her job duty. As one becomes a career manager, one may gradually lose the drive and instinct for serious academic contributions (Nason, 1980; Smart & McLaughlin, 1985; Gmelch & Burns, 1993; Seagren, Creswell, & Wheeler,

1993; Kinnick, 1994; Lucas, 2000). Still another route toward solidification is through a practice or service. One capitalizes on one's training, knowledge, or invention. One establishes a business or opens a practice and may become more and more concerned with profit and eventually abandon one's aspiration and effort for scholarly breakthroughs. Or as a practitioner, one may gather first-hand data from clients and gradually develop a theory, a method, or a treatment of scientific importance. No matter what route one takes, one needs to make a significant effort to emerge from the mundane, the secular, or the transient to become solidified in the unusual, the exceptional, and the eternal.

A career academician who reaches the stage of solidification can also make and act under false assumptions. The two general assumptions shared by many solidified scholars are: "I am special" and "I represent it all." By the first assumption, one feels one is a genius, blessed with the special talent, insight, skill, or opportunity to discover what others are not able to find, write about what others fail to see, control what others are incapable of handling, or profit from what others are scared or unaware of. With the second assumption, one feels one is the sovereign of one's discipline, institution, or profession. One may declare that one's discipline is in a theoretical or methodological crisis, calling for a general reform or revolution. One may proclaim that one's institution must commit to a particular philosophy, standard, or ideal, forcing all its members into a set track of thought or a fixed mode of endeavour. Specific to different routes of solidification, one is likely to assume that "knowledge is power" if one gains influence through scholarship. As one is cheered or admired, one may intuit that it is possible to conquer the whole world just by knowledge. Similarly, one is likely to assume that "power is everything" when one sits at the helm of an academic institution. And one is likely to assume that "money speaks" if one runs a knowledge-based corporation. One sets rules, gives orders, and applies rewards and penalties. One sees clearly how one can manipulate the mass of secular academicians, even the stars of scholarship, by the power one holds. One also keeps a development team of scientists and engineers within one's company. One hires, fires, shuffles, promotes, or demotes those scientists and engineers. One feels one can easily boss them around, no matter how much knowledge they have, as long as one has money to employ them.

Graduation

The last stage is graduation. Although some academicians vow that they will never graduate from their lifelong commitment to scientific inquiry, others admit that they cannot wait to retreat back to the wishes of their childhood, the excitement of their hobby, or the comfort of their family life. As far as employment and job duty are concerned, graduation is indeed an inevitable and important phase of a complete academic career (Dannefer, 1987). The major tasks one is faced with at the stage of graduation include: reviewing, winding-up, repositioning, adjusting, and slowing-down. Review involves both a retrospective examination of past work and an objective evaluation of current projects. In examining work throughout one's entire career, one sees ups and downs, gains and losses, or pride and regret. By evaluating one's ongoing projects, one can responsibly decide what to terminate, what to hand over, what to leave behind, and what to carry on. One wraps up an experiment, an analysis, a report, or a manuscript. Here, one may be able to clear a critical hurdle in the effort to establish a theory or method, unravel a puzzle or problem one has long dreamed of solving, or conclude a capstone project pursued for years. Repositioning is to evaluate oneself and identify a proper niche for oneself during retirement in the whole knowledge enterprise. Depending upon one's experience, reputation, visibility, network, age, energy level, time commitment, and other factors, one may deliver guest lectures from place to place, take short-term residency with a research center, participate in an issue-specific project, engage in writing, or volunteer in a service or educational organization. Adjustment is needed as a new line of activity in retirement requires a different approach, schedule, investment, expectation, or perspective than one is used to through the pre-retirement career. There is a time of ease and joy when one acts upon the wealth of lifelong learning. There is a time of frustration and sadness when one is confronted with challenges previously unheard of. Finally, slowing-down is to admit the declining mental and physical power available for academic activities as one draws close to the end of life. One keeps an eye on the academic world, reads scholarly articles, and may occasionally come up with some critical ideas. But overall, one knows that one is on the back stage of the knowledge enterprise, moving closer and closer to the absolute conclusion of one's academic career.

In a mood of graduation, a career academician can easily make false assumptions, specifically about his or her academic contribution, career, and discipline, and generally about life and science. One may assume: "Nobody really understands what I put forth in my theory or method." One may criticize the disciplinary establishment, lament the prevailing sentiment in academe, and admonish the mass of scholars for their inattention, obtuseness, mundaneness, and carelessness. One may also feel: "I have nothing to be proud of." Feeling a lack of self-worth in scholarship, one avoids talking about one's past work or the institution where one was employed, does not explore academic markets for research and teaching opportunities during retirement nor discuss any scholarly issues, and takes joy only in nonacademic hobbies or volunteering activities. With respect to one's discipline, one is likely to assume: "It is no longer my world." One has some ideas and thoughts about one's discipline and profession but buries them in one's mind, feeling that nobody would care to know about those ideas. At the same time, one may assume: "No life ever exists beyond my academic career." Here, one refuses to retire from academic work but continues a long habituated daily routine, with or without scholarly productivity. One retreats to life only when challenged by an accident, a disabling disease, or other drastic event. Finally, a graduating academician may cap his or her scholarly career with some general thoughts about science and the knowledge enterprise. One may assume: "Science is a game," and complacently aligns oneself with the large army of smart players in academia or assume: "Science is the world of geniuses and giants," and sentimentally characterizes oneself, along with the vast mass of ordinary academicians, as simple materials used by, or as little dwarfs in the service of, a few founders, pioneers, and leaders in the production of knowledge and domination.

INSTITUTIONAL ACCOMMODATIONS

What can an academic institution do to facilitate its employees in their lifelong drive toward success? Is a university willing and ready to change itself or some of its standard practices so that it can fully embrace an innovative idea to the benefit of its faculty and their career movement (Nason, 1980; Kinnick, 1994; Tierney & Bensimon, 1996; Coiner & George, 1998; Greenwood & Levin, 2001; Baez, 2002)?

One area for reform is the reward system. The existing reward system builds upon the principle of seniority and operates under the assumption of demonstrated performance. It does not reserve sufficient room for potentiality and needs. Some simple yet ubiquitous ironies result from this system. First, financial rewards are typically greatest at the end of one's career when one needs it least. Second, encouragement is often lowest during the early career when it is most wanted or desirable and greatest at the end of one's career, through honor, resource, and power, when one might not need much external support. To change such practices, a university may study faculty salaries and identify an average yearly salary for a normally progressing faculty member over his or her whole employment career with the university. A university may set salaries for its entry level and junior faculty high enough so that they can start a family, raise children, or support their elderly parents. The level of increase may be adjusted in response to age progression, life events, and career mobility. A faculty member receives a higher yearly salary, against an assumed total of his or her lifelong salaries, in the heyday of his or her career as well as when he or she has to support school-going children. He or she receives a lower yearly salary, against the assumed total of his or her lifelong salaries, when he or she pays off a home loan, sees his or her adult children leave, and lives in relative health and affluence. A faculty member who leaves before the assumed duration of employment is required to pay back to the institution the overdrawn portion of his or her lifelong salaries. Payback, however, automatically becomes unnecessary when all academic institutions follow the same practice. That is, all academic employers even up compensations to academic employees so that individual salaries and benefits are sensibly and rationally distributed throughout employee career pathways in terms of both performance and need, both contribution and promise.

Another area for reform lies in the evaluation system (Mortimer, Bagshaw, & Masland, 1985; Kaplowitz, 1986; Licata, 1986; Long, McGinnis, & Allison, 1993). The current evaluation system is controlled by the "old guard." The original rationale for such an evaluation system is that senior academicians have internalized academic norms, no longer need tight control, and can productively use their autonomy and freedom for creative work. The reality, however, is that many senior academics do not conform to this rational model, while at the same time insulated from accountability review. The likely outcome of any systematic study of scholarly productivity over individual career spans is that the majority of scholars complete most of their work in years when they are junior, relatively unknown, and subject to rigorous review and evaluation. In fact, as junior academicians strive for acceptance and recognition, they are more likely to do their best conscientiously and voluntarily. They will do the best they can in line with prevailing academic norms, standards, and practices. The evaluation system, therefore, might be revamped to loosen control over newcomers and tighten surveillance and monitoring of members of the old guard.

PERSONAL ADJUSTMENTS

There are always pride and regret, satisfaction and dissatisfaction, as well as reinforcement and modification over career pathways. Through the process, focus is put on performance, efficiency, effectiveness, and productivity. In the end, reflection is often transcendental about the consummation of academic career and the quality of mundane life. Specifically, are there any personal adjustments an academician can make in the life span to facilitate his or her navigation on the academic career (Maslow, 1954; Joughin, 1967; Finkelstein, 1984; Rodmann, 1995; Krau, 1997; Lewis, 1997; Hayes, 1998; Baldwin & Chronister, 2001; Silverman, 2001)?

Live a Simple Life

Life is a multidimensional sponge. It can easily absorb all your time and energy. As a scholar, one does not have to avoid or abandon life. One can live and enjoy every essential element of life while pursuing one's academic career. The key is to streamline one's life, free oneself from unnecessary life engagements, and live simply. For example, one may choose a married life but should select a spouse who understands the passion for knowledge and achievement. One may choose to have children, but one should not expect to enjoy a large family. One may choose to own a house but should not stretch budgets and time to maintain a huge estate with an extensive collection of personal "accoutrements." One may choose to invest in the market but should not attempt to chase stocks or mutual funds for best possible yields. One may choose to follow a hobby but should be aware of limitations of time and energy. There are big and small choices in life. Living a simple life often means that you choose a thing or a way of doing things in its basic form, without spending more than necessary time, money, and energy on it.

Enhance Your Work with Your Life

Depending upon one's discipline, one may benefit enormously from life observations, reflections, and experiences as fertile soil or motivation for serious philosophical contemplation, theoretical generalization, historical description, and policy analysis. Fighting as a soldier in Vietnam could not only motivate one to become a specialist in Vietnam, but also provide a real-world perspective in analysis of the country. Dealing with various moral dilemmas in life could make you a more thoughtful scholar of ethics. Being married and having children could offer you critical insights and motivation to delve deeply into the issues of marriage, family, childhood, adolescence, and personality development. Regardless of discipline, one can always develop a reciprocal relationship between work and life. For example, while deadlocked in the search for a concept, a formula, or a line of reasoning, enlightenment may emerge while walking with a spouse under the moonlight, playing with children in the yard, or chatting with family members at the dinner table. When disappointed, frustrated, or tired with work, one may be refreshed, recharged, and rejuvenated with courage, insight, and energy after a visit to a hometown or a vacation with family. It is in living that one finds the real meaning of success.

Enrich Your Life with Your Work

Divisions of scholarship loosely correspond to areas of life. One may take an informed, balanced, or holistic view of life, society, and history if one's major is in a discipline in the humanities and social sciences. Those educated and habituated to emphasize evidence, logical analysis, and rationality in exploring nature and to apply imagination, empathy, and sympathy in studying human dynamics may use education and career experience to help deal with real-life issues.

Be the Turtle, Not the Rabbit

In the folk story about the race between a turtle and a rabbit, the turtle moves slowly but keeps going and ultimately wins the race. The rabbit starts off with a huge lead but opts for a nap in the middle of the competition and eventually ends up as a loser in the game. As an academic, one may think that they are brilliant. But the best bet for success is to be as humble and persistent. Be content with a little progress each day in work. But also feel bad or even guilty when distractions prevail. Feel frustrated when tasks are not finished in a reasonable time frame. Such inner pressure provides the basic forces for devotion and dedication. Diligence towards one's subject puts one on track toward important findings and major breakthroughs. One may never find a complete chunk of time for academic work if one looks for a perfect time. The key is to put your mind on your scholarly pursuit and make use of every bit of usable time for it in your life.

Respect the Convention but Do Not be Conventional

The world, the academic world in particular, is well set up. One needs to follow rules and conventions to be recognized and accepted by the academic community. For example, there is a standard language for papers. There are established formats for presentations, styles for references, and codes for conduct. It is not wise to irritate insiders with a form or an appearance that is completely alien to them. An unusual form may alienate members of the "old guard" to the extent that they will not accept even conventional content. On the other hand, if ideas are packaged in conventional form, even unconventional substance can be sold. Respect conventions. Yet, also embrace innovative ideas, methods, and ways of analysis. Strive for new findings and breakthroughs. Think differently and do differently with regard to subject matter on a day-to-day basis. In the end, a successful career entails contribution. One contributes with something new, different, or perhaps even unique. One turns out something new, different, or unique only when you are unconventional in your way of thinking and acting throughout the whole journey of your academic career.

NOTE

I. Depending upon institutional context, establishing a research program could be equal or even more important than establishing a teaching portfolio. In some cases, it may even be the defining aspect of routinization.

REFERENCES

Alstete, J. (2000). *Posttenure faculty development: Building a system for faculty improvement and appreciation.* San Francisco, CA: Jossey-Bass.

Baez, B. (2002). *Affirmative action, hate speech, and tenure: Narratives about race, law, and the academy.* New York: Routledge Falmer.

Baldwin, R., & Chronister, J. (2001). *Teaching without tenure: Policies and practices for a new era.* Baltimore, MD: Johns Hopkins University Press.

Bianco-Mathis, V., & Chalofsky, N. (1999). *The full-time faculty handbook.* Thousand Oaks, CA: Sage Publications.

Blaxter, L., Hughes, C., & Tight, M. (1998). *The academic career handbook.* Philadelphia, PA: Open University Press.

Brown, D., & Brooks, L. (1996). *Career choice and development.* San Francisco, CA: Jossey-Bass Publishers.

Clark, B. (1987). *The academic profession: National, disciplinary, and institutional settings.* Berkeley and Los Angeles: University of California Press.

Clark, M., & Centra, J. (1985). Influences on the career accomplishments of Ph.D.'s. *Research in Higher Education, 23*, 256–269.

Clark, S., & Lewis, D. (1985). *Faculty vitality and institutional productivity: Critical perspectives for higher education.* New York: Teachers College Press.

Coiner, C., & George, D. (1998). *The family track: Keeping your faculties while you mentor, nurture, teach, and serve*. Urbana and Chicago: University of Illinois Press.

Dannefer, D. (1984). Adult development and social theory: A paradigmatic reappraisal. *American Sociological Review*, *49*, 100–116.

Dannefer, D. (1987). Aging as intracohort differentiation: Accentuation, the Matthew effect, and the life course. *Sociological Forum*, *2*, 211–236.

Elder, G. (1995). The life-course paradigm: Social change and individual development. In: P. Moen, G. Elder & K. Luscher (Eds), *Examining lives in context: Perspectives on the ecology of human development* (pp. 101–139). Washington, DC: American Psychological Association.

Elder, G. (2003). The life course in time and place. In: W. Heinz & V. Marshall (Eds), *Social dynamics of the life course: Transitions, institutions, and interrelations* (pp. 57–71). Hawthorne, NY: Aldine De Gruyter.

Etzkowitz, H., Kemelgor, C., Neuschatz, M., & Uzzi, B. (1992). Athena unbound: Barriers to women in academic science and engineering. *Science and Public Policy*, *19*, 157–179.

Finkelstein, M. (1984). *The American academic profession: A synthesis of social scientific inquiry since World War II*. Columbus, OH: Ohio State University Press.

Glassner, B., & Hertz, R. (2003). *Our studies, ourselves: Sociologists' lives and work*. New York: Oxford University Press.

Gmelch, W., & Burns, J. (1993). The cost of academic leadership: Department chair. *Innovative Higher Education*, *17*, 259–270.

Gould, R. (1978). *Transformations: Growth and change in adult life*. New York: Simon and Schuster.

Greenwood, D., & Levin, M. (2001). Reorganizing universities and "knowing how": University restructuring and knowledge creation for the 21st Century. *Organization*, *8*, 433–440.

Hayes, C. (1998). *Beyond the American dream: Lifelong learning and the search for meaning in a postmodern world*. Wasilla, AK: Autodidactic Press.

Heinz, W. (1991). Status passages, social risks and the life course: A conceptual framework. In: W. Heinz (Ed.), *Theoretical advances in life course research* (pp. 9–22). Weinheim, Germany: Deutscher Studien Verlag.

Heinz, W., & Kruger, H. (2001). Life course: Innovations and challenges for social research. *Current Sociology*, *49*, 29–57.

Hermanowicz, J. (2002). In the shadows of giants: Identity and institution building in the American academic profession. In: R. Settersten & T. Owens (Eds), *Advances in life course research: New frontiers in socialization* (pp. 133–162). Oxford, UK: JAI.

Huer, J. (1991). *Tenure for Socrates: A study in the betrayal of the American professor*. New York: Bergin and Garvey.

Joughin, L. (1967). *Academic freedom and tenure: A handbook of the American association of university professors*. Madison, WI: University of Wisconsin Press.

Kaplowitz, R. (1986). *Selecting college and university personnel: The quest and the questions*. Washington, DC: Association for the Study of Higher Education.

Kinnick, M. (1994). *Providing useful information for deans and department chairs*. San Francisco, CA: Jossey-Bass.

Krau, E. (1997). *The realization of life aspirations through vocational careers*. Westport, CT: Praeger.

Levinson, D. (1978). *The seasons of a man's life*. New York: Ballantine Books.

Lewis, L. (1975). *Scaling the ivory tower: Merit and its limits in academic careers*. Baltimore, MD: Johns Hopkins University Press.

Lewis, M. (1997). *Poisoning the ivy: The seven deadly sins and other vices of higher education in America*. Armonk, NY: M. E. Sharpe.

Licata, C. (1986). *Posttenure faculty evaluation: Threat or opportunity?* Washington, DC: Association for the Study of Higher Education.

Long, J., McGinnis, R., & Allison, P. (1993). Rank advancement in academic careers: Sex differences and the effects of productivity. *American Sociological Review*, *58*, 703–722.

Lucas, A. (2000). *Leading academic change: Essential roles for department chairs*. San Francisco, CA: Jossey-Bass.

MacDonald, K. (1995). *The sociology of the professions*. Thousand Oaks, CA: Sage Publications.

Marshall, V., Heinz, W., Kruger, H., & Verma, A. (2001). *Restructuring work and the life course*. Toronto: University of Toronto Press.

Maslow, A. (1954). *Motivation and personality*. New York: Harper.

Mayer, K., & Tuma, N. (1990). Life-course research and event history analysis. In: K. Mayer & N. Tuma (Eds), *Event history analysis in life course research* (pp. 3–20). Madison, WI: University of Wisconsin Press.

Moen, P., & Orrange, R. (2002). Careers and lives: Socialization, structural lag, and gendered ambivalence. *Advances in Life Course Research*, *7*, 231–260.

Mortimer, K., Bagshaw, M., & Masland, A. (1985). *Flexibility in academic staffing: Effective policies and practices*. Washington, DC: Association for the Study of Higher Education.

Nason, J. W. (1980). *Presidential assessment: A challenge to college and university leadership*. Washington, DC: Association of Governing Boards of Universities and Colleges.

Piper, D. (1992). Are professors professional? *Higher Education Quarterly*, *46*, 145–156.

Rajagopal, I., & Lin, Z. (1996). Hidden careerists in Canadian Universities. *Higher Education*, *32*, 247–266.

Rodmann, D. (1995). *Career transitions for chemists*. Washington, DC: American Chemical Society.

Rossides, D. (1998). *Professions and disciplines: Functional and conflict perspectives*. Upper Saddle River, NJ: Prentice-Hall.

Seagren, A., Creswell, J., & Wheeler, D. (1993). *The department chair: New roles, responsibilities, and challenges*. Washington, DC: Association for the Study of Higher Education.

Shaw, V. (2002). *Substance use and abuse: Sociological perspectives*. Westport and London: Greenwood Publishing Group.

Silverman, F. (2001). *Publishing for tenure and beyond: Strategies for maximizing your student ratings*. Westport, CT: Bergin and Garvey.

Smart, J., & McLaughlin, G. (1985). Administrative service and research performance: A study of chemistry department heads. *Research in Higher Education*, *22*, 31–42.

Tierney, W. (1997). Organizational socialization in higher education. *Journal of Higher Education*, *68*, 1–16.

Tierney, W., & Bensimon, E. (1996). *Promotion and tenure: Community and socialization in academe*. Albany, NY: State University of New York Press.

Zollinger, J. (2002). Life careers and the theory of action. In: R. Settersten & T. Owens (Eds), *Advances in life course research: New frontiers in socialization* (pp. 65–88). Oxford, UK: JAI.

AUTHOR INDEX

SUBJECT INDEX

2